Electrospinning for Proton Exchange Membrane Fuel Cells

Electrospinning for Proton Exchange Membrane Fuel Cells

Yong Liu and Mohideen Meerasahib Mohamedazeem

WILEY-VCH

Authors

Prof. Yong Liu
Beijing University of Chemical Technology
15, Beisanhuan East Road
Chaoyang District
Beijing
100029 China

Dr. Mohideen Meerasahib Mohamedazeem
Beijing University of Chemical Technology
15, Beisanhuan East Road
Chaoyang District
Beijing
100029 China

Cover Image: Courtesy of the authors

Typesetting: Straive, Chennai, India

Printed and bound: CPI Group (UK) Ltd, Croydon CR0 4YY

Library of Congress Card No.: applied for

British Library Cataloguing-in-Publication Data A catalogue record for this book is available from the British Library.

Bibliographic information published by the Deutsche Nationalbibliothek
The Deutsche Nationalbibliothek lists this publication in the Deutsche Nationalbibliografie; detailed bibliographic data are available on the Internet at <http://dnb.d-nb.de>.

Print ISBN: 978-3-527-35462-7
ePDF ISBN: 978-3-527-85020-4
ePub ISBN: 978-3-527-85019-8
oBook ISBN: 978-3-527-85021-1

C9783527354627_280825

Contents

About the Authors

Prof. Yong Liu is a full professor in the College of Materials Science and Engineering at Beijing University of Chemical Technology, China. His research spans a wide array of fields, including the development and application of special high-performance plastics and rubber products, electrospinning of ultrafine fibers for formaldehyde purification, PM2.5 air filtration, as well as the fabrication of fiber composite for fuel cell, solar cell, and biomedical devices. Prof. Liu's significant contributions to research have earned him prestigious accolades, including a National Science and Technology Progress Award, Beijing Science and Technology Award, two awards in Technological Invention at the provincial/ministerial level, and two Patent Excellence Awards. Recognized as an exceptional scientific and technological worker in the National Petroleum and Chemical Industry, he was honored as a Wiley Distinguished Author in 2022. Prof. Liu has an impressive publication record of 215 journal articles. He has published six academic books in Chinese or English. Prof. Liu been authorized 68 China or international patents, including nine transferred to companies.

Dr. Mohideen Meerasahib Mohamedazeem is an expert in Materials Science, with a deep focus on sustainable energy solutions. He completed his Master's degree in Materials Science from Anna University, India, in 2018. He later pursued his doctoral research and is currently working as a Postdoctoral Researcher at the College of Materials Science and Engineering at Beijing University of Chemical Technology (BUCT), China. His research primarily focused on the development of novel non-precious electrocatalysts for oxygen reduction reactions, using biomass, MXene, metal-organic frameworks (MOF) for proton exchange membrane (PEM) fuel cells. Dr. Mohamedazeem is passionate about advancing research aligned with the Sustainable Development Goals (SDGs) and carbon neutrality. His work aims to contribute to the global shift toward more sustainable energy solutions. He has received an International Scientist funding award from the Beijing Natural Science Foundation, where he will lead a project from 2024 to 2026. To date, Dr. Mohamedazeem has authored over 20 scientific articles, both as a first and co-author, in top Q1 SCI journals, and has published one book.

Preface

"Thinking is an endless flow of energy and understanding that connects everything to the universe. The universe is always speaking to us; all we need to do is pause and truly listen. When we do, we catch pieces of its story, woven seamlessly into our own thoughts. In those moments, we realize we're not just thinking, we're sharing in something infinite."

Tackling the ever-growing challenge of climate change requires immediate and innovative approaches to transitioning energy systems toward a low-carbon future. Transportation, which accounts for nearly one-fifth of global CO_2 emissions, remains a key focus area. Passenger vehicles alone contribute approximately 75% of the sector's emissions, further exacerbating environmental and public health concerns due to particulate matter (PM2.5) and other harmful pollutants, as noted by the World Health Organization. Transitioning to clean, zero-emission transportation technologies is no longer optional, and it is essential.

Proton exchange membrane fuel cells (PEMFCs) have emerged as a leading solution for net-zero transportation. However, their widespread deployment is constrained by the high costs associated with manufacturing key fuel cell components, particularly the catalyst layers, gas diffusion layers (GDL), and proton exchange membranes. Overcoming these barriers demands innovative, scalable, and cost-effective fabrication methods. Electrospinning—a highly versatile and relatively low-cost nanofiber production technique has demonstrated significant potential in addressing these challenges. Recent research highlights the application of electrospinning to create critical PEMFC components, achieving noteworthy performance enhancements and paving the way for more affordable, efficient systems.

This book, *Electrospinning for Proton Exchange Membrane Fuel Cells*, provides a detailed exploration of how electrospinning can drive innovation in PEMFC technology. Drawing on extensive research and practical expertise, the book offers a comprehensive examination of PEMFC components fabricated through electrospinning. Key topics include:

- **Catalyst Layers:** The fabrication of electrospun catalyst layers with unique morphological structures, their electrochemical performance under various operating conditions, and strategies for maximizing catalyst utilization and reducing costs.
- **Gas Diffusion Layers:** An in-depth discussion on the water management, gas transport mechanisms, and structural properties of electrospun GDLs, focusing on enhancing overall fuel cell efficiency and durability.
- **Proton Exchange Membranes:** Insights into the development of electrospun proton exchange membranes, including advances in proton conductivity, water uptake, swelling behavior, and thermal stability. The text highlights the benefits of solution and centrifugal electrospinning techniques in creating high-performance, cost-effective membranes.

This book is designed to serve as a valuable resource for researchers, engineers, and professionals in the energy and materials fields. By delving into the most recent advancements and practical applications of electrospinning technology, this volume aims to inspire innovation, encourage collaboration, and ultimately accelerate the transition to sustainable energy solutions.

Beijing
13 May 2025

Prof. Yong Liu
Dr. Mohideen Meerasahib Mohamedazeem

Acknowledgment

Embarking on this book was an exciting challenge, taking years of dedicated research, countless experiments, and hard-earned insights and shaping them into a clear and compelling narrative. Our aim has always been to translate complex discoveries into chapters that not only tell the story of our journey but also make a meaningful impact in the field. In this process, many of the insights and findings presented in this book are grounded in the innovative work of students passed through our lab: Shaopeng Zhang, Han Ding, Chaoming Li, Xin Li, Tianya Li, Dingbo Han, Caihan Zhu, Jingyi Sun, Jihao Wang, and Jinghui Song, whose research and experiments contributed significantly to our understanding and approach. We're also immensely grateful to Professor Jianbo Zhang of Tsinghua University, who support several postgraduate students studied in his lab. Thanks the Beijing University of Chemical Technology for the unwavering support, which has allowed us to push boundaries, pursue bold ideas, and bring this vision to life.

Prof. Yong Liu
Dr. Mohideen Meerasahib Mohamedazeem

1

Introduction to Proton Exchange Membrane Fuel Cells

1.1 Overview of Proton Exchange Membrane Fuel Cells Technology

Proton exchange membrane fuel cells (PEMFCs) are at the forefront of clean energy innovation, offering a sustainable and efficient means of power generation that is pivotal in addressing the challenges of global energy demands and environmental sustainability. PEMFCs work by converting the chemical energy of hydrogen directly into electrical energy through an electrochemical process, with water and heat as the only by-products. PEMFCs have widespread attention among clean energy technologies, renowned for their high efficiency, low operating temperatures, and rapid start-up times, making them suitable for a wide range of applications, from automotive and stationary power generation to portable power devices [1, 2]. As the world increasingly shifts toward renewable energy sources, PEMFCs provide a compelling alternative to traditional fossil fuel–based energy systems. Their ability to generate power with zero-greenhouse-gas emissions makes them an essential component in the global efforts to combat climate change and reduce the carbon footprint of energy production.

1.1.1 Brief History and Development of Proton Exchange Membrane Fuel Cell

The concept of fuel cells dates back to the nineteenth century when Sir William Grove first demonstrated the "gaseous voltaic battery" in 1839. However, it wasn't until the mid-twentieth century that significant advancements were made in fuel cell technology, primarily driven by the need for efficient power sources in space missions. While PEMFCs are a distinct type of fuel cell, their development is intrinsically linked to hydrogen as the primary fuel source. The exploration of hydrogen as a clean energy carrier has paralleled the advancement of PEMFC technology [3]. This synergy is crucial for the widespread adoption of fuel cell technologies, as hydrogen provides the necessary energy source that powers PEMFC systems. However H_2 as a low-carbon-energy system is not new; the wave of enthusiasm

Electrospinning for Proton Exchange Membrane Fuel Cells, First Edition.
Yong Liu and Mohideen Meerasahib Mohamedazeem.

for hydrogen began in the early 1970s due to the first global energy demand and environmental crisis; thus, many hydrogen energy-based research programs were launched. For instance, the International Energy Agency (IEA) was established in 1974, the *International Journal of Hydrogen Energy* launched in 1976, and the IEA Hydrogen and Fuel-Cell Technology Collaboration Programme in 1977. Since then, a growing number of researchers, international organizations, and companies have supported the hydrogen-based economy to address the fossil fuel demand and control greenhouse gas (GHG) emissions [4]. The historical key milestone of hydrogen and fuel cells is depicted in Figure 1.1.

Today, hydrogen and fuel cell technologies offer a promising path toward sustainable energy. Ongoing research and development aim to overcome current challenges, such as reducing the reliance on precious metals and improving system efficiency and durability. Continued investment in research and infrastructure is essential to realizing the full potential of hydrogen as a clean energy carrier. Moreover, hydrogen is a versatile energy carrier, not an energy source, having potential applications to provide energy services across all sectors: transportation, power, building, and industry. Hydrogen is itself a carbon-free carrier, but a significant amount of carbon footprint occurs during its production [1, 4]. Presently, hydrogen is produced from various primary energy sources such as fossil fuels, biomass, and renewables. Low-carbon technological options like carbon capture and storage (CCS) and electrolyzers require extensive development to reduce carbon dioxide (CO_2) emissions during hydrogen production. As an energy carrier, hydrogen can overcome the variable renewable energy flexibility issue on the energy supply-and-demand side by enabling the linkage between them by connecting various transmission and distribution networks [5]. Hydrogen and fuel cell technologies are an excellent solution to decarbonize the transport sector. Green hydrogen produced from renewable energy sources such as solar photovoltaic (PV) and wind power could be integrated into fuel cell vehicles as a promising alternative to internal combustion engines [1, 5].

1.1.2 Need for H_2-Powered Fuel Cell Technology in Today's World

In the face of addressing global environmental challenges and an urgent need for sustainable energy solutions, hydrogen-powered fuel cell technology has emerged as a critical component in the global energy transition. The pressing issues of climate change, air pollution, and energy security have highlighted the necessity of moving away from fossil fuel dependency and toward cleaner, renewable energy sources [6]. The transport sector is one of the fast-growing and key driving forces of anthropogenic environmental pressure, contributing nearly 20% of GHG emissions, with fossil fuels accounting for almost 90% of total energy consumption. One-quarter of all energy-related GHG emissions are related to transportation, with around 72% of these overall emissions generated from road transport. Recently, the COVID-19 pandemic decreased global transport emissions by 10%, or 7.2

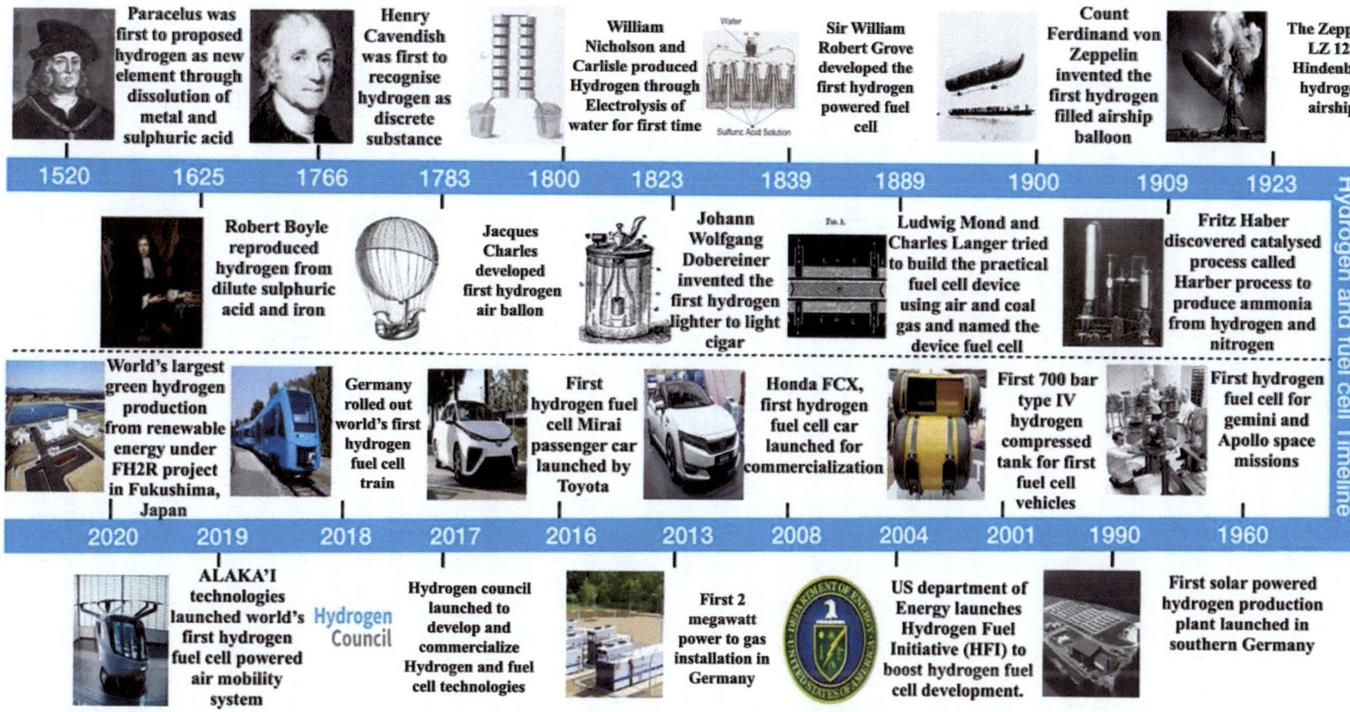

Figure 1.1 Hydrogen and fuel cell technology timeline. Source: Mohideen et al. [4]/with permission of Elsevier.

gigatonne (Gt) of carbon dioxide (CO_2) in 2020, compared to 8.5 Gt of CO_2 in 2019, due to the pandemic-related restrictions on domestic and international travel. However, in 2022, the CO_2 emissions associated with the transport sector began to rebound with the lifting of travel restrictions around the world [1, 4].

Today, a well-to-wheel passenger vehicle generates an average of 300 g of CO_2 per kilometer. According to the IEA, achieving a net-zero-emission transport sector requires a further drop in carbon emissions by 20% to 5.7 Gt by 2030 from the current level. Achieving such a drastic emission drop highly depends on the policies driving the revolutionary shift from fossil-fuel-based vehicles to clean and green mobility. Today, several low-carbon road transport technologies are readily available in the market, including plug-in hybrid vehicles, battery electric vehicles, and fuel cell electric vehicles (FCEVs). However, all these technologies are at various commercialization phases, and the significance of each in the sustainable future transport sector is a topic of discussion. Hydrogen-powered FCEV has the potential to decarbonize the transport sector compared to other low-carbon transport technologies. The proton exchange membrane (PEM) fuel cell stack is mostly used in FCEVs, offering high power density, high efficiency, and cold-start capabilities. In addition, hydrogen-powered FCEVs have widespread advantages, including high range and short refueling time (~500 km and 3 min), high well-to-wheel efficiency, smooth operation (low noise), and quick start-up. Owing to such overwhelming benefits, hydrogen-powered FCEVs will be a competitive edge in future transportation, especially for heavy-duty vehicles, trucks, buses, maritime shipping, and aviation sectors. However, compared to gasoline-powered and electric-powered vehicles, the number of FCEVs on the road is comparatively small, due to its high production cost of materials and infrastructure.

On the road to net-zero-emission scenario, it is anticipated that the hydrogen demand will reach almost 2.6% of the total transport sector energy demand by 2030 and over one-quarter by 2050. Nevertheless, in the current scenario, the use of hydrogen in the transport sector is much lower than that in other sectors, accounting for <0.01% of the energy consumed. For instance, if all the 1 billion cars, 25 million buses, and 190 million trucks on the road are replaced by FCEVs in the future, then the demand for global green hydrogen will be fourfold higher than the current level. From 2017 to 2020, the global share of FCEVs stock skyrocketed by an annual average of 70%, and as of June 2021, more than 40,000 FCEVs were on the road, mostly in the United States and Japan. The United States accounts for the second largest FCEV fleet, with more than 9,200 vehicles sold at the end of 2020. Japan presently has 4100 FCEVs on the road and is targeted to manufacture 800,000 passenger light-duty vehicles (PLDVs) by 2030. It was followed by Korea, which took the lead in the FCEVs market between 2019 and 2020, and by the end of June 2020 alone, 4400 PLDVs had been registered, and the government targeted 200,000 and 2.9 million light-duty FCEVs by 2025 and 2040, respectively.

Although FCEVs hold tremendous advantages in decarbonization transport sector, the cost of fuel cell components is a key barrier that hinders the widespread commercialization of FCEVs. For instance, the high cost of

Table 1.1 Approximate cost of platinum content in PEMFC vehicles based on current target platinum loading.

Stack power (kW)	Application	Example	Total Pt (g)	Total cost of Pt (USD)
1–25	Portable power and scooters		0.3–8	10–240
25–75	Range extenders, buses, and trucks		8–25	240–750
>75	Passenger vehicles		25–35	750–1050

Source: Banham et al. [7]/Elsevier/CC BY 4.0.

the expensive platinum metal used as a cathode catalyst in a fuel cell stack itself is responsible for 56% of its total cost of ownership. As shown in Table 1.1, for mid-sized passenger FCEVs, 25–35 g of Pt is required, which is 10-fold higher than a diesel autocatalyst. Therefore, reducing the platinum loading of the catalyst in the fuel cell stack is critical for the economic viability of FCEVs. In this direction, Daimler has significantly reduced platinum usage in their FCEVs by nearly 90% of platinum by ~8 g to just 5 g per vehicle [8]. Similarly, Toyota's Mirai FCEV cuts platinum utilization by 50% through alloying with cobalt (Pt/Co), an approach that plays.

In this context, electrospun nanofibers, prepared via the electrospinning technique, have garnered considerable attention to reduce the cost of the key components as well as improve the performance of fuel cell. The concept of electrospinning experienced a resurgence in the 1990s as nanomaterials became a research priority. The technique has since found diverse applications in biotechnology, filtration, energy storage, medicine, aerospace, and electronics [9]. As illustrated in Figure 1.2, a typical electrospinning setup includes a high-voltage power supply, a spinning apparatus (composed of a feed pump and syringe), and a collector. The process involves generating a strong electrostatic field between the spinning apparatus and the collector. When the electrostatic force acting on the polymer solution overcomes its surface tension and viscosity, a charged jet forms at the apex of the Taylor cone. This jet stretches, undergoes solvent evaporation and solidification, and finally deposits on the collector as nanofibers. The resulting fiber diameters typically range from tens of nanometers to several microns [11].

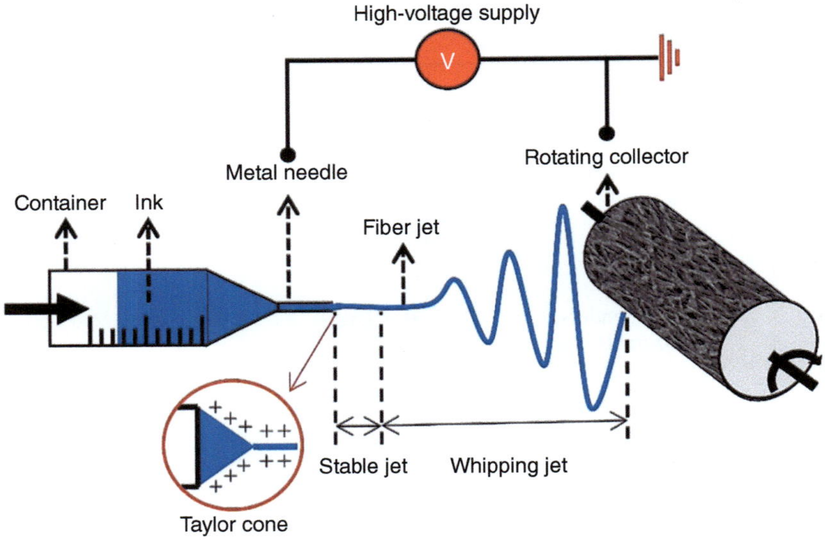

Figure 1.2 Schematic diagram of electrospinning device. Source: Yong et al. [10]/with permission of Elsevier.

1.2 Key Components of Proton Exchange Membrane Fuel Cell

PEMFCs are complex systems that rely on the seamless integration of several critical components to convert hydrogen into electrical energy efficiently. Each component plays a vital role in ensuring the effective operation of the fuel cell, with maximum efficiency and durability. A deeper understanding of these components is crucial for advancing PEMFC technology and optimizing its performance across various applications. The core components of a PEMFC include the PEM, catalyst layer (CL), gas diffusion layer (GDL), bipolar plates, and the flow field. Together, these elements form a cohesive system that facilitates the electrochemical reactions necessary for power generation in FCEVs, as illustrated in Figure 1.3.

1.2.1 Catalyst Layer

The CL is typically a thin, porous structure consisting of catalyst particles, an ionomer, and a carbon support. The platinum or platinum alloy catalyst particles are dispersed on a high-surface-area carbon support to maximize the available catalytic surface area. The ionomer, often Nafion, is interspersed throughout the layer to aid proton conduction and ensure effective contact with the membrane [12–15]. The primary function of the CL is to accelerate the electrochemical reactions within the fuel cell, ensuring efficient conversion of hydrogen and oxygen into water and electrical energy. The efficiency and performance of the PEMFC are heavily dependent on the activity and stability of the CL. The catalysts must exhibit high

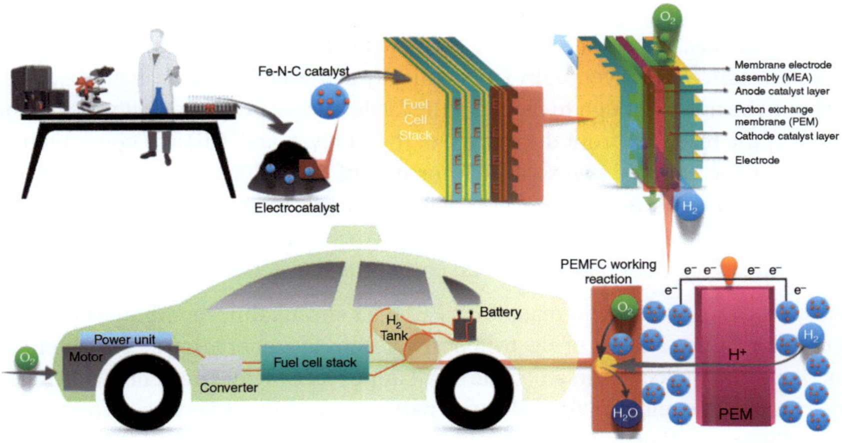

Figure 1.3 Schematic illustration of key components in FCEVs. Source: Mohideen et al. [8]/with permission of Elsevier.

catalytic activity to facilitate fast reaction rates while maintaining stability under the operating conditions of the fuel cell [16].

The thickness of the CL is a crucial factor influencing its performance. Thicker CLs provide a higher total catalytic surface area per geometric area, which can enhance performance by increasing reaction rates. However, they also introduce higher resistance to the transport of electrons, protons, reactants, and products, potentially negatively affecting overall fuel cell performance [17–19]. This trade-off between surface area and transport resistance necessitates careful optimization of the CL thickness. The porosity of the CL is approximately 30%, with a hierarchical pore structure that includes primary pores less than 10 nm and secondary pores less than 100 nm. This porosity facilitates the effective transport of reactants and products, as well as the distribution of gases across the catalyst surface. Achieving the optimal balance of porosity and pore size is critical for maintaining efficient mass transport and minimizing diffusion losses [20].

At the CL, two critical electrochemical reactions occur that convert chemical energy into electrical energy:

(i) **Hydrogen Oxidation Reaction (HOR)**: At the anode, hydrogen molecules (H_2) are split into protons (H^+) and electrons (e^-) by the catalyst. The protons migrate through the PEM, whereas the electrons are directed through an external circuit, providing electric power. HOR reaction can be represented by the following equation:

$$H_2 \rightarrow 2H^+ + 2e^-$$

(ii) **Oxygen Reduction Reaction (ORR)**: ORR occurs at the cathode of the PEMFC, where oxygen molecules are reduced to form water. The ORR kinetics at the cathode is slower than that at the anode even though the Pt catalyst loading at the cathode was 10 times higher than those at the anode. The reason for sluggishness and high catalyst loading of the cathode is the grand challenges of ORR, namely, that (i) the cathode catalyst should withstand extreme corrosive circumstances and eventually be chemically active enough to activate O_2, (ii) it should be noble enough to facilitate the facile release of by-product water at the end of the reactions from the surface of the catalyst, and, (iii) as a consequence, ORR at the cathode is responsible for half of the voltage loss in the PEMFC system, whereas the voltage loss of HOR is considerably small even with a very low Pt loading of 0.05 mg/cm^2. As a consequence, the high loading of Pt accounts more than half of the total cost of the fuel cell [8].

ORR is the most prominent reaction in PEMFCs and alkaline membrane exchange fuel cells, involving a series of electron transfer steps by undergoing electrochemical reduction of O_2 molecules in an acidic or alkaline electrolyte. The general ORR catalytic reactions are either a four-electron pathway ($4e^-$) or a two-step, two-electron ($2e^-$) pathway, as shown in the following equations:

(i) The direct $4e^-$ pathway: It is a universally accepted reaction process that directly reduces the O_2 molecules to H_2O in acid media, and to OH^- in alkaline media.
$4e^-$ pathway in acid medium:

$$O_2 + 4H^+ + 4e^- \rightarrow 2H_2O$$

$4e^-$ pathway in alkaline medium:

$$O_2 + 2H_2O + 4e^- \rightarrow 4OH^-$$

(ii) The two-step, two-electron ($2e^-$) pathway: The reaction process includes a series of steps by producing H_2O_2 and then reduce to H_2O.
$2e^-$ pathway in acid medium:

$$O_2 + 2H^+ + 2e^- \rightarrow H_2O_2$$

$$H_2O_2 + 2H^+ + 2e^- \rightarrow 2H_2O$$

$2e^-$ pathway in alkaline medium:

$$O_2 + H_2O + 2e^- \rightarrow HO_2{}^- + OH^-$$

$$HO_2{}^- + H_2O + 2e^- \rightarrow 3OH^-$$

However, in the $4e^-$ pathway, two types of possible reaction kinetics were proposed, such as the dissociative and associative reaction pathways, as shown in Table 1.2.

It is essential to note that the dissociating energy barrier of the initial O_2 adsorption molecule plays a crucial role in deciding whether the catalyst undergoes an associative or dissociative pathway. However, due to the high O_2 dissociation barriers on metal-N_x moieties, the associative mechanism is more suitable for nonprecious metal catalysts in both acidic and alkaline environments [21]. Both mechanisms involve various intermediate species during the reaction pathways, including hydroxyl (OH^*), oxygenated (O^*), and super hydroxyl (OOH^*), and their corresponding transformations during the reaction make the ORR mechanism more complicated. Initially, O_2 molecule adsorption takes place, followed by weakening of the O–O bond and cleaving it to form two atomic O_2^* species that leads to the direct formation of OH^- without generating OOH^- [22–25].

Table 1.2 Associative and dissociative mechanisms of ORR in acid and alkaline electrolytes.

Acid	Alkaline
$4e^-$ associative mechanism	$4e^-$ associative mechanism
$O_2(g) + H^+ + e^- + ^* \rightarrow OOH^*$	$O_2(g) + H_2O + e^- + ^* \rightarrow OOH^* + OH^-$
$OOH^* + H^+ + e^- \rightarrow O^* + H_2O$	$OOH^* + e^- \rightarrow O^* + OH^-$
$O^* + H^+ + e^- \rightarrow OH^*$	$O^* + H_2O + e^- \rightarrow OH^* + OH^-$
$OH^* + H^+ + e^- \rightarrow ^* + H_2O$	$OH^* + e^- \rightarrow ^* + OH^-$
$4e^-$ dissociative mechanism	$4e^-$ dissociative mechanism
$O_2(g) + ^* \rightarrow O_2^*$	$O_2(g) + ^* \rightarrow O_2^*$
$O_2^* \rightarrow O^* + O^*$	$O_2^* \rightarrow O^* + O^*$
$O^* + H^+ + e^- \rightarrow OH^*$	$O^* + H_2O + e^- \rightarrow OH^* + OH^-$
$OH^* + H^+ + e^- \rightarrow H_2O + ^*$	$OH^* + e^- \rightarrow ^* + OH^-$
$2e^-$ associative mechanism	$2e^-$ associative mechanism
$O_2(g) + H_+ + e^- + ^* \rightarrow OOH^*$	$O_2(g) + H_2O + e^- + ^* \rightarrow OOH^* + OH^-$
$OOH^* + H^+ + e^- + ^* \rightarrow H_2O_2 + ^*$	$OOH^* + e^- \rightarrow HO_2^- + ^*$

Concerning the above-mentioned ORR mechanisms, it is clear that the performance of the ORR pathway depends on the nature of the electrocatalyst and electrolyte. Therefore, to reduce Pt usage without compromising the stability, durability, and ORR performance of fuel cell [26], the appropriate electrocatalysts can be developed using the following strategies: (i) alloying Pt with other transition metal, (ii) replacing Pt with nonprecious transition metals, and (iii) metal-free carbon-based electrocatalyst.

1.2.2 Gas Diffusion Layer

GDL is a vital component of PEMFCs, ensuring effective transport of gases and water between the CL and the flow field channels. Situated between the CL and the bipolar plates, the GDL facilitates uniform reactant distribution, provides electrical conductivity, and optimizes water management—factors essential for reliable fuel cell operation (Figure 1.4) [27–30].

Typically made from porous carbon materials such as carbon paper or carbon cloth, the GDL offers high permeability and excellent electrical conductivity. Carbon cloth, being more porous and less tortuous than carbon paper, exhibits greater compressibility. However, its higher compressibility can cause material to intrude into flow field channels, leading to uneven gas flow and performance variations [31, 32]. In contrast, carbon fiber paper is often preferred to the macroporous layer due to its superior mechanical strength. Its natural brittleness ensures higher compressive strength, supporting uniform gas flow and stable mechanical integrity within the fuel cell.

Water management is a critical function of the GDL, as it must maintain the membrane's hydration while preventing flooding at the CL. Effective water handling is key to achieving both high performance and long-term durability [33]. To this end,

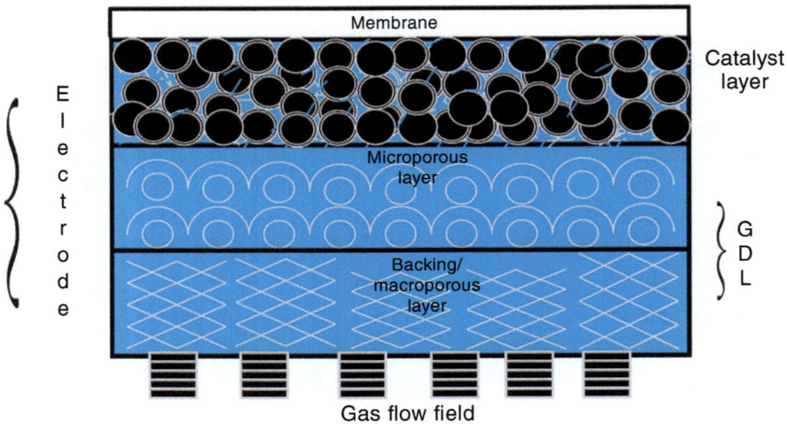

Figure 1.4 Schematic illustration of GDL.

GDLs are frequently treated with hydrophobic coatings—commonly polytetrafluoroethylene (PTFE)—to enhance water removal and prevent water accumulation. The GDL's porous structure not only facilitates the diffusion of hydrogen and oxygen to the CL but also aids in removing water vapor and any excess liquid water. Properly optimized porosity is essential to strike a balance between efficient gas transport and effective water management [34].

The thickness of the GDL typically falls within the range of 200–400 μm. While a thicker GDL can improve mechanical support and enhance water management, it can also increase resistance to gas diffusion. Thus, achieving the right balance of thickness and porosity is crucial for maximizing PEMFC performance [35].

1.2.3 Proton Exchange Membrane

The PEM is the core component of a PEMFC, serving as the electrolyte that facilitates proton transport while preventing the crossover of gases. The primary role of PEMs is to isolate gases and insulate electrons, allowing protons (H^+) to be transferred from the anode to the cathode, where they recombine with reduced oxygen atoms (O_2^-) to form water [36, 37]. If oxygen atoms on the cathode are not fully reduced or react with hydrogen permeating from the anode, hydrogen peroxide (H_2O_2) can form, which then decomposes into corrosive peroxide radicals (•OOH) and hydroxyl radicals (•OH). Under high temperatures and pressures within the fuel cell, these generated radicals accelerate the degradation of polymer materials. Given these circumstances, PEMs must meet strict requirements to ensure high efficiency and durability of the device. An ideal PEM should have high proton conductivity, good mechanical strength, high thermal stability, chemical stability, and low cost [38, 39].

Beyond conducting protons, PEMs also prevent electron transfer and maintain gas separation, ensuring that protons reach the cathode where they combine with oxygen to form water. However, if oxygen does not fully reduce or reacts with hydrogen crossing the membrane, it can generate hydrogen peroxide (H_2O_2). At high temperatures and pressures, H_2O_2 decomposes into highly reactive radicals (•OH and •OOH) that degrade the membrane material. Thus, PEMs must be highly durable and meet rigorous standards for chemical, thermal, and mechanical stability to ensure long-term fuel cell performance.

In practical applications, PEMs need to deliver not only exceptional performance but also cost-effectiveness. They should have proton conductivity above 100 mS/cm, excellent electrical insulation, and minimal gas permeability, with hydrogen leakage currents under 2 mA/cm^2. Additionally, they must remain stable under hydrolytic, thermal, and oxidative conditions while maintaining sufficient water content to support conductivity even at low humidity. Robust mechanical properties in the hydrated state are crucial for reliable membrane electrode assembly operation, and keeping production costs below \$20/m^2 is a key economic target. By meeting these stringent requirements, PEMs can ensure high efficiency, stability, and affordability in PEM fuel cell systems.

1.2.3.1 Proton Transport Mechanisms

Proton conductivity is a critical parameter for evaluating PEMs. To enhance proton transport within the membrane, it is essential to first understand the underlying transport mechanisms. As illustrated in Figure 1.5, proton transport in PEMs primarily occurs through two mechanisms: the hopping mechanism (also known as the Grotthuss mechanism) [40] and the vehicle mechanism [41].

In the hopping mechanism (see Figure 1.5a), proton transport is achieved through the breaking and forming of hydrogen bonds as protons move from one active site to another. Initially, protons generated at the anode via oxidation reactions combine with water molecules to form hydronium ions. When a new proton appears nearby, the hydronium ion becomes unstable, causing a hydrogen bond to break. This releases the proton, which then "hops" to a neighboring ion site, forming a new hydrogen bond. While water molecules serve as carriers for transient hydrogen bonding, this mechanism can still function under high-temperature, low-humidity conditions. In fact, even in anhydrous environments, protons can bind to active carriers such as $-SO_3H$, $-NH_2$, $-COOH$, and $-PO_3H$, migrating to another active site via the same bond-breaking and bond-forming processes [42].

In the vehicle mechanism (see Figure 1.5b), protons associate with water molecules to form hydrated hydrogen ions (e.g. H_3O^+ or $H_5O_2{}^+$) that diffuse as a

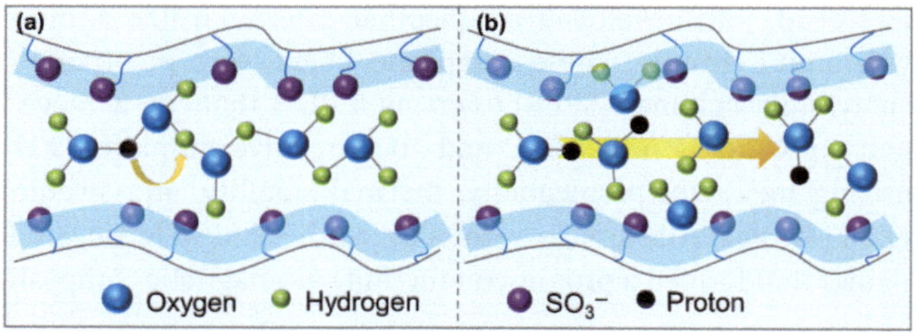

Figure 1.5 (a) Hopping mechanism and (b) vehicle mechanism.

single entity, without relying entirely on hydrogen bonds [43]. During this process, the proton and its associated water molecules move together, while unbound water molecules diffuse in the opposite direction. This results in a concentration gradient across the membrane, which facilitates the diffusion of both hydrated hydrogen ions and water molecules. For proton transport via the vehicle mechanism, the membrane's microstructure must contain continuous clusters of hydrated ions to provide pathways. However, under low-water-content conditions, the insufficient quantity and connectivity of these ion clusters hinder continuous proton transport, significantly reducing proton conductivity [44].

Proton transport is a complex process, typically occurring through both hopping and vehicle mechanisms. To enhance proton transport, three main approaches can be taken: increasing water content, optimizing proton transport channels, and increasing the number of proton carriers. Although increasing water content may appear to be the simplest method, doing so indiscriminately can lead to fragmented hydrophobic domains and decreased continuity of hydrophilic domains. This limitation not only reduces the extent of improvement but also compromises the mechanical stability of the membrane. Consequently, optimizing proton transport channels and increasing the number of proton carriers are more practical strategies.

i) **Optimizing Proton Transport Channels**: Improving proton transport channels can enhance the utilization of acid–water interactions, thereby increasing proton conductivity while maintaining appropriate water content [45]. Broadly speaking, proton transport channels refer to hydrophilic domains formed by acidic groups in the polymer chains. Under hydration, these acidic groups dissociate, generating proton carriers that migrate with the aid of hydrodynamic forces. In fact, proton transport channels can take various forms. When certain segments of a polymer's main chain, side chain, or functional groups exhibit differing hydrophilic and hydrophobic properties, microphase separation can occur, creating nanoscale interpenetrating regions [46, 47].

   However, traditional PEM materials like perfluorosulfonic acids or sulfonated hydrocarbon polymers tend to form "dead zones" that interrupt channel continuity, impeding smooth proton migration. Adding fillers—such as nanofibers, nanotubes, graphene, or metal-organic frameworks (MOFs)—can help eliminate these dead zones, transforming the convoluted channels into streamlined highways for proton transport.

   For example, Li et al. [48] developed bio-inspired caterpillar-like alumina fibers loaded with amine groups. These composite fibers were incorporated into sulfonated poly(ether sulfone) (SPES) to create a PEM that can achieve a proton conductivity of 0.263 S/cm at 80 °C and 100% relative humidity (RH), while also improving methanol permeability, thermal stability, and swelling resistance. Liu et al. [49] achieved similar results by synthesizing a ferricyanide coordination polymer that formed a proton-conducting paramagnetic composite with phosphotungstic acid. By applying a magnetic field during the polymer casting process, they aligned the proton transport channels vertically, overcoming phosphotungstic acid loss. This design, supported by redox cycling of cyanide, produced a PEM with a maximum power density of 1107 mW/cm^2 and exceptional durability.

ii) **Increasing Proton Carriers**: Adding more proton carriers is another effective way to improve proton conductivity. Proton carriers generally fall into three categories: acidic, basic, and acid–base pair carriers. Acidic carriers, the earliest developed proton carriers, offer relatively high proton conductivity (10^{-2}–10^{-1} S/cm). They use acidic functional groups like $-SO_3H$ or $-PO_3H_2$ to donate protons, thereby facilitating proton transport. However, excessive acidic groups can lead to excessive water absorption and swelling, which reduces acid concentration and dimensional stability, ultimately affecting the overall performance of the membrane [50]. Basic carriers, on the other hand, act as proton acceptors (e.g. $-NH_2$ or $-NH^-$), initially capturing free H^+ ions to form $-NH_3^+$. These then form intermediates such as $-NH_3^+\ldots OH^-\ldots H^+$, allowing H^+ ions to dissociate and transfer. Although basic carriers can transport protons, the strong electrostatic interactions between the basic groups and H^+ ions make proton dissociation more difficult, resulting in proton conductivities of less than 10^{-2} S/cm [51].

In contrast, acid–base pair carriers, a newer class of proton carriers, contain both acidic and basic groups. The electrostatic interactions between these groups facilitate proton hopping from acidic donor sites to basic acceptor sites. The electrostatic pairing also lowers the energy barrier for proton migration, thereby accelerating the process [52]. For instance, Sun et al. [53] synthesized double-shelled nanotubes with a carboxylate inner shell and an imidazole outer shell, which were then incorporated into sulfonated poly(ether ether ketone) (SPEEK) membranes. The imidazole outer shell formed acid–base pairs with both the inner shell and the SPEEK matrix, significantly enhancing proton hopping. The resulting membrane exhibited a proton conductivity of 0.336 S/cm at 80 °C and 100% RH, twice that of the original SPEEK membrane.

Similarly, Chen's team [54] combined sulfonic acid–functionalized graphitic carbon nitride nanosheets with sulfonated aromatic polymers. The acid–base pairs between the nanosheets and the matrix promoted sulfonate group dissociation and hydrogen-bond network formation. This "zipper-like" interface interaction induced phase separation and created long-range ionic pathways, resulting in excellent fuel cell performance of 717 mW/cm at 80 °C and 100% RH. Qu et al. [55] developed an acid–base composite membrane made from a carboxyl-containing polyimide and a basic polybenzimidazole for high-temperature PEM applications. Through electrostatic interactions and a continuous hydrogen-bond network, this membrane exhibited improved proton conductivity and enhanced phosphoric acid retention.

1.2.3.2 Classification of PEM

PEMs are a critical component of PEMFCs, directly determining the fuel cell performance. Consequently, PEMs must meet the following criteria [56, 57]:

1) High proton conductivity without electron conduction.
2) Excellent mechanical strength and dimensional stability, maintaining integrity under operational conditions.
3) Outstanding chemical and thermal stability, resisting degradation in harsh environments.
4) Low fuel crossover, effectively blocking reactants.
5) Low cost.

PEMs can be broadly categorized into homogeneous membranes and composite membranes based on their composition.

i) **Homogeneous Membranes**: Homogeneous membranes consist of a single type of material and can be further divided into fully fluorinated, partially fluorinated, and nonfluorinated membranes based on their polymer fluorine content. Nafion, the most common fully fluorinated sulfonic acid membrane, has a structure as shown in Figure 1.6a. Its hydrophobic PTFE backbone provides mechanical strength, while the hydrophilic side chains, terminated with sulfonic acid groups, offer proton transport sites. The distinct hydrophilic–hydrophobic nature of its main and side chains leads to phase separation in the microstructure, forming proton transport channels approximately 5–10 nm in diameter. High sulfonic acid group content endows Nafion with excellent proton conductivity, exceeding 100 mS/cm at 80 °C [59]. Thanks to its PTFE backbone, Nafion also boasts good mechanical, chemical, and thermal stability. However, Nafion's performance is highly sensitive to temperature and water content. Its low glass transition temperature limits its operating temperature to

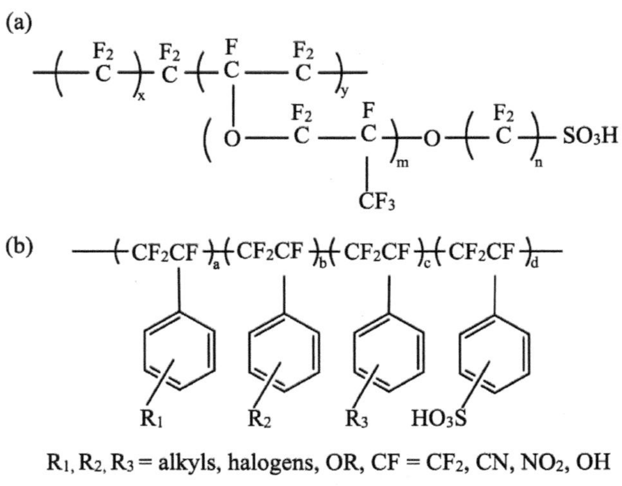

Figure 1.6 Structural formula of (a) Nafion and (b) BAM-3G. Source: Adapted from Basura et al. [58].

below 80 °C, and its proton conductivity drops significantly under low-humidity conditions due to reduced phase separation. Additionally, its high cost and significant fuel crossover impede further application [60–62].

To lower costs and simplify fabrication, researchers have increasingly turned to non-perfluorinated sulfonic acid PEMs. Partially fluorinated sulfonic acid PEMs are one example. These materials are similar to fully fluorinated membranes, with PTFE or polyvinylidene fluoride (PVDF) backbones and non-fluorinated aromatic groups in the side chains. For instance, the BAM-3G series PEMs from Canada's Ballard Power Systems are synthesized by copolymerizing substituted trifluorostyrenes and sulfonated trifluorostyrenes, as shown in Figure 1.6b. These membranes feature low sulfonic acid content, high efficiency, and extended lifetimes (up to 15,000 hours for single cells) at a fraction of Nafion's cost, making them more accessible [58].

Another major category of homogeneous membranes is non-fluorinated sulfonic acid PEMs, which mainly include sulfonated aromatic polymers such as SPEEK, SPES, sulfonated polyimides, and sulfonated polybenzimidazoles [63–66]. Compared to aliphatic polymers, these polymers contain a higher proportion of rigid aromatic groups in their molecular backbones, providing the membranes with good mechanical properties and thermal stability. They can easily be sulfonated to introduce sulfonic acid groups, are derived from widely available materials, and have relatively simple preparation methods. Moreover, the absence of costly fluorination steps significantly reduces their production costs. However, due to shorter side chains, less-pronounced phase separation, and a lower sulfonic acid group content compared to perfluorosulfonic acid membranes, nonfluorinated sulfonic acid PEMs generally exhibit lower proton conductivity [67]. Additionally, they tend to have high humidity requirements and limited durability, which remain challenges to overcome.

ii) **Composite Membranes**: Whether fully fluorinated, partially fluorinated, or nonfluorinated sulfonic acid membranes, each type has its own strengths and weaknesses, making it difficult to find a homogeneous PEM that excels in all aspects. Modifying a homogeneous PEM—such as by incorporating fillers—can produce composite membranes that address these limitations. Composite membranes offer unique advantages over homogeneous membranes, including the ability to integrate the benefits of the fillers and to tailor performance improvements for specific needs. Depending on the nature of the filler, modification methods can be categorized as either organic–inorganic or organic–organic approaches.

a) **Organic–Inorganic Modification**: Introducing inorganic components can effectively enhance proton conductivity, particularly for sulfonated polymers under low-humidity conditions. Common inorganic fillers include silica [68], heteropoly acids [69], carbon materials [34], and MOFs [70]. These fillers can improve water retention, dimensional stability, mechanical strength, and methanol permeability to varying degrees. Inorganic fillers can be introduced by methods such as doping, sol-gel processes, or in situ synthesis.

For example, Yan et al. [71] prepared sulfonated mesoporous silica frameworks with controlled channel sizes of approximately 0.5 nm, slightly larger than the diameter of H_3O+ ions (0.28 nm) and slightly smaller than methanol molecules (0.4 nm). These molecular sieve-like membranes, when combined with Nafion, showed methanol permeability three orders of magnitude lower than Nafion. Pei et al. [72] introduced CeO_2-based core–shell micro/nanostructures rich in tourmaline into perfluorosulfonic resin, resulting in a composite membrane with excellent proton conductivity and chemical stability. At 80 °C and 100% RH, the composite membrane achieved a maximum power density of 1006 mW/cm^2, higher than Nafion's 906 mW/cm^2 under the same conditions. Additionally, CeO_2 helps scavenge hydroxyl radicals, slowing degradation and improving the membrane's durability.

b) **Organic–Organic Modification**: Organic modifications involve incorporating functional organic materials into the polymer matrix. This approach can compensate for the inherent shortcomings of the polymer matrix while selectively enhancing specific properties. Organic fillers can be divided into small organic molecules and macromolecular organic substances [73–75].

Small organic molecules include covalent-organic frameworks, hydrogen-bonded organic frameworks, and ionic liquids (ILs), while macromolecular organic fillers include PVDF, polyvinyl alcohol, and sulfonated polymers. These fillers interact with the polymer matrix through various mechanisms, supplementing missing functions and enhancing existing performance. For instance, Sun's group [41] hydrogen-bonded Nafion with polyvinyl alcohol and then post-modified it with 4-carboxybenzaldehyde to produce a repairable PEM. This PEM demonstrated superior mechanical stability (with a tensile strength of 20.3 MPa and strain of 380%), improved proton conductivity (0.11 S/cm at 80 °C, 1.2 times that of Nafion), and resistance to methanol permeation. Under methanol fuel cell conditions, the membrane could self-heal mechanical damage up to tens of microns in size.

Zhao and co-workers [76] embedded Nafion nanofibers into a polybenzimidazole matrix to create a Nafion/polybenzimidazole composite membrane. The polybenzimidazole provided high ion selectivity, while the fibrous Nafion ensured high proton conductivity and low cost. Compared to Nafion 212, the composite membrane, with only 40 wt% Nafion, exhibited a 58-fold reduction in vanadium crossover and achieved 99.8% coulombic efficiency and 80.0% energy efficiency at a high current density of 210 mA/cm^2 in vanadium redox flow batteries.

References

1 Mohideen, M.M., Subramanian, B., Sun, J. et al. (2023). Techno-economic analysis of different shades of renewable and non-renewable energy-based hydrogen for fuel cell electric vehicles. *Renewable and Sustainable Energy Reviews* 174: 113153.

2 Cano, Z.P., Banham, D., Ye, S. et al. (2018). Batteries and fuel cells for emerging electric vehicle markets. *Nature Energy* 3: 279–289.

3 Pollet, B.G., Kocha, S.S., and Staffell, I. (2019). Current status of automotive fuel cells for sustainable transport. *Current Opinion in Electrochemistry* 16: 90–95.
4 Mohideen, M.M., Ramakrishna, S., Prabu, S. et al. (2021). Advancing green energy solution with the impetus of COVID-19 pandemic. *Journal of Energy Chemistry* 59: 688–705.
5 Staffell, I., Scamman, D., Abad, A.V. et al. (2019). The role of hydrogen and fuel cells in the global energy system. *Energy & Environmental Science* 12: 463–491.
6 Mohideen, M.M., Liu, Y., and Ramakrishna, S. (2020). Recent progress of carbon dots and carbon nanotubes applied in oxygen reduction reaction of fuel cell for transportation. *Applied Energy* 257: 114027.
7 Banham, D., Zou, J., Mukerjee, S. et al. (2021). Ultralow platinum loading proton exchange membrane fuel cells: performance losses and solutions. *Journal of Power Sources* 490: 229515.
8 Mohideen, M.M., Radhamani, A.V., Ramakrishna, S. et al. (2022). Recent insights on iron based nanostructured electrocatalyst and current status of proton exchange membrane fuel cell for sustainable transport. *Journal of Energy Chemistry* 69: 466–489.
9 Wang, Y. (2018). A brief history and application of electrospinning technology development. *China Synthetic Fiber Industry* 41 (04): 52–57.
10 Liu, Y., Ding, H., Ding, D. et al. (2018). Overview of electrospinning preparation of proton exchange membrane fuel cell catalyst layer. *Journal of Electrochemistry* 24 (06): 639–654.
11 Lv, Y., Sang, X., Tian, Z. et al. (2022). Electrospun hydroxyapatite loaded L-polylactic acid aligned nanofibrous membrane patch for rotator cuff repair. *International Journal of Biological Macromolecules* 217: 180–187.
12 Wang, Y., Chen, K.S., Mishler, J. et al. (2011). A review of polymer electrolyte membrane fuel cells: technology, applications, and needs on fundamental research. *Applied Energy* 88 (4): 981–1007.
13 Larminie, J. and Dicks, A. (2003). *Fuel Cell Systems Explained*, 2e, 1–43. Antony Rowe Ltd: Chippenham.
14 Shui, J., Wang, M., Du, F. et al. (2015). N-doped carbon nanomaterials are durable catalysts for oxygen reduction reaction in acidic fuel cells. *Science Advances* 1 (1): e1400129.
15 Qin, C., Wang, J., Yang, D. et al. (2016). Proton exchange membrane fuel cell reversal: a review. *Catalysts* 6 (12): 197.
16 Tamura, T. and Kawakami, H. (2010). Aligned electrospun nanofiber composite membranes for fuel cell electrolytes. *Nano Letters* 10 (4): 1324–1328.
17 Fofana, D., Hamelin, J., and Bénard, P. (2013). Modelling and experimental validation of high performance low platinum multilayer cathode for polymer electrolyte membrane fuel cells (PEMFCs). *International Journal of Hydrogen Energy* 38 (24): 10050–10062.
18 Tian, Z.Q., Lim, S.H., Poh, C.K. et al. (2011). A highly order-structured membrane electrode assembly with vertically aligned carbon nanotubes for ultra-low pt loading PEM fuel cells. *Advanced Energy Materials* 1 (6): 1205–1214.

19 Manikandan, T. and Ramalingam, S. (2016). A review of optimization algorithms for the modeling of proton exchange membrane fuel cell. *Journal of Renewable and Sustainable Energy* 8 (3): 1103–1112.

20 Li, Y., Li, Q., Wang, H. et al. (2019). Recent progresses in oxygen reduction reaction electrocatalysts for electrochemical energy applications. *Electrochemical Energy Reviews* 2: 518–538.

21 Zhong, H.-X., Zhang, Y., and Zhang, X.-B. (2018). Superior oxygen reduction electrocatalyst: hollow porous spinel microsphere. *Chem* 4: 196–198.

22 Thompson, S.T., Wilson, A.R., Zelenay, P. et al. (2018). ElectroCat: DOE's approach to PGM-free catalyst and electrode R&D. *Solid State Ionics* 319: 68–76.

23 Wang, X., Li, Z., Qu, Y. et al. (2019). Review of metal catalysts for oxygen reduction reaction: from nanoscale engineering to atomic design. *Chem* 5: 1486–1511.

24 Primbs, M., Sun, Y., Roy, A. et al. (2020). Establishing reactivity descriptors for platinum group metal (PGM)-free Fe–N–C catalysts for PEM fuel cells. *Energy & Environmental Science* 13: 2480–2500.

25 Huang, Z.-F., Song, J., Dou, S. et al. (2019). Strategies to break the scaling relation toward enhanced oxygen electrocatalysis. *Matter* 1: 1494–1518.

26 Ganesan, A. and Narayanasamy, M. (2019). Ultra-low loading of platinum in proton exchange membrane-based fuel cells: a brief review. *Materials for Renewable and Sustainable Energy* 8: 18.

27 Okonkwo, P.C. and Otor, C. (2021). A review of gas diffusion layer properties and water management in proton exchange membrane fuel cell system. *International Journal of Energy Research* 45 (3): 3780–3800.

28 Omrani, R. and Shabani, B. (2017). Gas diffusion layer modifications and treatments for improving the performance of proton exchange membrane fuel cells and electrolysers: a review. *International Journal of Hydrogen Energy* 42 (47): 28515–28536.

29 Cindrella, L., Kannan, A.M., Lin, J.F. et al. (2009). Gas diffusion layer for proton exchange membrane fuel cells—a review. *Journal of Power Sources* 194 (1): 146–160.

30 Athanasaki, G., Jayakumar, A., and Kannan, A.M. (2023). Gas diffusion layers for PEM fuel cells: materials, properties and manufacturing—a review. *International Journal of Hydrogen Energy* 48 (6): 2294–2313.

31 Jayakumar, A., Sethu, S.P., Ramos, M. et al. (2015). A technical review on gas diffusion, mechanism and medium of PEM fuel cell. *Ionics* 21: 1–18.

32 Zhang, G., Qu, Z., Tao, W.Q. et al. (2022). Porous flow field for next-generation proton exchange membrane fuel cells: materials, characterization, design, and challenges. *Chemical Reviews* 123 (3): 989–1039.

33 Ji, M. and Wei, Z. (2009). A review of water management in polymer electrolyte membrane fuel cells. *Energies* 2 (4): 1057–1106.

34 Madheswaran, D.K., Thangavelu, P., Krishna, R. et al. (2023). Carbon-based materials in proton exchange membrane fuel cells: a critical review on performance and application. *Carbon Letters* 33 (6): 1495–1518.

35 Zhang, B., Hua, Y., and Gao, Z. (2022). Strategies to optimize water management in anion exchange membrane fuel cells. *Journal of Power Sources* 525: 231141.

36 Wang, Y., Pang, Y., Xu, H. et al. (2022). PEM fuel cell and electrolysis cell technologies and hydrogen infrastructure development—a review. *Energy & Environmental Science* 15 (6): 2288–2328.

37 Qasem, N.A. (2024). A recent overview of proton exchange membrane fuel cells: fundamentals, applications, and advances. *Applied Thermal Engineering* 252: 123746.

38 Xie, Z., Wang, J., Zhao, G. et al. (2024). A review on durability of key components of PEM fuel cells. *Catalysis Science & Technology* 252.

39 Zhang, K., Liang, X., Wang, L. et al. (2022). Status and perspectives of key materials for PEM electrolyzer. *Nano Research Energy* 1 (3): e9120032.

40 Li, J., Lou, J., Wang, Z. et al. (2020). Facilitating proton transport with enhanced water conservation membranes for direct methanol fuel cells. *ACS Sustainable Chemistry & Engineering* 8 (15): 5880–5890.

41 Li, Y., Liang, L., Liu, C. et al. (2018). Self-healing proton-exchange membranes composed of nafion–poly(vinyl alcohol) complexes for durable direct methanol fuel cells. *Advanced Materials* 30 (25): 1707146.

42 Oka, T., Oshima, A., and Washio, M. (2024). Free volume study of sulfonated FEP proton exchange membrane. *Radiation Physics and Chemistry* 215: 111364.

43 Xu, G., Ke, A., Xu, G. et al. (2024). Enabling high-temperature application of Nafion membrane via imitating ionic clusters in proton conduction channels. *International Journal of Hydrogen Energy* 56: 330–337.

44 Guan, P., Lei, J., Liu, X. et al. (2022). Origins of water state and ionic cluster morphology for high proton conductivity of short side-chain perfluorinated sulfonic acid membranes. *Chemistry of Materials* 34 (17): 7845–7857.

45 Guan, P., Zou, Y., Zhang, M. et al. (2024). High-temperature low-humidity proton exchange membrane with "stream-reservoir" ionic channels for high-power-density fuel cells. Science. *Advances* 9 (17): eadh1386.

46 Golubenko, D.V., Shaydullin, R.R., and Yaroslavtsev, A.B. (2019). Improving the conductivity and permselectivity of ion-exchange membranes by introduction of inorganic oxide nanoparticles: impact of acid–base properties. *Colloid and Polymer Science* 297 (5): 741–748.

47 Chen, L., Ren, Y., Fan, F. et al. (2023). Artificial frameworks towards ion-channel construction in proton exchange membranes. *Journal of Power Sources* 574: 233081.

48 Li, L., Liu, X., Guo, Y. et al. (2022). Reasonable construction of proton conducting channel via biomimetic caterpillar-like alumina fiber to improve the properties of its composite proton exchange membrane. *International Journal of Hydrogen Energy* 47 (69): 29915–29924.

49 Liu, X., Li, Y., Xue, J. et al. (2019). Magnetic field alignment of stable proton-conducting channels in an electrolyte membrane. *Nature Communications* 10 (1): 842.

50 Wang, G., Kang, J., Yang, S. et al. (2024). Influence of structure construction on water uptake, swelling, and oxidation stability of proton exchange membranes. *International Journal of Hydrogen Energy* 50: 279–311.

51 He, Y., Wang, J., Zhang, H. et al. (2014). Polydopamine-modified graphene oxide nanocomposite membrane for proton exchange membrane fuel cell under anhydrous conditions. *Journal of Materials Chemistry A* 2 (25): 9548–9558.

52 Li, G., Shen, R., Hu, S. et al. (2022). Norbornene-based acid–base blended polymer membranes with low ion exchange capacity for proton exchange membrane fuel cell. *Advanced Composites and Hybrid Materials* 5 (3): 2131–2137.

53 Sun, X., Zhang, T., Liu, X. et al. (2020). Multi-functionalized acid-base double-shell nanotubes are incorporated into the proton exchange membrane to cope with low humidity conditions. *International Journal of Hydrogen Energy* 45 (55): 30673–30688.

54 Lu, Y., Liu, Y., Li, N. et al. (2020). Sulfonated graphitic carbon nitride nanosheets as proton conductor for constructing long-range ionic channels proton exchange membrane. *Journal of Membrane Science* 601: 117908.

55 Qu, E., Cheng, G., Xiao, M. et al. (2023). Composite membranes consisting of acidic carboxyl-containing polyimide and basic polybenzimidazole for high-temperature proton exchange membrane fuel cells. *Journal of Materials Chemistry A* 11 (24): 12885–12895.

56 Zeng, L., Lu, X., Yuan, C. et al. (2024). Self-enhancement of perfluorinated sulfonic acid proton exchange membrane with its own nanofibers. *Advanced Materials* 36: 2305711.

57 Xin, Z., Yanyi, Z., and Xiaobing, W. (2023). Research on testing and evaluation technology of proton exchange membrane for fuel cell. *Energy Reports* 10: 1943–1950.

58 Basura, V.I., Chuy, C., Beattie, P.D. et al. (2001). Effect of equivalent weight on electrochemical mass transport properties of oxygen in proton exchange membranes based on sulfonated α,β,β-trifluorostyrene (BAM®) and sulfonated styrene-(ethylene-butylene)-styrene triblock (DAIS-analytical) copolymers. *Journal of Electroanalytical Chemistry* 501 (1): 77–88.

59 Pourzare, K., Mansourpanah, Y., Farhadi, S. et al. (2020). Improving the efficiency of Nafion-based proton exchange membranes embedded with magnetically aligned silica-coated Co_3O_4 nanoparticles. *Solid State Ionics* 351: 115343.

60 Teixeira, F.C., De Sá, A.I., Teixeira, A.P.S. et al. (2019). Nafion phosphonic acid composite membranes for proton exchange membranes fuel cells. *Applied Surface Science* 487: 889–897.

61 Li, J., Xu, G., Cai, W. et al. (2018). Non-destructive modification on Nafion membrane via in-situ inserting of sheared graphene oxide for direct methanol fuel cell applications. *Electrochimica Acta* 282: 362–368.

62 Wang, H., Li, X., Zhuang, X. et al. (2017). Modification of Nafion membrane with biofunctional SiO_2 nanofiber for proton exchange membrane fuel cells. *Journal of Power Sources* 340: 201–209.

63 Li, J., Cui, N., Liu, D. et al. (2024). SPEEK-co-PEK-x proton exchange membranes with controllable sulfonation degree for proton exchange membrane fuel cells. *International Journal of Hydrogen Energy* 50: 606–617.

64 Justin Jose Sheela, A.S., Moorthy, S., Maria Mahimai, B. et al. (2023). Sulfonated poly ether sulfone membrane reinforced with bismuth-based organic and inorganic additives for fuel cells. *ACS Omega* 8 (30): 27510–27518.

65 Balasubramanian, A., Gunasekaran, M., and Kannan, T. (2022). Photo crosslinked stilbene-containing sulfonated polyimide membranes as proton exchange membranes in fuel cell. *European Polymer Journal* 176: 111418.

66 Liang, J., Ge, J., Wu, K. et al. (2020). Sulfonated polyaryletherketone with pendant benzimidazole groups for proton exchange membranes. *Journal of Membrane Science* 597: 117626.

67 Liu, L., Wang, C., He, Z. et al. (2021). Bi-functional side chain architecture tuned amphoteric ion exchange membranes for high-performance vanadium redox flow batteries. *Journal of Membrane Science* 624: 119118.

68 Zheng, C., Xie, N., Liu, X. et al. (2024). Durability improvement of proton exchange membrane fuel cells by doping silica–ferrocyanide antioxidant. *Journal of Membrane Science* 690: 122195.

69 Ryu, G.Y., Jae, H., Kim, K.J. et al. (2023). Hollow heteropoly acid-functionalized ZIF composite membrane for proton exchange membrane fuel cells. *ACS Applied Energy Materials* 6 (8): 4283–4296.

70 Devrim, Y. and Colpan, C.O. (2024). Assessment of polybenzimidazole/MOF composite membranes for the improvement of high-temperature PEM fuel cell performance. *International Journal of Hydrogen Energy* 58: 470–478.

71 Yan, X., Li, H., Lin, C. et al. (2020). An inorganic-framework proton exchange membrane for direct methanol fuel cells with increased energy density. *Sustainable Energy & Fuels* 4 (2): 772–778.

72 Pei, S., Xu, K., Xu, X. et al. (2023). CeO_2 stabilized by tourmaline as a novel inorganic filler to simultaneously increase the conductivity and durability of proton exchange membranes. *Inorganic Chemistry Frontiers* 10 (11): 3335–3344.

73 Yang, X., Zhu, H., Jiang, F. et al. (2020). Notably enhanced proton conductivity by thermally-induced phase-separation transition of Nafion/Poly (vinylidene fluoride) blend membranes. *Journal of Power Sources* 473: 228586.

74 Xu, X.Q., Cao, L.H., Yang, Y. et al. (2021). Hybrid Nafion membranes of ionic hydrogen-bonded organic framework materials for proton conduction and PEMFC applications. *ACS Applied Materials & Interfaces* 13 (47): 56566–56574.

75 Xu, F., Han, Y., Huang, K. et al. (2024). Enhanced performance of high-temperature proton exchange membranes via polyfluorene cross-linked zwitterionic liquid. *Journal of Membrane Science* 694: 122434.

76 Wan, Y.H., Sun, J., Jian, Q.P. et al. (2022). A Nafion/polybenzimidazole composite membrane with consecutive proton-conducting pathways for aqueous redox flow batteries. *Journal of Materials Chemistry A* 10 (24): 13021–13030.

2

Classification of Catalyst for Proton Exchange Membrane Fuel Cell

2.1 State-of-the-Art Electrocatalysts

As discussed in Chapter 1, Section 1.2.1, the performance of the oxygen reduction reaction (ORR) is heavily influenced by the nature of the electrocatalyst and the electrolyte. Therefore, reducing platinum (Pt) usage while maintaining stability, durability, and ORR efficiency is a key research goal. Several strategies have been developed to address this challenge, including [1–4]:

1) Alloying Pt with other transition metals.
2) Replacing Pt with nonprecious transition metals.
3) Utilizing metal-free, carbon-based electrocatalysts.

The following sections from 2.1.1 to 2.15 provide a detailed discussion of these types of electrocatalysts, highlighting their advantages and barriers in the context of ORR performance.

2.1.1 Pt-Alloy Electrocatalysts

Pt-based materials are among the most efficient catalysts for ORR. However, the high cost and limited availability of Pt present significant barriers to widespread adoption in both proton exchange membrane fuel cells (PEMFCs) and anion exchange membrane fuel cells (AEMFCs). According to the U.S. Department of Energy (DOE), reducing Pt loading in the cathode layer is a critical step in lowering the overall cost of fuel cell electric vehicles (FCEVs). Since 2002, Pt loading has decreased from 1.0 to 0.125 mg/cm^2 [5, 6], resulting in a reduction of system costs from \$50 to \$30 kW^{-1}. For example, Daimler has reduced its Pt usage by 90% since 2009 [7, 8]. Nonetheless, current Pt loading levels still fall short of DOE targets, as illustrated in Figure 2.1.

Another significant challenge for Pt-based electrocatalysts is their limited stability. At low Pt loadings, catalysts are prone to CO poisoning and dissolution, leading to reduced efficiency and durability [9, 10]. Additionally, the size of Pt nanoparticles plays a critical role in their performance. Smaller Pt nanoparticles have higher

Electrospinning for Proton Exchange Membrane Fuel Cells, First Edition.
Yong Liu and Mohideen Meerasahib Mohamedazeem.

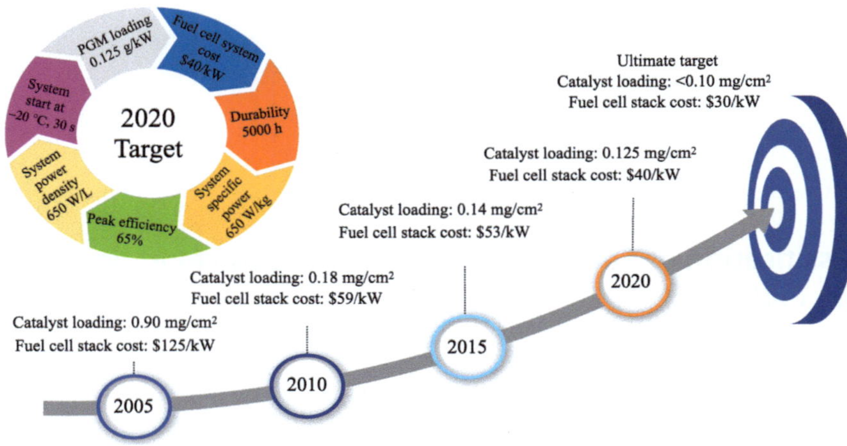

Figure 2.1 Current status and target of Pt group metal catalyst on PEMFC. Source: Mohideen et al. [8]/with permission of Elsevier.

surface energy, which can lead to diffusion, dissolution, and coalescence with larger particles—a process known as Ostwald ripening—during ORR. Corrosion of the carbon support further exacerbates this issue. Under operating conditions, the structure of the carbon material can degrade, causing Pt particles to detach or agglomerate, ultimately hindering mass transport and reducing ORR efficiency.

To address these issues, alloying Pt with transition metals on suitable carbon supports has emerged as a promising strategy. This approach has been studied for decades, with early predictions suggesting that Pt alloys with cobalt (Co), iron (Fe), and nickel (Ni) could achieve up to three times the ORR performance of pure Pt [11]. However, alloying introduces its own challenges. Transition metal ions can oxidize or dissolve due to their lower dissolution potentials compared to Pt. For instance, Pt has a dissolution potential of approximately 0.8 V, whereas Co, Ni, and Fe dissolve at potentials around −0.2 V, 0.1 V, and −0.5 V, respectively, versus the normal hydrogen electrode. The resulting dissolved metal cations can damage the membrane electrode assembly (MEA) by exchanging protons with the membrane [12–15].

Limitations: Carbon corrosion remains a major concern for Pt-based electrocatalysts. Selecting carbon supports with high porosity, high specific surface area, and excellent electronic conductivity is crucial for maintaining efficient mass transport. The interaction between Pt and the carbon support strongly influences the structure, size, and growth of Pt nanoparticles, which in turn affects overall ORR performance in both acidic and alkaline environments [16, 17]. However, when the cathode potential exceeds approximately 1.4 V—such as during frequent start-up or shutdown cycles—the carbon support can become thermodynamically unstable and corrode [18]. This degradation of the carbon material compromises the performance and durability of Pt-based catalysts, underscoring the importance of developing more robust carbon supports and optimized Pt alloys.

2.1.2 Carbon-Supported Transition Metal Catalysts (M-N-C)

Carbon-supported transition metal catalysts, commonly referred to as M-N-C, have emerged as promising candidates for PEMFCs and anion exchange membrane fuel cells. These catalysts leverage abundant, cost-effective precursors and demonstrate excellent electrocatalytic activity. M-N-C catalysts are typically synthesized through high-temperature pyrolysis (above 800 °C) under inert conditions, producing numerous active sites that facilitate the ORR. In such catalysts, carbon serves as a primary contributor to the active sites, while heteroatoms such as nitrogen (N) or sulfur (S) and a small fraction of transition metals (approximately 1–3 atomic percent) provide additional activity [19].

Despite the small amount of metal incorporated into M-N-C catalysts, identifying the precise nature of the active sites responsible for ORR remains a matter of ongoing debate. Among the various M-N-C catalysts, Fe-N-C systems have demonstrated the highest electrocatalytic activity for ORR [20]. However, the specific structure of active sites often varies between studies. For example, Ohsaka et al. observed that Fe-N-C catalysts exhibit both four-electron ($4e^-$) and two-electron ($2e^-$) reduction pathways, with their occurrence depending on the catalyst loading and the electrolyte used [21].

Wang et al. [22] reported that square planar Fe-N_4 active sites favor a $4e^-$ pathway in acidic electrolytes, while Tylus et al. [23] proposed (see Figure 2.2) that Fe-N_4 catalyzes a single-site $2 \times 2e^-$ reaction pathway in alkaline electrolytes and a double-site $2e^- + 2e^-$ mechanism in acidic media, where Fe/Fe_xO_y nanoparticles act as secondary active sites. According to density functional theory (DFT) calculations, Kattel et al. suggested that the dissociative $4e^-$ OOH pathway is more likely to occur on FeN_4 [24]. Nevertheless, achieving a definitive understanding of the FeN_x active sites for ORR remains challenging, as different research groups report varying findings even when following similar preparation methods [25]. This complexity continues to fuel discussions among researchers regarding the underlying ORR mechanism in M-N-C catalysts [26].

Another significant barrier to the practical application of M-N-C catalysts is their difficulty in competing with Pt-based electrocatalysts in terms of performance stability and reproducibility [27]. Some researchers have argued that catalysts exhibiting high ORR activity at low Fe content (>0.1 at%) suggest that Fe is not the sole active site component [28]. Other studies proposed that Fe contributes indirectly by forming specific nitrogen-based active sites [29], while more recent investigations point to FeNx moieties themselves as the primary ORR-active species [30]. From these debates, two leading hypotheses for M-N-C active sites have emerged: (i) the interaction of nitrogen atoms with metal dopants (M-Nx), and (ii) the interaction of nitrogen-doped carbon (N-C) alone. In the former case, nitrogen heteroatoms embedded within the carbon framework create a conducive environment for metal ions to form catalytically active M-Nx moieties [31].

Limitations: One of the major limitations of M-N-C electrocatalysts lies in the complex identification of active sites. Fe-N-C catalysts, for example, exhibit multiple coexisting FeNx moieties, such as FeN_2, FeN_3, FeN_4, FeN_5, and FeN_6 [32]. Altering the nitrogen coordination number changes the electronic structure of the Fe

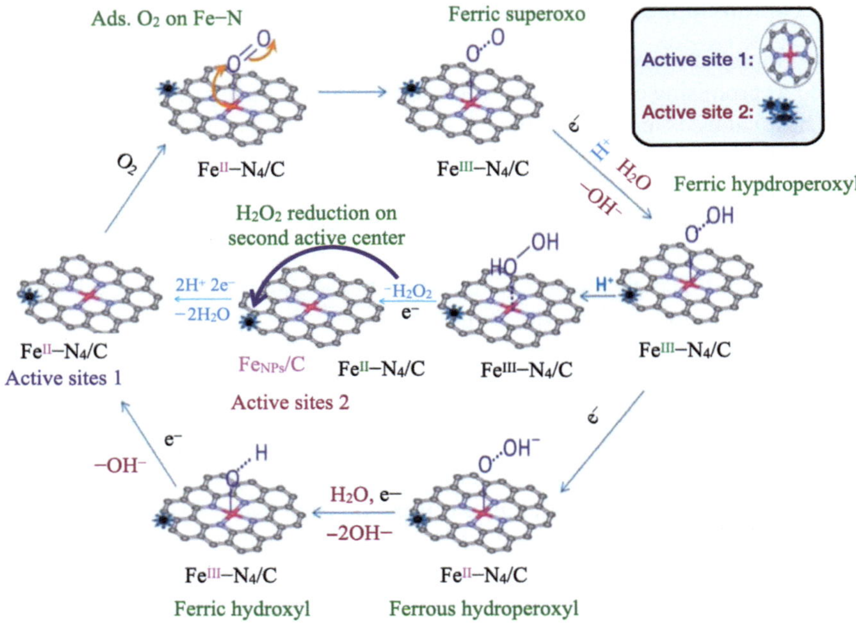

Figure 2.2 Proposed ORR mechanism pathway on FeN_4/C and adjacent Fe_{NPS}/C in acidic (H^+) and alkaline (H_2O/OH^-) electrolytes. Source: Tylus et al. et al. [23]/with permission of American Chemical Society.

atom, thereby influencing ORR activity. Further categorization of FeNx configurations based on the surrounding carbon atoms has led to structures like FeN_4C_8, FeN_4C_{10}, and FeN_4C_{12} [33]. Using time-of-flight secondary ion mass spectrometry (TOF-SIMS), Dodelet et al. identified two distinct catalytic active sites in Fe-N-C, detecting FeN_4C and FeN_2C moieties. Their results suggested that FeN_2C is more active toward ORR than is FeN_4C [34]. Kabir et al., through X-ray photoelectron spectroscopy (XPS) and DFT calculations, identified FeN_3C moieties as more stable than FeN_2C [35]. Despite the debates surrounding the specific FeNxC coordination structure, FeN_4 configurations remain widely accepted as key contributors to Fe-N-C catalysts' high ORR performance.

2.1.3 MXene-Based ORR Catalysts

MXenes, a class of two-dimensional nonprecious metal carbide and nitride materials, have attracted significant interest as potential ORR catalysts due to their excellent electronic conductivity, layered structure, and favorable surface hydrophilicity [36]. Over 20 types of MXenes have been successfully synthesized to date, with computational studies predicting a wide range of structural and property variations [37]. MXenes are generally described by the formula $M_{n+1}X_nT_x$, where "M" represents a transition metal, "X" stands for carbon or nitrogen, "T" denotes surface functional groups (such as –O, –F, and –OH), and "n" ranges from 1 to 3. The most commonly studied MXene, $Ti_3C_2T_x$, is typically derived from the MAX phase Ti_3AlC_2 by chemically etching the Al layer using strong acids like hydrofluoric acid [38].

$Ti_3C_2T_x$ serves as an excellent support material for both Pt-based and Pt-free catalysts. Its Ti layer contributes to durability under acidic and oxidative conditions, while the carbon layer provides an effective conductive pathway, enabling a four-electron ($4e^-$) ORR mechanism [39]. The interaction between the Ti and C layers also modifies the electronic properties of deposited metal nanoparticles, thereby enhancing electrocatalytic performance for applications such as methanol oxidation and ORR. However, despite its promise, research on MXene-based ORR catalysts remains limited compared to other nonprecious electrocatalysts [40].

Recent studies have demonstrated the potential of MXene-supported catalysts. Zhang et al. reported that uniformly dispersing Pt nanoparticles on MXene surfaces reduced agglomeration and created stable active sites for ORR in alkaline media [41]. Similarly, Wang et al. found that MXene offered numerous active sites for anchoring Pt nanowires, leading to a fourfold increase in mass activity compared to commercial Pt/C catalysts [42]. Moreover, the integration of MXenes with nonprecious metals has also shown promise. For example, combining Fe-N-C with MXene significantly lowered charge transfer and intrinsic resistances, enhancing ORR activity and cyclic stability in both acidic and alkaline environments [43].

Limitations: Despite these advancements, the experimental understanding of MXene-based ORR kinetics remains limited, particularly under practical fuel cell conditions. Additionally, MXene synthesis poses challenges related to material stability and the tendency of layers to restack, which can diminish catalytic activity. Consequently, the catalytic performance of MXenes still lags behind that of many established nonprecious ORR catalysts.

2.1.4 Metal-Organic Framework Electrocatalysts

Metal-organic frameworks (MOFs), a class of porous carbon-based materials, have garnered attention as ORR electrocatalysts due to their high electrical conductivity, extensive surface area, and versatile chemical structures. Incorporating heteroatoms such as nitrogen and sulfur into the carbon framework alters the local charge distribution and creates positively charged sites on adjacent carbon atoms, facilitating oxygen adsorption and charge transfer and thereby improving ORR activity [44]. MOFs are constructed by linking metal ions or clusters with organic ligands, resulting in tunable structures and functionalities that are ideal for electrocatalysis.

Transition metals such as Fe, Co, and Ni are commonly incorporated into MOFs to create active sites due to their favorable electrochemical properties, abundance, and cost-effectiveness [45]. For example, Agarwal et al. synthesized pseudo-carbon nanotubes by combining ZIF-8 (an MOF) with a sacrificial tellurium template and a ternary metal blend (Zn, Co, Fe), yielding an ORR catalyst that outperformed commercial Pt/C. They also developed mixed-metal sulfide oxygen evolution reaction

(OER) catalysts by incorporating Co, Fe, and Ni into ZIF-67, achieving an onset potential of 1.65 V versus reversible hydrogen electrode (RHE) at 10 mA/cm^2 with better stability than IrO_2 [46].

Other studies have explored strategies such as host–guest chemistry to optimize MOF-derived catalysts. Lai et al. developed ZIF-8-derived Fe-N-C catalysts with well-defined Fe-Nx active sites that significantly improved ORR kinetics in acidic media [47]. Chen et al. synthesized ZIF-67-derived Co-MOFs hybridized with copper, forming Co-Cu-MOF oxides that demonstrated an overpotential of 327 mV and retained 96.8% of their initial current density. This performance was attributed to the enhanced conductivity of Cu^{2+} doping and the synergistic effects of Cu^{+}/Cu^{2+} and Co^{2+}/Co^{3+} redox pairs [48].

Limitations: One of the main challenges for MOF-derived electrocatalysts is the high-temperature pyrolysis process, which often collapses the original MOF structure. This makes it difficult to control the newly formed pore structure and maintain the structural integrity of the resulting material. Additionally, the role of heteroatoms in preserving the stability of MOF-derived carbon frameworks remains poorly understood, presenting another barrier to their widespread adoption.

2.1.5 Carbon-Based Metal-Free Electrocatalyst

On the road to development of nonprecious electrocatalyst, a new class of carbon-based metal-free electrocatalyst was discovered for ORR in 2009. For the first time, Dai et al. found N-doped carbon nanotube (CNT) exhibited threefold higher ORR activity, long-term durability, and stability than did commercial Pt/C [49]. As discussed in Section 1.4.1, CO poisoning is one of the significant barrier to precious Pt-alloy nanoparticles catalyst. But the reported metal-free N-doped CNT showed negligible CO poisoning, and good methanol crossover pave the way for widespread research reporting on metal-free electrocatalyst. Gazing through atomic-scale, the enhanced activity of such metal-free electrocatalyst is due to the level of heteroatoms doping in the carbon matrix. At a desired nitrogen doping level, the nitrogen atoms alter electronic properties by creating a net positive charge to the adjacent carbon atom. Such modification improves the chemisorption behavior of oxygen molecules, facilitating electron transfer toward ORR [50].

Recently, the possible ORR mechanism of N-doped carbon material was proposed [51]. In that, the presence of pyridinic-N initiates the formation of Lewis basicity, which facilitates the adsorption of oxygen molecules first on to the neighbor carbon atom to the pyridinic-N, leading to the protonation of the adsorbed oxygen molecules. As shown in Figure 2.3, N-doped carbon favors $4e^-$ pathway through single active sites through A→B→C→D→E→A, and $2e^-$ pathways by more than one site through A→B→C→F→E→A, respectively. Besides mono-doping, co-doping of heteroatoms such as nitrogen, sulfur, and phosphorous with carbon material also showed to enhance the electrocatalytic activity better than many M-N-C and Pt-alloy catalyst toward ORR [52].

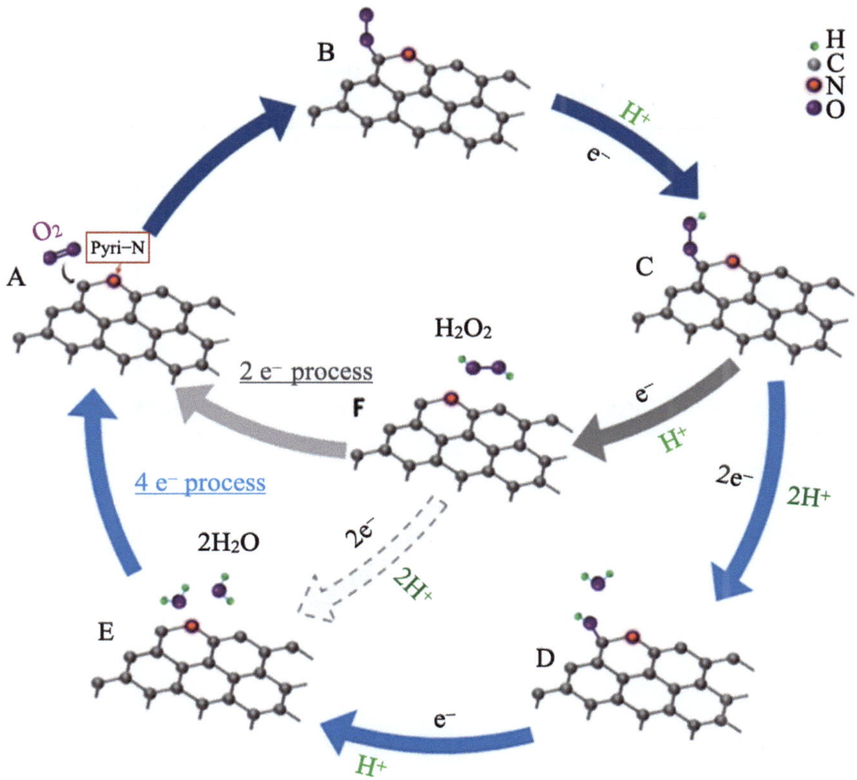

Figure 2.3 ORR pathway of N-doped carbon electrocatalyst. Source: Shen et al. [52]/with permission of Elsevier.

2.2 Fabrication of Catalyst Layers (CL) in Proton Exchange Membrane Fuel Cell

Over the past three decades, automotive PEMFC catalyst layer preparation has evolved through five significant stages: the polytetrafluoroethylene (PTFE) bonding method [16], the Nafion film method [17], the deposition method [18], the gradient catalyst layer [53], and the ordered nanofiber catalyst layer [54]. Concurrently, MEA technology has progressed through three generations: gas diffusion electrodes (GDEs), catalyst-coated membranes (CCMs), and ordered electrodes. Table 2.1 outlines the parameters for each generation of MEA [55].

2.2.1 GDE Membrane Electrode

The GDE membrane electrode involves a two-step process: first, the catalyst layer is deposited onto the gas diffusion layer (GDL), and then this GDL with

Table 2.1 The related parameters of three-generation MEA.

Parameter	GDE MEA	CCM MEA	Ordered MEA	2020 DOE target
PEM thickness (μm)	50–125	10–25	10–25	120 °C High-temperature film (0.02 Ω/cm^2)
Catalyst type		Platinum carbon, Platinum alloy	Platinum-gold alloy, core-shell metal, nonplatinum alloy	—
Carrying capacity (mg/cm^2)	>0.8	<0.8	<0.15	0.125
Catalyst utilization rate (%)	~30	>40	>80	—
Catalyst support material	C	C, Fe graphite/ Au, Ni, W, Co	CNT, TiO_2 nanotubes	—
Catalyst layer thickness (μm)	20~40	~10	~1	~1
Power density (W/cm^2)	0.4	0.78	0.845	1
Manufacturing cost ($/kW)	~100	~85	~17	~14

CCM, catalyst-coated membrane; CNT, carbon nanotube; DOE, U.S. Department of Energy; GDE, gas diffusion electrode; MEA, membrane electrode assembly; PEM, proton exchange membrane.

the catalyst layer is hot-pressed onto the PEM. Early methods employed PTFE emulsions as binders to fix the catalyst onto the GDL. Although the PTFE bonding method helped stabilize the catalyst [16], proton transport within the catalyst layer depended heavily on water films. This reliance on water not only limited proton conduction efficiency but also increased the risk of water accumulation ("flooding"), which could impair fuel cell performance [56].

Another major limitation was the mismatch in expansion coefficients between the catalyst layer and the PEM. Over extended operational periods, these differences led to delamination, significantly affecting the fuel cell stack's durability. To address these issues, the next generation of catalyst layer preparation methods introduced Nafion as a binder instead of PTFE. This innovation improved the adhesion between the catalyst layer and the PEM [17]. However, despite these advancements, GDE-based catalyst layers retained certain inherent drawbacks. For instance, they were often relatively thick, which extended the transport path for gases and water within the fuel cell. These longer paths increased the transmission losses of reactants and products, ultimately hindering fuel cell efficiency.

2.2.2 Catalyst-Coated Membrane Electrode

CCM electrodes represent an advanced approach to PEMFC catalyst layer fabrication, including methods such as deposition and gradient layering. In CCM electrodes, the catalyst layer is directly applied onto the PEM, which is then covered by the GDL. The deposition method [18] involves applying the catalyst and proton-conducting medium onto a Nafion membrane using techniques such as chemical vapor deposition, physical vapor deposition, spraying, and electrodeposition. After deposition, the membrane is dried. Currently, this is a mainstream method for producing automotive PEMFC catalyst layers. However, a key limitation is the uneven distribution of reactant gases within the catalyst layer. This non-uniformity can cause variations in critical parameters such as temperature, current density, and voltage [57–59], which in turn affect the consistency of the electrochemical reaction.

Additionally, the lack of continuous three-phase channels and interfaces means that some catalyst sites are not fully used, lowering overall effectiveness. GDL, by contrast, aims to address these inherent non-uniformities by tailoring the distribution of components such as Nafion content and Pt loading throughout the catalyst layer. This method builds on third-generation catalyst layer technologies, as illustrated in Figure 2.4. By fine-tuning the internal composition, gradient layers improve Pt utilization and enhance the overall reaction efficiency [53, 60, 61]. However, despite these advancements, CCM-based catalyst layers still face several critical challenges:

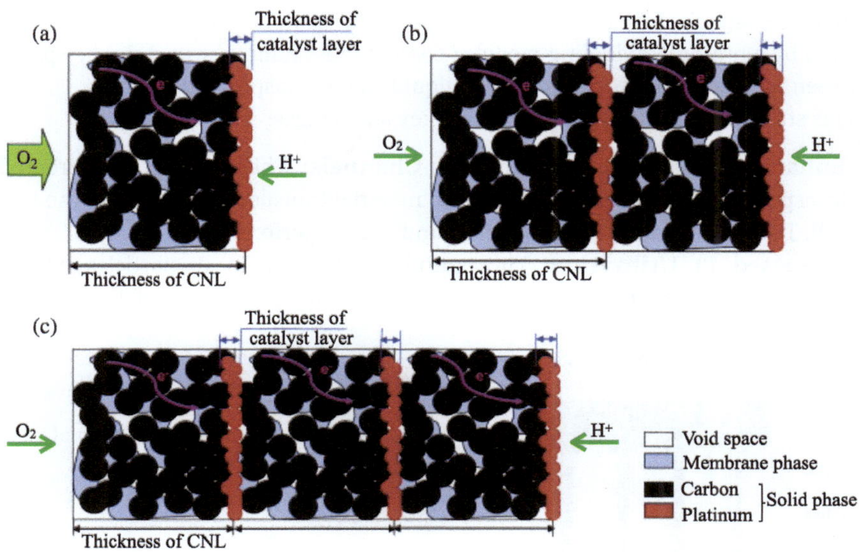

Figure 2.4 Schematic of the CNL with the sputtered Pt particles for the three studied configurations: (a) one Pt sputtered layer, (b) two Pt sputtered layers and (c) three Pt sputtered layers. Source: Dong et al. [58]/with permission of Elsevier.

1) **Low Pt Utilization:** A significant fraction of Pt resides within the porous carbon structure, preventing adequate proton-conducting medium coverage. As a result, this portion of Pt remains inactive and does not participate fully in the electrochemical reaction.
2) **High Three-Phase Transport Losses:** Traditional catalyst layers comprise porous media with complex, tortuous transport pathways. These hinder efficient movement of reactants, products, and charge carriers, leading to increased material transport losses and reduced overall efficiency.
3) **Durability Issues:** Pt particles are supported on carbon structures that are prone to corrosion under high operating potentials. This leads to the dissolution, migration, and agglomeration of Pt particles over time, reducing the active surface area and causing long-term performance degradation [62].

Overall, while CCM electrodes, especially those using gradient layers, have significantly advanced PEMFC catalyst technology, challenges such as low Pt utilization, transport losses, and durability remain critical areas for further research and optimization.

2.2.3 Ordered Membrane Electrode

The concept of an ordered catalyst layer was first introduced by Middelman in 2002 [63], envisioning a design where separate, well-defined pathways exist for proton, electron, and reactant transport. This "ideal" electrode layer (illustrated in Figure 2.5) is constructed from vertically aligned carbon nanotubes or rods, with Pt particles attached to their outer surfaces. The proton transport channels are formed by a proton-conducting medium coating these nanotubes, while the spaces between the nanotubes serve as gas and liquid water transport pathways.

This structured arrangement offers two key advantages:

1) **Enhanced Three-Phase Transport Channels:** The distinct, well-ordered transport pathways significantly reduce material transport losses within the fuel cell. This leads to improved efficiency and overall performance.
2) **Increased Pt Utilization:** The optimized coverage of Pt particles by the proton-conducting medium ensures that a greater proportion of the catalyst

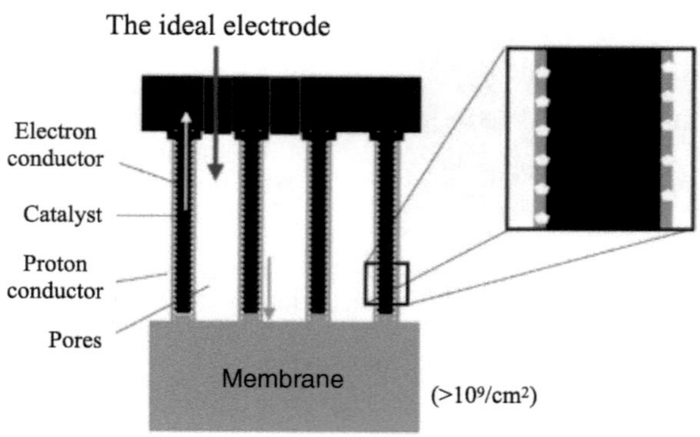

Figure 2.5 The ideal electrode of fuel cell. Source: Middelman [63]/with permission of Elsevier.

is actively participating in the electrochemical reactions. This results in better catalyst efficiency and a reduction in the amount of precious metal needed.

Over the past decade, the development of ordered catalyst layers has emerged as a cutting-edge research area in automotive PEMFC technology. By enabling higher catalyst efficiency and more effective transport, these ordered layers hold promise for advancing PEMFC performance and reducing costs.

2.3 Conclusion

In summary, we explored the classification of catalysts and fabrication methods for PEMFC catalyst layers. Pt-alloy catalysts, while effective for ORR, remain limited by high cost and stability concerns. M-N-C catalysts present a cost-efficient, nonprecious alternative, though the nature of their active sites remains unclear. MXene-based catalysts show promise due to their stability and conductivity, but their ORR activity still lags behind more established materials. MOF-derived catalysts offer tunable structures and high versatility, yet challenges such as structural integrity loss and unclear heteroatom roles persist. Regarding catalyst layer fabrication, GDE layers are simple but suffer from poor water management and limited durability, while CCM approaches improve Pt utilization but encounter high transport losses. Emerging ordered catalyst layers offer structured transport channels and greater efficiency, paving the way for advanced PEMFC designs. Moving forward, research should focus on improving the identification of active sites, enhancing structural stability, and developing catalyst layers that boost performance, lower costs, and ensure long-term durability.

References

1 Wang, X., Li, Z., Qu, Y. et al. (2019). Review of metal catalysts for oxygen reduction reaction: From nanoscale engineering to atomic design. *Chem* 5: 1486–1511.

2 Primbs, M., Sun, Y., Roy, A. et al. (2020). Establishing reactivity descriptors for platinum group metal (PGM)-free Fe–N–C catalysts for PEM fuel cells. *Energy & Environmental Science* 13: 2480–2500.

3 Huang, Z.-F., Song, J., Dou, S. et al. (2019). Strategies to break the scaling relation toward enhanced oxygen electrocatalysis. *Matter* 1: 1494–1518.

4 Lu, Z., Chen, G., Siahrostami, S. et al. (2018). High-efficiency oxygen reduction to hydrogen peroxide catalysed by oxidized carbon materials. *Nature Catalysis* 1: 156–162.

5 Ostroverkh, A., Johánek, V., Dubau, M. et al. (2019). Optimization of ionomer-free ultra-low loading Pt catalyst for anode/cathode of PEMFC via magnetron sputtering. *International Journal of Hydrogen Energy* 44: 19344–19356.

6 Parra, D., Valverde, L., Pino, F.J. et al. (2019). A review on the role, cost and value of hydrogen energy systems for deep decarbonisation. *Renewable and Sustainable Energy Reviews* 101: 279–294.

7 Wang, Y., Chen, K.S., Mishler, J. et al. (2011). A review of polymer electrolyte membrane fuel cells: Technology, applications, and needs on fundamental research. *Applied Energy* 88: 981–1007.

8 Mohideen, M.M., Radhamani, A.V., Ramakrishna, S. et al. (2022). Recent insights on iron based nanostructured electrocatalyst and current status of proton exchange membrane fuel cell for sustainable transport. *Journal of Energy Chemistry* 69: 466–489.

9 Xu, H., Ci, S., Ding, Y. et al. (2019). Recent advances in precious metal-free bifunctional catalysts for electrochemical conversion systems. *Journal of Materials Chemistry A* 7: 8006–8029.

10 Osgood, H., Devaguptapu, S.V., Xu, H. et al. (2016). Transition metal (Fe, Co, Ni, and Mn) oxides for oxygen reduction and evolution bifunctional catalysts in alkaline media. *Nano Today* 11: 601–625.

11 Gao, G., Zhu, G., Chen, X. et al. (2023). Optimizing Pt-based alloy electrocatalysts for improved hydrogen evolution performance in alkaline electrolytes: A comprehensive review. *ACS Nano* 17: 20804–20824.

12 Huang, C., Zhang, W., Li, J. et al. (2023). Pt alloy nanocrystal catalysts: Diminishing sizes and improving stability. *The Journal of Physical Chemistry C* 127: 12827–12840.

13 Janssen, M., Weber, P., and Oezaslan, M. (2023). Recent advances of various Pt-based catalysts for oxygen reduction reaction (ORR) in polymer electrolyte membrane fuel cells (PEMFCs). *Current Opinion in Electrochemistry* 40: 101337.

14 Kim, J., Yang, W., Chung, J.Y. et al. (2024). Synthesis and characterization of Pt_3Co alloy magnetic nanoparticles in multidomain region for oxygen reduction reaction catalysts in PEMFCs. *International Journal of Hydrogen Energy* 51: 1484–1495.

15 Chen, Y., Zhao, X., Yan, H. et al. (2023). Manipulating Pt-skin of porous network Pt-Cu alloy nanospheres toward efficient oxygen reduction. *Journal of Colloid and Interface Science* 652: 1006–1015.

16 Ticianelli, E.A., Derouin, C.R., Redondo, A. et al. (1988). Methods to advance technology of proton exchange membrane fuel cells. *Journal of the Electrochemical Society* 135 (9): 2209–2214.

17 Paganin, V.A., Ticianelli, E.A., and Gonzalez, E.R. (1996). Development and electrochemical studies of gas diffusion electrodes for polymer electrolyte fuel cells. *Journal of Applied Electrochemistry* 26 (3): 297–304.

18 O'Hayre, R., Lee, S.J., Cha, S.W. et al. (2002). A sharp peak in the performance of sputtered platinum fuel cells at ultra-low platinum loading. *Journal of Power Sources* 109 (2): 483–493.

19 Das, S., Kundu, A., Kuila, T. et al. (2023). Recent advancements on designing transition metal-based carbon-supported single atom catalysts for oxygen electrocatalysis: Miles to go for sustainable Zn-air batteries. *Energy Storage Materials* 61: 102890.

20 Liu, B., Huang, B., Lin, C. et al. (2017). Porous carbon supported Fe-NC composite as an efficient electrocatalyst for oxygen reduction reaction in alkaline and acidic media. *Applied Surface Science* 411: 487–493.

21 Muthukrishnan, A., Nabae, Y., Okajima, T. et al. (2015). Kinetic approach to investigate the mechanistic pathways of oxygen reduction reaction on Fe-containing N-doped carbon catalysts. *ACS Catalysis* 5: 5194–5202.

22 Wang, W., Jia, Q., Mukerjee, S. et al. (2019). Recent insights into the oxygen-reduction electrocatalysis of Fe/N/C materials. *ACS Catalysis* 9: 10126–10141.

23 Tylus, U., Jia, Q., Strickland, K. et al. (2014). Elucidating oxygen reduction active sites in pyrolyzed metal–nitrogen coordinated non-precious-metal electrocatalyst systems. *The Journal of Physical Chemistry C* 118: 8999–9008.

24 Kattel, S. and Wang, G. (2014). Reaction pathway for oxygen reduction on FeN_4 embedded graphene. *The Journal of Physical Chemistry Letters* 5: 452–456.

25 Wang, H.-Y., Weng, C.-C., and Yuan, Z.-Y. (2021). Insights into efficient transition metal-nitrogen/carbon oxygen reduction electrocatalysts. *Journal of Energy Chemistry* 56: 470–485.

26 Asset, T. and Atanassov, P. (2020). Iron-nitrogen-carbon catalysts for proton exchange membrane fuel cells. *Joule* 4: 33–44.

27 Zhu, C., Li, H., Fu, S. et al. (2016). Highly efficient nonprecious metal catalysts towards oxygen reduction reaction based on three-dimensional porous carbon nanostructures. *Chemical Society Reviews* 45: 517–531.

28 Matter, P.H., Wang, E., Millet, J.-M.M. et al. (2007). Characterization of the iron phase in CN x-based oxygen reduction reaction catalysts. *The Journal of Physical Chemistry C* 111: 1444–1450.

29 Sougrati, M.T., Goellner, V., Schuppert, A.K. et al. (2016). Probing active sites in iron-based catalysts for oxygen electro-reduction: A temperature-dependent 57Fe Mössbauer spectroscopy study. *Catalysis Today* 262: 110–120.

30 Sharifi, T., Gracia-Espino, E., Chen, A. et al. (2020). Oxygen reduction reactions on single-or few-atom discrete active sites for heterogeneous catalysis. *Advanced Energy Materials* 10: 1902084.

31 Qu, L., Wang, Z., Guo, X. et al. (2019). Effect of electrode Pt-loading and cathode flow-field plate type on the degradation of PEMFC. *Journal of Energy Chemistry* 35: 95–103.

32 Zitolo, A., Ranjbar-Sahraie, N., Mineva, T. et al. (2017). Identification of catalytic sites in cobalt-nitrogen-carbon materials for the oxygen reduction reaction. *Nature Communications* 8: 957.

33 Shen, H., Thomas, T., Rasaki, S.A. et al. (2019). Oxygen reduction reactions of Fe-NC catalysts: Current status and the way forward. *Electrochemical Energy Reviews* 2: 252–276.

34 Lefèvre, M., Dodelet, J., and Bertrand, P. (2002). Molecular oxygen reduction in PEM fuel cells: Evidence for the simultaneous presence of two active sites in Fe-based catalysts. *The Journal of Physical Chemistry B* 106: 8705–8713.

35 Kabir, S., Artyushkova, K., Kiefer, B. et al. (2015). Computational and experimental evidence for a new $TM–N_3/C$ moiety family in non-PGM electrocatalysts. *Physical Chemistry Chemical Physics* 17: 17785–17789.

36 Xiong, D., Li, X., Bai, Z. et al. (2018). Recent advances in layered $Ti_3C_2T_x$ MXene for electrochemical energy storage. *Small* 14: 1703419.

37 Zhan, C., Sun, W., Xie, Y. et al. (2019). Computational discovery and design of MXenes for energy applications: Status, successes, and opportunities. *ACS Applied Materials & Interfaces* 11: 24885–24905.

38 Biswas, S. and Alegaonkar, P.S. (2021). MXene: Evolutions in chemical synthesis and recent advances in applications. *Surfaces* 5: 1–34.

39 Ahmad Junaidi, N.H., Wong, W.Y., Loh, K.S. et al. (2021). A comprehensive review of MXenes as catalyst supports for the oxygen reduction reaction in fuel cells. *International Journal of Energy Research* 45: 15760–15782.

40 Das, H.T., Balaji, T.E., Dutta, S. et al. (2022). Recent advances in MXene as electrocatalysts for sustainable energy generation: A review on surface engineering and compositing of MXene. *International Journal of Energy Research* 46: 8625–8656.

41 Zhang, C., Ma, B., Zhou, Y. et al. (2020). Highly active and durable Pt/MXene nanocatalysts for ORR in both alkaline and acidic conditions. *Journal of Electroanalytical Chemistry* 865: 114142.

42 Wang, R., Chang, Z., Fang, Z. et al. (2020). Pt nanowire/$Ti_3C_2T_x$-CNT hybrids catalysts for the high performance oxygen reduction reaction for high temperature PEMFC. *International Journal of Hydrogen Energy* 45: 28190–28195.

43 Wang, W.-T., Batool, N., Zhang, T.-H. et al. (2021). When MOFs meet MXenes: Superior ORR performance in both alkaline and acidic solutions. *Journal of Materials Chemistry A* 9: 3952–3960.

44 Pan, Y., Gao, J., Li, Y. et al. (2024). Constructing nitrogen-doped carbon hierarchy structure derived from metal-organic framework as high-performance ORR cathode material for Zn-air battery. *Small* 20: 2304594.

45 Nguyen, Q.H., Tinh, V.D.C., Oh, S. et al. (2024). Metal-organic framework-polymer complex-derived single-atomic oxygen reduction catalyst for anion exchange membrane fuel cells. *Chemical Engineering Journal* 481: 148508.

46 Agarwal, S., Yu, X., and Manthiram, A. (2020). A pair of metal organic framework (MOF)-derived oxygen reduction reaction (ORR) and oxygen evolution reaction (OER) catalysts for zinc-air batteries. *Materials Today Energy* 16: 100405.

47 Lai, Q., Zheng, L., Liang, Y. et al. (2017). Metal–organic-framework-derived Fe-N/C electrocatalyst with five-coordinated Fe-N x sites for advanced oxygen reduction in acid media. *ACS Catalysis* 7: 1655–1663.

48 Chen, T., Ren, H., Li, L. et al. (2024). Synergistic cobalt–copper metal–organic framework-derived oxide electrocatalyst for efficient oxygen evolution reaction. *Materials Chemistry and Physics* 313: 128689.

49 Dai, L., Xue, Y., Qu, L. et al. (2015). Metal-free catalysts for oxygen reduction reaction. *Chemical Reviews* 115: 4823–4892.

50 Wu, G., Santandreu, A., Kellogg, W. et al. (2016). Carbon nanocomposite catalysts for oxygen reduction and evolution reactions: From nitrogen doping to transition-metal addition. *Nano Energy* 29: 83–110.

51 Ma, R., Lin, G., Zhou, Y. et al. (2019). A review of oxygen reduction mechanisms for metal-free carbon-based electrocatalysts. *npj Computational Materials* 5: 78.

52 Shen, H., Yang, L., Wu, Y. et al. (2023). Influence of activating and supporting oxygen in M–N–C electrocatalysts for oxygen reduction. *Electrochimica Acta* 466, 143001.

53 Fofana, D., Hamelin, J., and Bénard, P. (2013). Modelling and experimental validation of high performance low platinum multilayer cathode for polymer electrolyte membrane fuel cells (PEMFCs). *International Journal of Hydrogen Energy* 38 (24): 10050–10062.

54 Tian, Z.Q., Lim, S.H., Poh, C.K. et al. (2011). A highly order-structured membrane electrode assembly with vertically aligned carbon nanotubes for ultra-low pt loading PEM fuel cells. *Advanced Energy Materials* 1 (6): 1205–1214.

55 Wang, C., Zhao, B., and Zhang, J. (2016). Key technologies of proton exchange membrane fuel cell membrane electrode. *Science and Technology Report* 34 (6): 62–68.

56 Dubau, L., Castanheira, L., Maillard, F. et al. (2014). A review of PEM fuel cell durability: Materials degradation, local heterogeneities of aging and possible mitigation strategies. *WIREs Energy and Environment* 3 (6): 540–560.

57 Hogarth, W.H.J., Steiner, J., Benziger, J.B. et al. (2007). Spatially-resolved current and impedance analysis of a stirred tank reactor and serpentine fuel cell flow-field at low relative humidity. *Journal of Power Sources* 164 (2): 464–471.

58 Dong, Q., Kull, J., and Mench, M.M. (2005). Real-time water distribution in a polymer electrolyte fuel cell. *Journal of Power Sources* 139 (1): 106–114.

59 Manikandan, T. and Ramalingam, S. (2016). A review of optimization algorithms for the modeling of proton exchange membrane fuel cell. *Journal of Renewable and Sustainable Energy* 8 (3): 1103–1112.

60 Guangying, C., Bo, Z., Cheng, W. et al. (2016). Gradient membrane electrode assembly of proton exchange membrane fuel cell. *Chemical Bulletin (Beijing, China)* 79 (1): 9–15.

61 Santis, M., Freunberger, S.A., Reiner, A. et al. (2006). Homogenization of the current density in polymer electrolyte fuel cells by in-plane cathode catalyst gradients. *Electrochimica Acta* 51 (25): 5383–5393.

62 Le Canut, J., Abouatallah, R.M., and Harrington, D.A. (2006). Detection of membrane drying, fuel cell flooding, and anode catalyst poisoning on PEMFC stacks by electrochemical impedance spectroscopy. *Journal of the Electrochemical Society* 153 (5): A857–A864.

63 Middelman, E. (2002). Improved PEM fuel cell electrodes by controlled self-assembly. *Fuel Cells Bulletin* 2002 (11): 9–12.

3

Electrospun Catalyst Layer for Proton Exchange Membrane Fuel Cell

3.1 Introduction

Newly assembled fuel cells often face unstable operating conditions due to poorly developed interfaces within the catalyst layer (CL). These inadequate interfaces hinder the redox reactions of the catalyst and disrupt the transport of protons, electrons, and reactant gases. To stabilize fuel cell performance and ensure consistent operation, an activation period is typically required prior to regular use [1]. Comparative studies between nanofiber CLs and traditional CLs reveal notable differences in activation behavior. Traditional CLs show only modest performance improvements during activation, whereas nanofiber CLs undergo significant changes. This chapter focuses on understanding the activation mechanisms of nanofiber CLs by examining the evolution of key electrochemical parameters—such as electrochemical impedance spectroscopy (EIS), electrochemically active surface area (ECSA), and exchange current density (j_0)—throughout the activation process.

3.2 Materials Preparation

3.2.1 Preparation of Nanofiber Catalyst Layers

This section outlines the process for creating nanofiber-based CLs through electrospinning. Key steps include determining the appropriate components and proportions in the electrospinning solution, preparing a uniform spinning solution, and fabricating the CL using the electrospinning technique. Figure 3.1 illustrates the overall preparation workflow.

1) **Selection of Electrospinning Components and Proportions**: Preparing a suitable electrospinning solution is a challenging yet crucial step. In our experiments, we used the established Pt/C catalyst combined with Nafion as the proton-conducting medium, following an optimal ratio (carbon in Pt/C: Nafion = 1:1) reported in the literature [2]. However, direct spinning of Nafion fibers is difficult due to its short molecular chains. To enhance chain entanglement and

Electrospinning for Proton Exchange Membrane Fuel Cells, First Edition.
Yong Liu and Mohideen Meerasahib Mohamedazeem.

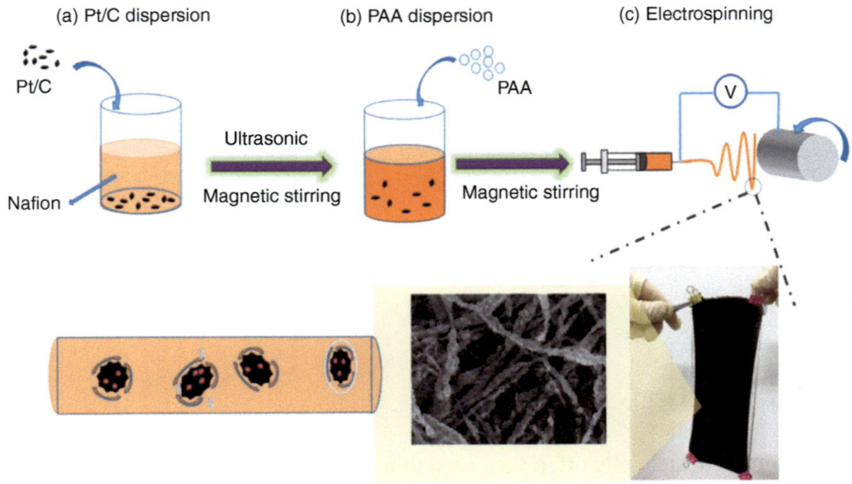

Figure 3.1 Schematic presentation for the fabrication mechanism of nanofibers catalyst layer for fuel cell.

improve spinnability, polyacrylic acid (PAA) was introduced as a carrier polymer. PAA was selected over poly(ethylene oxide) (PEO) due to its better dispersibility and ability to form uniform nanofibers. Although PAA lacks conductivity and can interfere with CL properties, its content was minimized while still enabling a stable electrospinning process.

Early observations revealed that the high activity of the Pt/C catalyst could cause sparking when mixed directly with Nafion solution. Prewetting the catalyst with water proved necessary. Moreover, because water evaporates slowly during spinning, isopropanol or ethanol was added to accelerate solvent evaporation. Through optimization, the final electrospinning solution consisted of Pt/C catalyst, Nafion, PAA, isopropanol or ethanol, and water. The exact component ratios are listed in Table 3.1.

2) **Preparation of a Uniform Electrospinning Solution**: The Pt/C catalyst, with particle diameters of 2–3 nm, requires careful dispersion to ensure uniformity in the final CL. In this experiment, a combination of magnetic stirring and ultrasonic treatment was used. Ultrasonication was performed in a thermostatic chamber to prevent overheating, which can lead to nanoparticle agglomeration. The optimal process involved three cycles of 5-minute magnetic stirring followed by 30-minute ultrasonic treatment. PAA was added after these steps and stirred for 24 hours to achieve a homogeneous spinning solution. Figure 3.2 outlines the sequence and duration of these steps.
3) **Fabrication of** CLs **via Electrospinning**: Electrospinning involves using a high-voltage electric field to charge the polymer solution, forming a jet stream that solidifies into nanofibers as the solvent evaporates. These nanofibers are collected on a rotating drum, creating a uniform CL. Optimal spinning conditions, including voltage, spinning distance, and feed rate, were derived from previous studies. The specific parameters are listed in Table 3.2.

Table 3.1 The composition of electrospinning solution.

Component	Pt/C	Nafion (5%)	Polyacrylic acid	Deionized water (D.I.)	Isopropanol
Mass	0.3 g	3.6 g	0.09 g	0.9 g	1.5 g

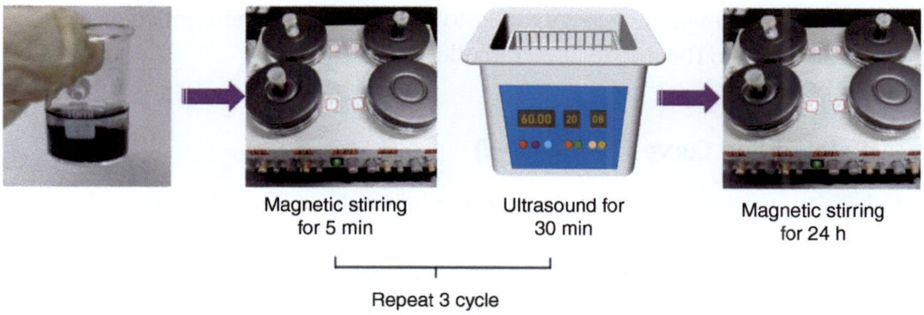

Figure 3.2 Schematic presentation for the fabrication mechanism of uniform electrospinning solution. Specific steps: (i) Place a magnetic stirring rotor in a sealed beaker and stir for 5 minutes; (ii) transfer the beaker to an ultrasonic cleaner at 10 °C for 30 minutes, ensuring the water level is higher than the solution level; (iii) repeat the above steps three times; and (iv) add PAA to the solution, reseal, and stir for 24 hours.

Table 3.2 Experimental parameters of electrospinning nanofibers catalyst layer.

Positive voltage (kV)	Negative voltage (kV)	Injection speed (mm/min)	Distance (mm)	Transmit speed (mm/min)	Transmit distance (mm)
9	−5	0.12	90	40	72

Specific steps: (i) Load approximately 2 ml of the prepared spinning solution into a syringe and set it in the electrospinning apparatus; (ii) attach the Nafion 211 membrane to the rotating collector drum with adhesive tape; (iii) set the spinning parameters (rotation speed, voltage, injection speed, translational speed, and distance) and begin electrospinning; and (iv) after the first layer, flip the Nafion membrane and repeat the spinning process.

3.3 Preparation and Electrochemical Characterization

3.3.1 Preparation of Single Cell

The $5 \times 5\,cm^2$ single cell used in this study was assembled by first cutting two pieces of gas diffusion layers (GDLs) from SGL and trimming them to $5.5 \times 5.5\,cm^2$. The bipolar plates and current collectors were cleaned with alcohol and allowed to air dry. During assembly, sealing rings and gaskets were placed on the bipolar plates, followed by two drops of water on the flow field side of each plate to ensure proper contact with the GDL. The membrane electrode assembly (MEA) was then sandwiched between the anode and cathode bipolar plates, with the GDLs aligned over their respective flow fields. Once all components were aligned, the assembly was tightened to a torque of 4 N·m. The final step involved connecting the anode and cathode inlets and outlets, as well as the appropriate electrical wiring, ensuring that the positive load wires were attached to the cathode current collector and the negative load wires to the anode current collector.

3.3.2 Polarization Curve Testing ($V-I$)

The polarization curve is a graph showing the relationship between electrode potential and current density, serving as a crucial method for fuel cell performance characterization. In this experiment, the voltage was measured under constant current conditions. The performance of the fuel cell was tested under hydrogen–oxygen and hydrogen–air conditions [3, 4]. The specific test conditions are shown in Table 3.3.

3.3.3 Cyclic Voltammetry Testing

Cyclic voltammetry (CV) was conducted to evaluate the electrochemically active surface area (ECSA), which is a key indicator of the number of active sites in the CL [5, 6]. With platinum as the catalyst, the adsorption–desorption charge of hydrogen on the Pt surface was measured via in situ CV, as illustrated in Figure 3.3. The ECSA was then determined using Eq. (3.1), where Q represents the measured charge, Lca is the cathode Pt loading, Ag is the electrode area, and 210 μC/cm(Pt)^2 corresponds to the single-layer hydrogen adsorption charge density.

$$A_{\mathrm{Pt}} = \frac{Q}{210\,\mu\mathrm{C/cm^2_{Pt}}(L_{\mathrm{ca}}A_{\mathrm{g}})}10^5\left[\mathrm{m^2_{Pt}/g_{Pt}}\right] \tag{3.1}$$

The CV experiments were carried out using a VMP3 potentiostat, configured with a four-terminal connection. Measurements were taken across a voltage range of 0.045–1.0 V, with a scan rate of 40 mV/s. Hydrogen was fed to the anode, while nitrogen was supplied to the cathode. Prior to testing, gases were introduced until the cell voltage fell below 0.13 V to ensure the removal of residual oxygen. Test parameters are detailed in Table 3.4.

Table 3.3 The measurement conditions of polarization curve.

Parameter	Cell temperature	Pressure	H_2 rate	Air rate	O_2 rate	Relative humidity
Value	70 °C	1 bar	600 sccm	3000 sccm	600 sccm	100%

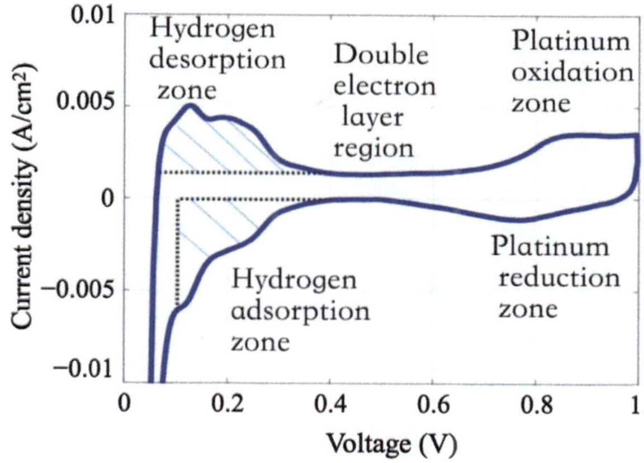

Figure 3.3 The typical CV curve for Pt.

Table 3.4 The measurement conditions of cyclic voltammetry.

Parameter	Cell temperature	H_2 rate	N_2 rate	Relative humidity
Value	30 °C	200 sccm	200 sccm	100%

3.3.4 Linear Sweep Voltammetry Testing

In fuel cells, hydrogen crossover refers to hydrogen that bypasses the reaction and directly passes through the PEM to the cathode. Severe hydrogen crossover can affect cell performance and even pose safety risks. In this experiment, in situ linear sweep voltammetry (LSV) testing was used to measure the hydrogen crossover current [7, 8]. The equipment used is the PEM Test 8900, with a four-terminal connection. The measurement range is 0.2–0.6 V, and the scan rate is 2 mV/s. Hydrogen gas (200 sccm) is supplied to the anode and nitrogen gas (200 sccm) to the cathode. (Note: Before the LSV test, the gases are introduced until the cell voltage drops below 0.13 V, ensuring complete oxygen removal from the cell.) As the hydrogen crossover current varies with humidity and temperature, specific testing conditions are not provided.

3.3.5 Electrochemical Impedance Spectroscopy Testing

Electrochemical impedance spectroscopy (EIS) applies a small alternating current (or voltage) signal to a working fuel cell and measures the corresponding voltage (or current) signal to calculate the impedance of different components within the fuel cell. The impedance spectrum in Figure 3.4 shows the following:

Ohmic Resistance (Rohm): The intersection of the high-frequency region with the real axis, representing resistance from electron transport (external circuits, GDL, and CL interfaces) and proton transport in the PEM [9, 10].

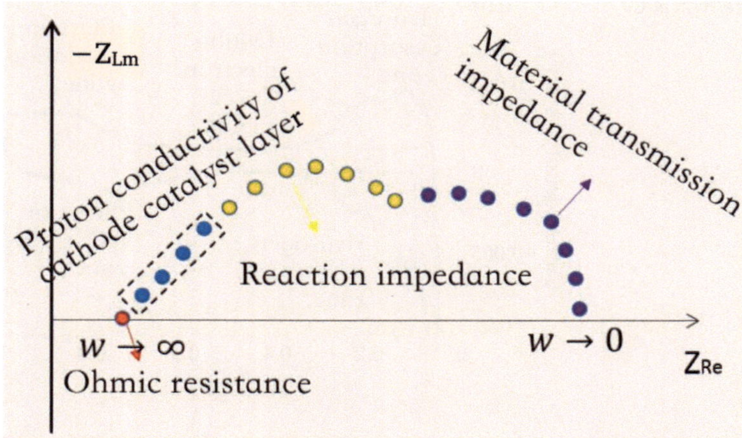

Figure 3.4 The typical EIS for fuel cell.

Proton Transport Resistance: Represented by a 45° sloped line at lower frequencies in the cathode CL.

Charge Transfer Resistance (R_{ct}): A semicircular curve in the mid-frequency region, reflecting the impedance of the oxygen reduction reaction (ORR).

Mass Transport Resistance (R_{mt}): A second semicircular curve in the low-frequency region, caused by limitations in gas transport.

The EIS testing equipment used in this experiment is the PEM Test 8900. The cell is connected in a four-terminal configuration, and the impedance spectrum is measured using a current excitation method. The measurement frequency range is 100–10,000 Hz, and the amplitude is set below 5% of the operating current to avoid affecting the fuel cell's performance.

3.4 Results and Discussion

3.4.1 Polarization Curve (*V*−*I*) of Preconditioning Process

Figure 3.5 shows the polarization curve test results for both traditional and nanofiber CLs over the course of the activation process. The seven polarization curves in each set represent performance data collected during seven consecutive tests, spanning a 2.5-hour activation period. Both catalyst types exhibit improved performance as activation progresses, but the magnitude of change differs significantly. Traditional CLs show a performance increase of less than 30%, stabilizing within roughly one hour, indicating a relatively rapid activation completion. Nanofiber CLs, on the other hand, demonstrate a more pronounced performance increase of over 200% during activation, with continued improvement even beyond the 2.5-hour mark, suggesting that the activation process is still ongoing (Figure 3.5).

To better understand these substantial changes in nanofiber catalyst performance, a series of experiments were designed to track their electrochemical and structural

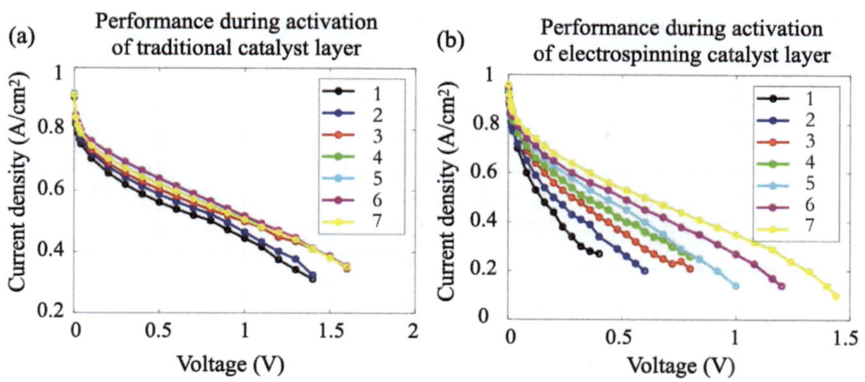

Figure 3.5 Polarization curve of two catalyst layers with a platinum loading of 0.1 mg/cm^2 during the activation process.

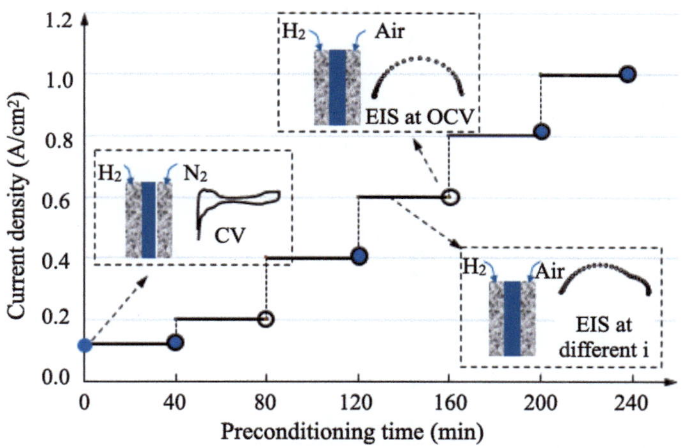

Figure 3.6 Schematic measurements during the preconditioning process of fuel cell.

evolution throughout activation. Initial tests established that current densities applied to nanofiber CLs must remain below 0.2 A/cm^2 at the early stages of activation. As the activation proceeds, performance improves, enabling the application of higher current densities. After approximately four hours, performance stabilizes with no further notable gains. For comparative testing, the activation protocol outlined in Figure 3.6 was used to assess both catalyst types. This protocol incorporated EIS and CV measurements.

The activation process was conducted in six steps. During each step, the fuel cell operated at a constant current for 40 minutes. Seven consecutive EIS measurements were taken during this time, each separated by a 3-minute interval. At the conclusion of each step, the cell was held at open-circuit voltage, followed by additional EIS measurements. Once the EIS tests were completed, the cathode gas was switched to nitrogen, and CV tests were performed.

3.4.2 Electrochemical Impedance Spectroscopy of Preconditioning Process

Figure 3.7 illustrates the EIS results for the conventional CL over six distinct activation periods. Each spectrum displays two semicircular features: the mid-frequency semicircle corresponds to electron transfer resistance, reflecting the ease of the redox reaction, while the low-frequency semicircle represents mass transfer resistance, indicative of the transport efficiency for protons, electrons, and gases [11].

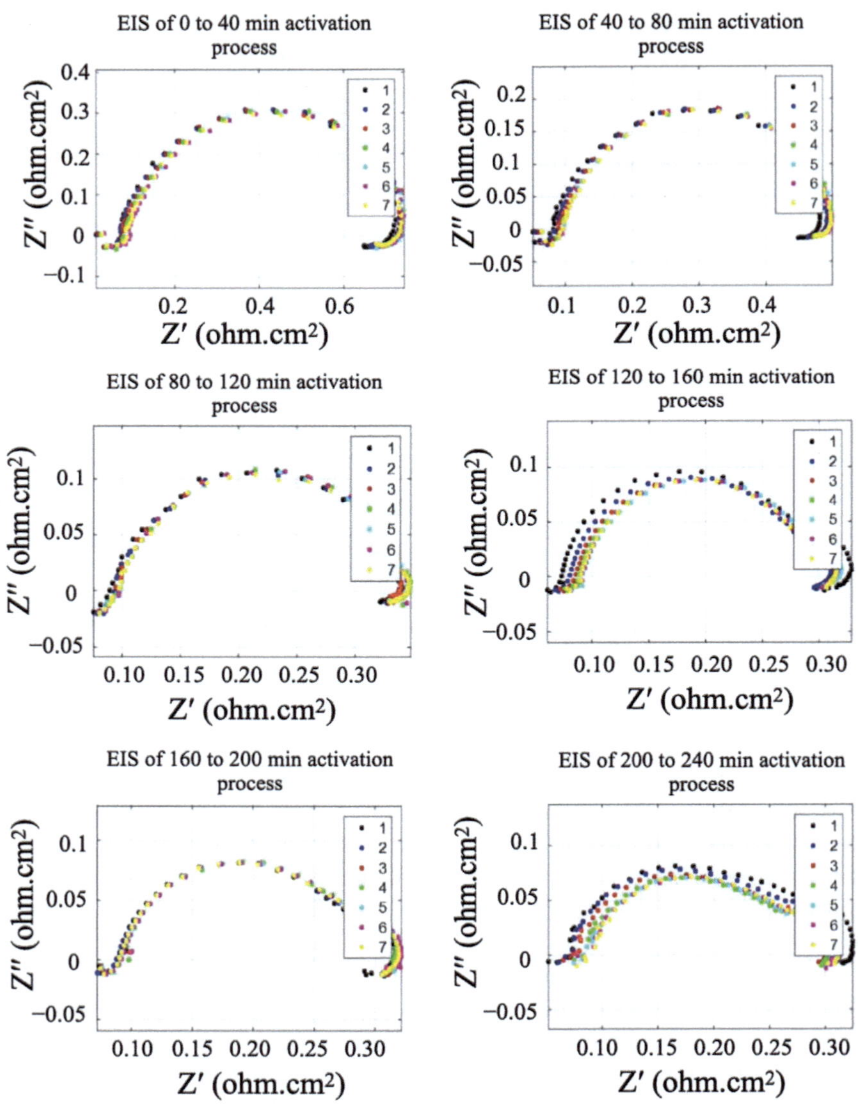

Figure 3.7 The EIS of conventional catalyst layer with a platinum loading of 0.1 mg/cm^2 during the preconditioning process.

Notably, the EIS spectra for each activation period show minimal variation, signifying that both electron transfer resistance and mass transfer resistance in the conventional CL remain largely stable throughout the activation process.

In contrast, Figure 3.8 presents the EIS data for the nanofiber CL across the same six activation periods. Within each time interval, the EIS spectra reveal notable changes. During the first four intervals—0–40 minutes, 40–80 minutes, 80–120 minutes, and 120–160 minutes—the semicircles progressively shrink, indicating reduced charge transfer and mass transfer resistances. This suggests that the redox reactions become easier and material transport resistance diminishes as activation progresses. By the final two intervals—160–200 minutes and

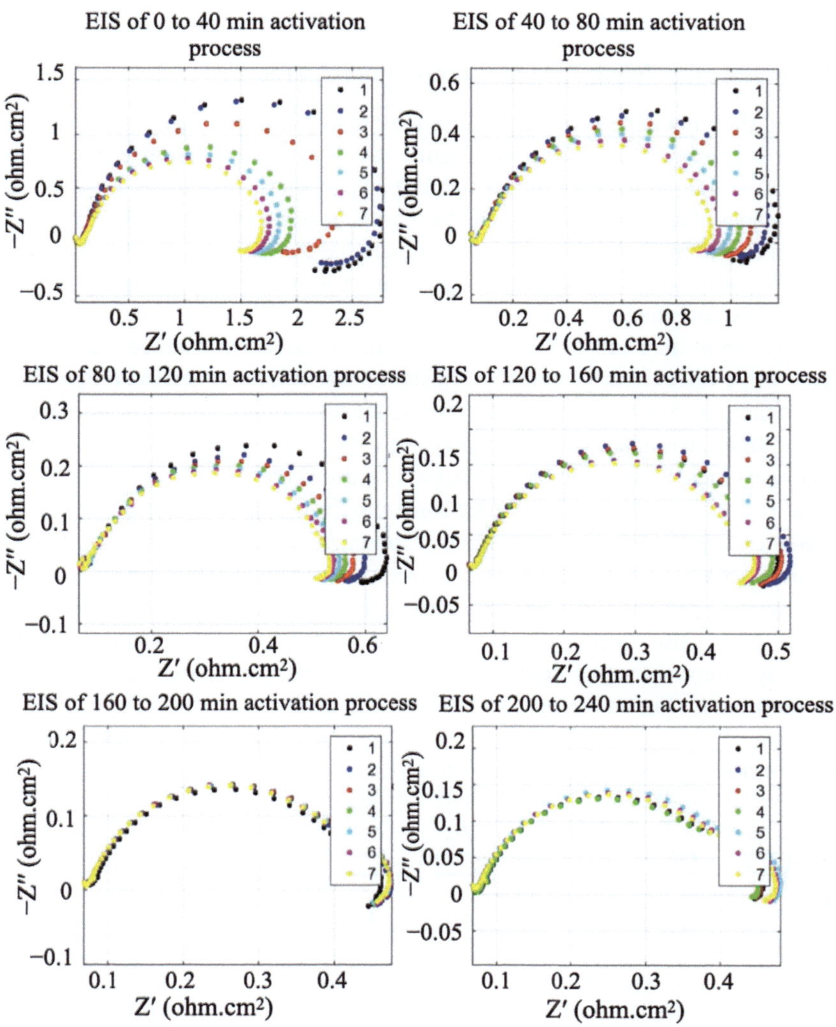

Figure 3.8 The EIS of electrospinning nanofibers catalyst layer with a platinum loading of 0.1 mg/cm^2 during the preconditioning process.

200–240 minutes—the semicircles stabilize, demonstrating that both electron transfer and mass transport resistances have reached steady levels.

A direct comparison of the EIS results from both CLs highlights significantly greater resistance reductions in the nanofiber CL during activation. This observation aligns with the performance curve improvements seen earlier, linking the enhanced performance of nanofiber CLs to the substantial decreases in both electron transfer and mass transfer resistances over time. However, by the end of activation (200–240 minutes), the conventional CL still exhibits slightly lower electron transfer and mass transfer resistances compared to the nanofiber CL, suggesting that under these operational conditions, the conventional CL maintains easier reaction pathways.

3.4.3 Impedance Fitting Data of Preconditioning Process

To analyze the impedance changes in the nanofiber CL during activation, equivalent electrical circuit fitting was used (Figure 3.9) [12, 13]. For low current densities (0.12, 0.2, 0.4, and 0.6 A/cm^2), mass transfer resistance was negligible. However, in the early stages of activation, anode electron transfer resistance was noticeable. The equivalent circuit for low-current activation included the charge transfer resistances of both the anode and cathode, as well as ohmic resistance, which mainly reflects the membrane resistance.

At high current densities (0.8 and 1 A/cm^2), mass transfer resistance became significant, while anode charge transfer resistance was relatively small compared to cathode charge transfer and mass transfer resistances. Consequently, the equivalent circuit for high-current activation included cathode charge transfer resistance, mass transfer resistance, and ohmic resistance (see Figure 3.4).

Fitting results revealed that the electron transfer resistance (R_{ct}) decreased by 42%, 28%, 18%, 9.4%, 2%, and 27% at activation currents of 0.12 A/cm^2, 0.2 A/cm^2, 0.4 A/cm^2, 0.6 A/cm^2, 0.8 A/cm^2, and 1 A/cm^2, respectively, indicating that the redox reactions progressively became more facile over time. The mass transfer resistance (R_{mt}) also decreased—by 15% and 12% at 0.8 A/cm^2 and 1 A/cm^2, respectively—demonstrating improved material transport channels during the activation process (Figure 3.10).

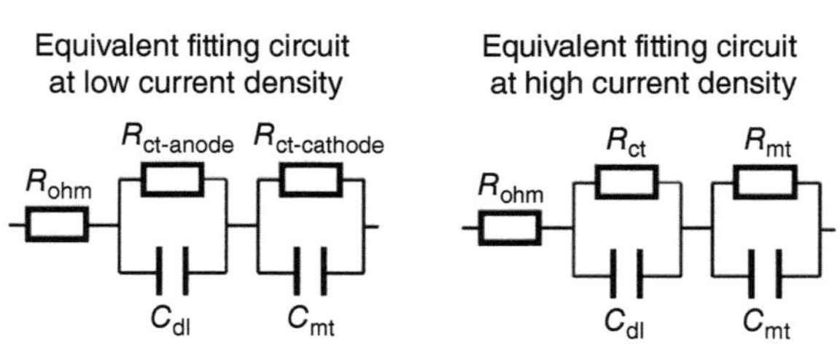

Figure 3.9 The equivalent electrical circuit for fitting the EIS data for electrospinning nanofibers catalyst layer.

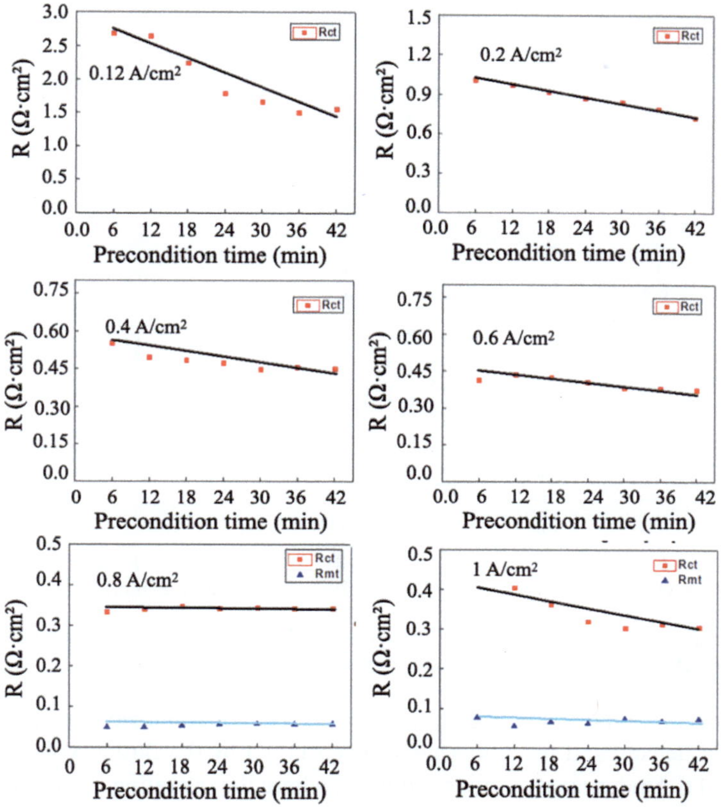

Figure 3.10 The equivalent fitting data for electrospinning nanofibers catalyst layer.

3.4.4 Electrochemically Active Surface Area of Preconditioning Process

As shown in Figure 3.11, the ECSA of the traditional CL ranged from 38.3 to 54.1 m^2/gPt during the activation process, ultimately increasing by 43.6% from the start to the end of activation. This gradual increase in ECSA corresponds to the continuous formation of more active reaction sites as the CL settles into its operating state. For the nanofiber CL, the ECSA ranged from 65.9 to 75.9 m^2/gPt, with an overall increase of 7.2% by the end of activation. Notably, the ECSA of the nanofiber CL was approximately 1.4 times higher than that of the traditional CL. This difference is attributed to the larger specific surface area of the nanofiber structure, which enhances the availability of active sites and improves the catalytic efficiency.

3.4.5 Exchange Current Density of Preconditioning Process

The exchange current density serves as a critical parameter for assessing the ease of electrochemical reactions at the electrode surface. A higher exchange current

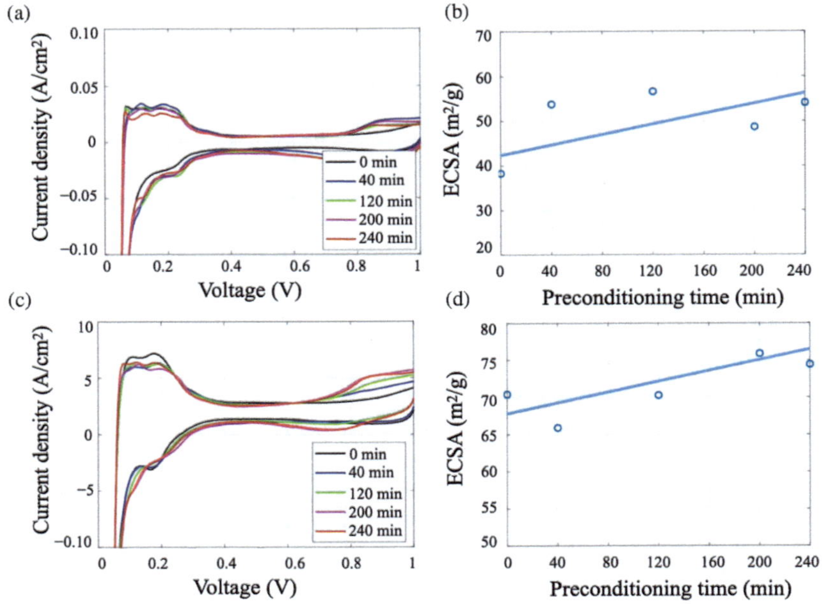

Figure 3.11 (a) and (c) represent CV curves of conventional CL and electropun CL, (b) and (d) represent the ECSA curves of conventional conventional CL and electropun CL.

density indicates more favorable reaction kinetics. The relationship between exchange current density and reaction kinetics is governed by the Butler–Volmer equation (Eq. (3.1)). Under open-circuit conditions, where the operating current is significantly lower than the exchange current density, the activation overpotential approaches zero. Through a Taylor expansion of the Butler–Volmer equation, Eq. (3.2) can be derived, enabling the calculation of R_{ct}^{ocv}. By measuring the impedance spectrum under open-circuit conditions and fitting it to an equivalent circuit model, the open-circuit charge transfer resistance (R_{ct}^{ocv}) can be obtained, allowing for the estimation of the exchange current density of the fuel cell [14].

From Figure 3.12b, the exchange current density of the traditional CL ranges from 6.5 to 8.5×10^{-4} A/cm^2, showing a continuous increase throughout activation, ultimately achieving a 30.2% overall improvement. Similarly, Figure 3.12d reveals that the exchange current density of the nanofiber CL ranges from 6.6 to 11.2×10^{-4} A/cm^2, with an overall increase of 84.9% during activation. Furthermore, a direct comparison at the end of activation shows that the exchange current density of the nanofiber CL is 1.32 times higher than that of the traditional CL. This indicates that the nanofiber CL exhibits significantly greater reaction activity than the traditional CL.

$$I_0 = i_{O_2}^0 \left(e^{\frac{n_{\alpha_0} \alpha_0 F \eta_c}{RT}} - e^{\frac{n_{\alpha_0} (1-\alpha_0) F \eta_c}{RT}} \right) \tag{3.2}$$

$$\eta_c = \frac{RT}{n_{\alpha_0} F i_{O_2}^0} I_0 \tag{3.3}$$

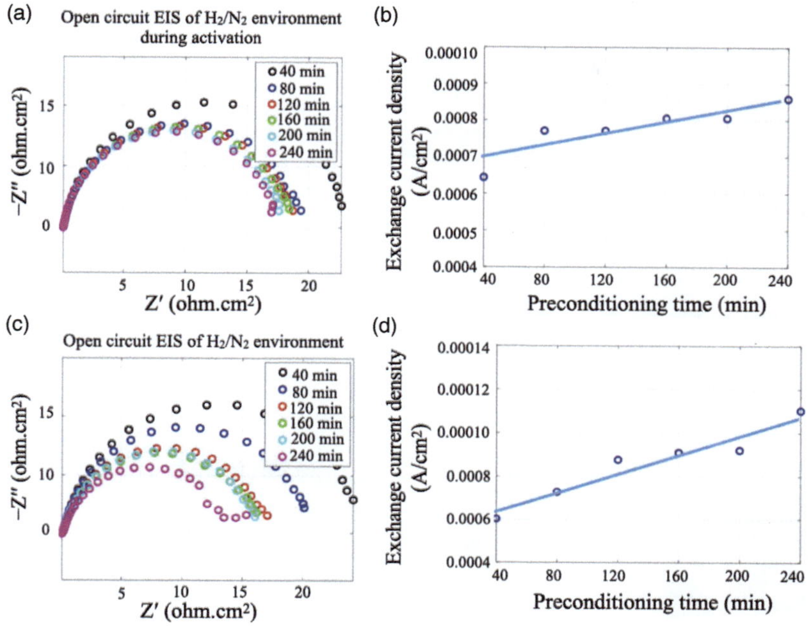

Figure 3.12 (a) and (c) represent EIS curves of conventional CL and electropun CL, (b) and (d) represent the power density curves of conventional conventional CL and electropun CL under hydrogen–nitrogen atmosphere during the preconditioning process.

where $i^0_{O_2}$ is the exchange current density; I_0 is the operating current; η_c is the activation potential; n_{α_0} is the electron transfer efficiency, typically taken as 2.0; F is the Faraday constant (96,485 C/mol); R is ideal gas (8.314 J/(mol·K)); and T is the operating temperature of the fuel cell.

3.4.6 Mechanism of Activation Process

Based on the results of the electrochemical characterization, it is evident that the performance improvement of the traditional CL during activation is primarily due to the increased number of active reaction sites within the CL. This leads to an increase in the ECSA, which, in turn, affects the exchange current density. For the nanofiber CL, the performance changes during activation are due to two main factors: (i) the decrease in electron transfer resistance and the increase in ECSA, which leads to an enhancement in the exchange current density; and (ii) the continuous reduction of mass transfer resistance during activation. Thus, the significant performance change in the nanofiber CL is mainly attributed to the dissolution of PAA (as shown in Figure 3.13). PAA itself is a water-soluble substance [15].

During the fuel cell reaction, water produced dissolves the PAA in the fibers, reducing the polymer thickness covering the Pt surface. This facilitates the easier arrival of reactant gases at the Pt surface. Additionally, since PAA is not a proton-conducting medium, its presence hinders the proton transport pathway [4].

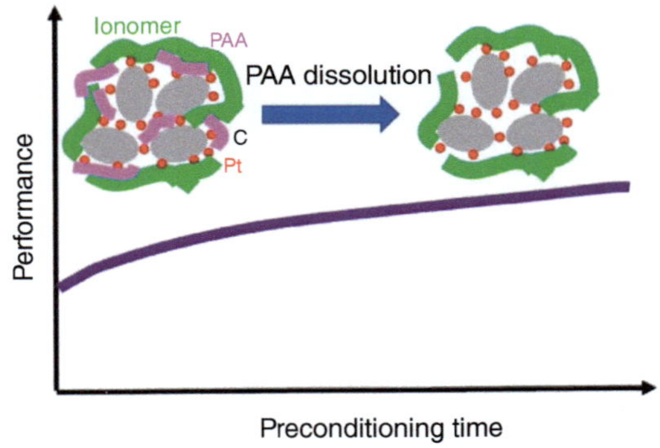

Figure 3.13 Experimental hypothesis.

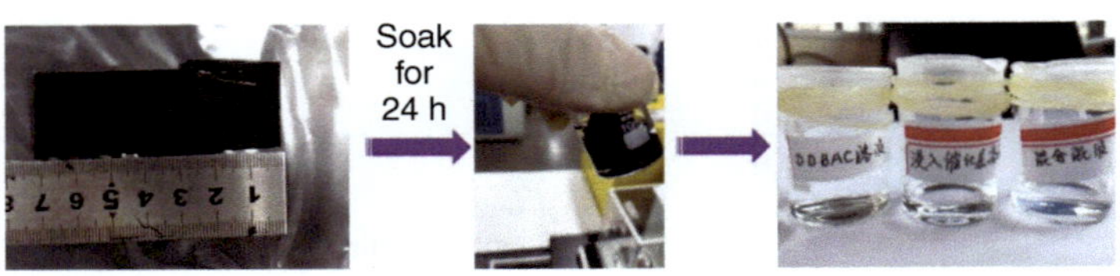

Figure 3.14 PAA detection experiment.

Finally, by comparing the impedance spectra after activation, it can be observed that the electron transfer resistance and mass transfer resistance of the nanofiber CL are still higher than those of the traditional CL. This may be because PAA is not completely dissolved, and some PAA remains within the fibers, maintaining their structural integrity.

To validate the experimental hypothesis, chemical detection and thermogravimetric analysis experiments were conducted to confirm PAA dissolution in the nanofiber CL. The chemical detection process, illustrated in Figure 3.14, involved immersing the prepared nanofiber CL in a known quantity of water at room temperature for 24 hours. This water was then poured into a solution containing dodecyl dimethyl benzyl ammonium chloride (DDBAC), causing it to become cloudy. As PAA was the only component capable of producing this effect, the results confirmed that PAA in the nanofiber structure is water soluble [16].

For further validation, a thermogravimetric analysis experiment was conducted. To simulate conditions before and after activation, the nanofiber CL on aluminum foil was immersed in water at 70 °C for 24 hours, mimicking the 70 °C and 100% humidity environment of the activation process. The fibers were subsequently dried in a vacuum oven at 50 °C for 8 hours. Once dried, they were carefully removed from the foil and analyzed thermogravimetrically (Figure 3.15).

The results, presented in Figure 3.16a, showed that the Pt content in the catalyst was approximately 41.2%, closely matching the manufacturer's stated 40%. Beyond 500 °C, Nafion, PAA, and the carbon components were completely oxidized, leaving only Pt. This indicates that the nanofiber's primary components—Nafion, PAA,

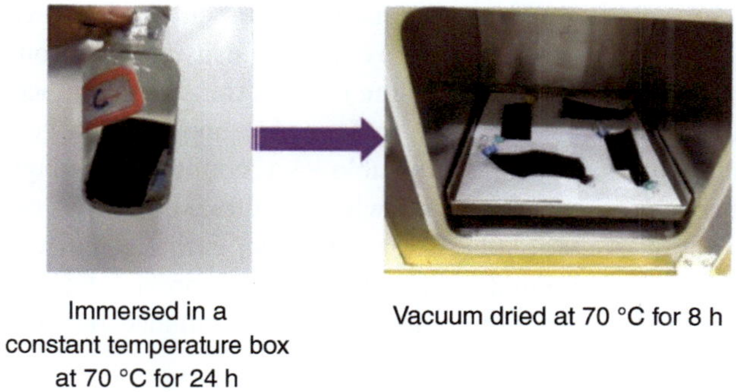

Figure 3.15 Water treatment process for thermal gravimetric sample.

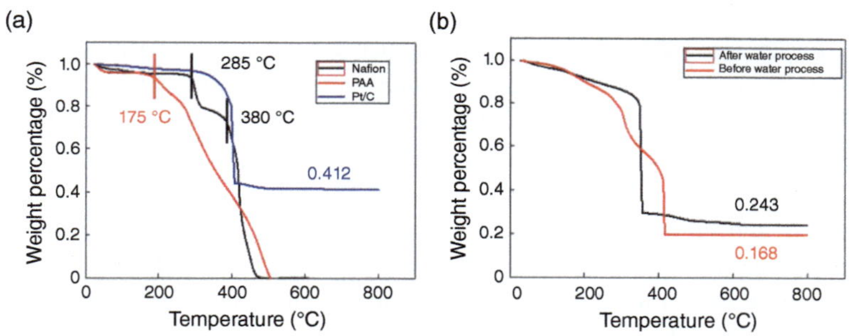

Figure 3.16 Thermal gravimetric analysis (a) PAA, Nafion, Pt/c (b) electrospinning nanofibers catalyst layer before and after water treatment.

and Pt/C—remained intact during the electrospinning process, apart from solvent evaporation. Figure 3.16b highlights that the Pt content in the fibers prior to water treatment was 19.8%, likely reflecting residual water or solvents within the material. After water treatment, this increased to 24.3%, representing a 4.5% rise, which corresponds to material loss during the water treatment process. Since neither Nafion nor Pt/C are water soluble, this material loss was attributed to the dissolution of PAA. Notably, complete PAA dissolution would have resulted in a Pt content of 25%. The observed 24.3% suggests that a small amount of PAA remained, maintaining the structural integrity of the nanofibers after the activation process.

3.5 Conclusion

In this chapter, we examined the activation process of traditional and nanofiber CLs. The traditional CL's performance improved by about 30% during activation, with its ECSA increasing by 43.6% and exchange current density rising by 30.2%. Impedance spectra revealed minimal changes in electron and mass transfer resistances, suggesting that improved performance was driven mainly by the exposure of more active reaction sites. The nanofiber CL, however, exhibited a greater performance enhancement, exceeding 200% during activation. This improvement was marked by a 7.3% increase in ECSA, an 84.9% rise in exchange current density, a continuous reduction in electron transfer resistance, and a decline in mass transfer resistance, resulting in smoother three-phase transport. The performance boost was largely attributed to the partial dissolution of the water-soluble PAA, confirmed through chemical detection and thermogravimetric analysis, which facilitated better access to Pt active sites. Post-activation, the nanofiber CL achieved an exchange current density 1.3 times higher than the traditional layer and an ECSA approximately 1.4 times greater, highlighting the nanofiber structure's ability to enhance reaction activity and fuel cell performance.

References

1 Luo, M., Luo, Z., and Pan, M. (2006). Study on activation process of a PEM fuel cell stack. *Journal of Wuhan University of Technology* 28 (2): 499–502.

2 Xie, J., Xu, F., Wood III, D.L. et al. (2010). Influence of ionomer content on the structure and performance of PEFC membrane electrode assemblies. *Electrochimica Acta* 55 (24): 7404–7412.

3 Ballengee, J.B. and Pintauro, P.N. (2011). Composite fuel cell membranes from dual-nanofiber electrospun mats. *Macromolecules* 44 (18): 7307–7314.

4 Tamura, T. and Kawakami, H. (2010). Aligned electrospun nanofiber composite membranes for fuel cell electrolytes. *Nano Letters* 10 (4): 1324–1328.

5 Lopez-Haro, M., Guétaz, L., Printemps, T. et al. (2014). Three-dimensional analysis of Nafion layers in fuel cell electrodes. *Nature Communications* 5: 5229.

6 Ticianelli, E.A., Derouin, C.R., Redondo, A. et al. (1988). Methods to advance technology of proton exchange membrane fuel cells. *Journal of the Electrochemical Society* 135 (9): 2209–2214.

7 Fofana, D., Hamelin, J., and Bénard, P. (2013). Modelling and experimental validation of high performance low platinum multilayer cathode for polymer electrolyte membrane fuel cells (PEMFCs). *International Journal of Hydrogen Energy* 38 (24): 10050–10062.

8 Tian, Z.Q., Lim, S.H., Poh, C.K. et al. (2011). A highly order-structured membrane electrode assembly with vertically aligned carbon nanotubes for ultra-low pt loading PEM fuel cells. *Advanced Energy Materials* 1 (6): 1205–1214.

9 Lee, J., Yoo, J.M., Ye, Y. et al. (2015). Development of highly stable and mass transfer-enhanced cathode catalysts: support-free electrospun intermetallic FePt nanotubes for polymer electrolyte membrane fuel cells. *Advanced Energy Materials* 5 (11): 1402093.

10 Wang, X., Richey, F.W., Wujcik, K.H. et al. (2014). Ultra-low platinum loadings in polymer electrolyte membrane fuel cell electrodes fabricated via, simultaneous electrospinning/electrospraying method. *Journal of Power Sources* 264 (264): 42–48.

11 Bard, A.J. and Faulkner, L.R. (1986). *Principles and Applications of Chemical Methods*. Beijing.: Chemical Industry Press.

12 Huang, J. and Zhang, J. (2016). Theory of impedance response of porous electrodes: simplifications, inhomogeneities, non-stationarities and applications. *Journal of the Electrochemical Society* 163 (9): A1983–A2000.

13 Huang, J., Li, Z., Liaw, B.Y. et al. (2016). Graphical analysis of electrochemical impedance spectroscopy data in Bode and Nyquist representations. *Journal of Power Sources* 309: 82–98.

14 Zhang, J., Tang, Y., Song, C. et al. (2006). PEM fuel cell open circuit voltage (OCV) in the temperature range of 23°C to 120°C. *Journal of Power Sources* 163 (1): 532–537.

15 Buchholz, F.L. (2000). *Polyacrylamides and Poly(Acrylic acids)*. England: Wiley-VCH Verlag GmbH & Co. KGaA.

16 Zhaoyan, L., Yue, Z., Zhongfa, C. et al. (2014). Determination of polyacrylic acid in circulating water by turbidity. *Industrial Water Treatment* 34 (4): 76–78.

4

Analysis of Nanofiber Catalyst Layers Performance Under Various Temperature and Humidity Conditions

4.1 Introduction

Operating temperature and humidity are critical factors influencing the performance of fuel cells [1]. To better understand the performance of both the traditional and nanofiber catalyst layers (CLs), this chapter investigates polarization curves under hydrogen–air conditions at varying temperatures (60, 70, 80 °C) and relative humidities (RH) of 40%, 60%, 80%, and 100%. These tests aim to identify the optimal operating conditions for each CL. Additionally, loss separation analysis is conducted on the polarization curves obtained under different conditions, offering insight into the underlying reasons for performance differences observed between the two CLs.

4.2 Materials Preparation

All CLs are prepared as same as reported in Section 3.2.

4.3 Result and Discussion

4.3.1 Polarization Curves (*V*−*I*) at Constant Temperature and Different Humidity Levels

Figure 4.1 presents the polarization curves of nanofiber and traditional CLs, both with a platinum loading of 0.1 mg/cm^2, at a fixed temperature while varying humidity levels. It is evident that the performance of both CLs improves as humidity increases. This improvement is attributed to enhanced water content in the proton-conducting medium, which optimizes proton transport channels. Consequently, this reduces ohmic resistance—primarily governed by the proton exchange membrane (PEM)—and mass transport resistance, both of which contribute to better fuel cell performance [2].

Electrospinning for Proton Exchange Membrane Fuel Cells, First Edition.
Yong Liu and Mohideen Meerasahib Mohamedazeem.

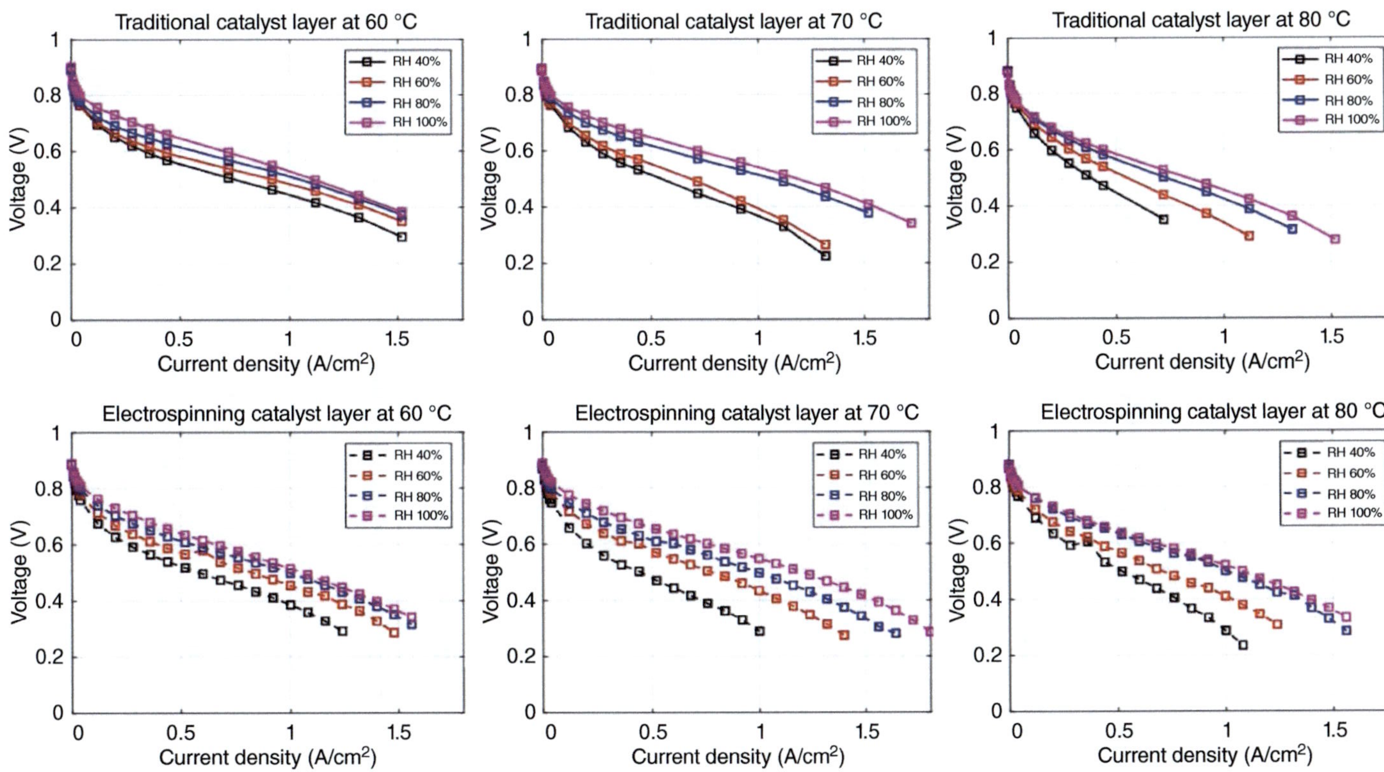

Figure 4.1 The polarization curve of two catalyst layer with a platinum loading of 0.1 mg/cm^2 under same temperature and different humidity.

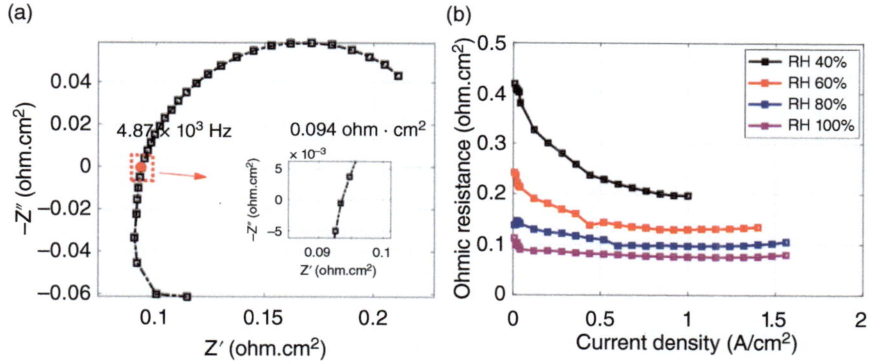

Figure 4.2 (a) The electrochemical impedance spectroscopy curve of electrospinning nanofibers catalyst layer with a platinum loading of 0.1 mg/cm^2. (b) High-frequency resistance and current density diagram of electrospinning nanofibers catalyst layer with a platinum loading of 0.1 mg/cm^2 under 70 °C and different humidity.

To further examine this effect, electrochemical impedance spectroscopy (EIS) was performed at different current densities and humidity levels to evaluate the cell's impedance. The ohmic resistance, determined by the intersection of the EIS curve with the real axis, is shown in Figure 4.2a. The extracted ohmic resistance values under various current densities and humidity conditions are summarized in Figure 4.2b. Key findings include the following: (i) As the current density rises, the ohmic resistance decreases, but the rate of decline diminishes. This is because increased current densities generate more water, and once a certain threshold is reached, proton transport resistance becomes less sensitive to changes in membrane hydration. (ii) When RH increases from 40% to 100%, the ohmic resistance at a given current density decreases consistently. This is due to the higher membrane water content at elevated humidity, which improves proton transport efficiency.

4.3.2 Polarization Curves (*V*–*I*) at Different Temperatures and Constant Humidity

Figure 4.3 shows the polarization curves of nanofiber and traditional CLs with a platinum loading of 0.1 mg/cm^2 at varying temperatures under identical humidity conditions. Under low and moderate humidity levels (RH 40% to RH 80%), the performance of both CLs decreases as temperature rises. This decline is mainly due to higher temperatures reducing the water content in the proton-conducting medium, which disrupts continuous proton transport channels, increases proton transport resistance, and ultimately lowers fuel cell performance [3].

However, under fully humidified conditions (RH 100%), the performance of both fuel cell types initially improves and then diminishes as temperature increases. The highest performance is observed at 70 °C. At this temperature, the increased reaction rates and improved proton conduction outweigh the negative impact of water evaporation, resulting in better overall performance. At 80 °C, however, the rapid evaporation of water in the proton-conducting medium leads to a significant loss of water content, which increases resistance and reduces performance.

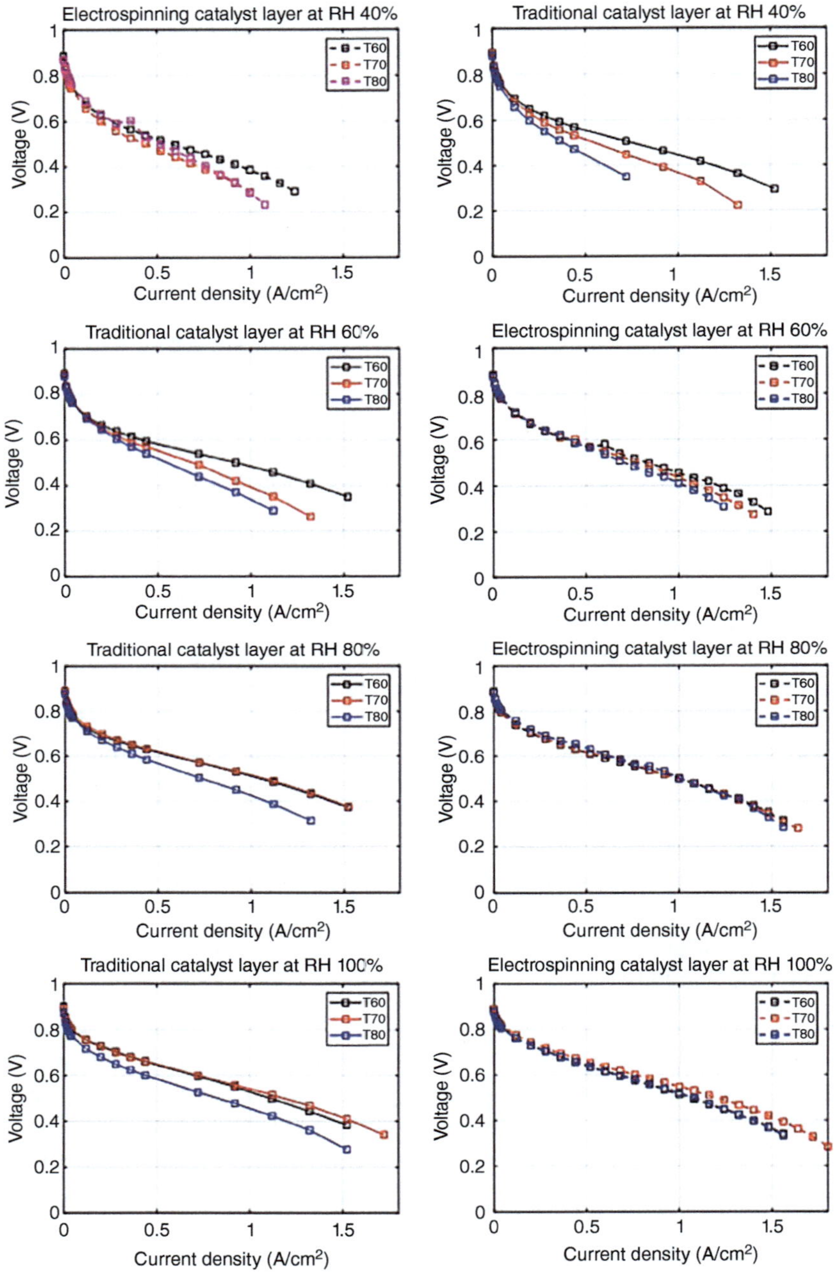

Figure 4.3 The polarization curve of two catalyst layer with a platinum loading of 0.1 mg/cm^2 under same humidity and different temperature.

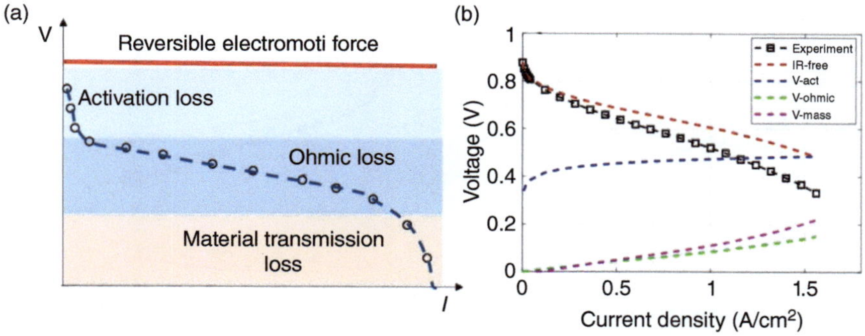

Figure 4.4 (a) The typical polarization curve. (b) Voltage losses of fuel cell.

4.3.3 Polarization Curves and Loss Separation at Different Temperatures and Humidity

In fuel cells, internal resistance leads to a voltage drop between the anode and cathode, resulting in a cell voltage lower than the reversible electromotive force. On a polarization curve, these voltage losses can be categorized into three main regions (Figure 4.4a). At low current densities, activation loss (V_{act}) dominates, reflecting the oxygen reduction reaction (ORR) voltage loss. As the current density increases, the voltage decline becomes linear due to ohmic resistance, and this ohmic loss (V_{ohmic}) becomes the primary factor. At higher current densities, limitations in oxygen transport at the cathode cause a nonlinear voltage drop, resulting in mass transport loss (V_{mass}) and a limiting current density [4, 5].

To evaluate the influence of the nanofiber and traditional CLs under various temperature and humidity conditions, a loss separation analysis was performed. The separation process involved the following:

1) **Ohmic Loss Separation (V-ohmic):** EIS was used to measure ohmic resistance across different current densities and conditions. The measured voltage was corrected by adding the ohmic loss component, producing an IR-free V–I curve.
2) **Activation Loss Separation (V-act):** At low current densities (0–0.1 A/cm^2), fuel cell performance is primarily influenced by activation and ohmic losses. Using the Tafel equation ($\eta = a + b \log j$), the IR-free V–I curve in this range was linearly fitted, isolating the activation loss.
3) **Mass Transport Loss Separation (V-mass):** The reversible electromotive force (E_{rev}) was calculated according to Eq. (4.1), and mass transport loss (V-mass) was determined using Eq. (4.2).

$$E_{rev} = 1.23 - 0.9 \times 10^{-3}(T - 298) + \frac{2.303RT}{4F} \times \log\left[\left(\frac{P_{O_2}}{P^*_{O_2}}\right)\left(\frac{P_{H_2}}{P^*_{H_2}}\right)^2\right] \tag{4.1}$$

$$V_{mass} = E_{rev} - V_{cell} - V_{ohmic} - V_{act} \tag{4.2}$$

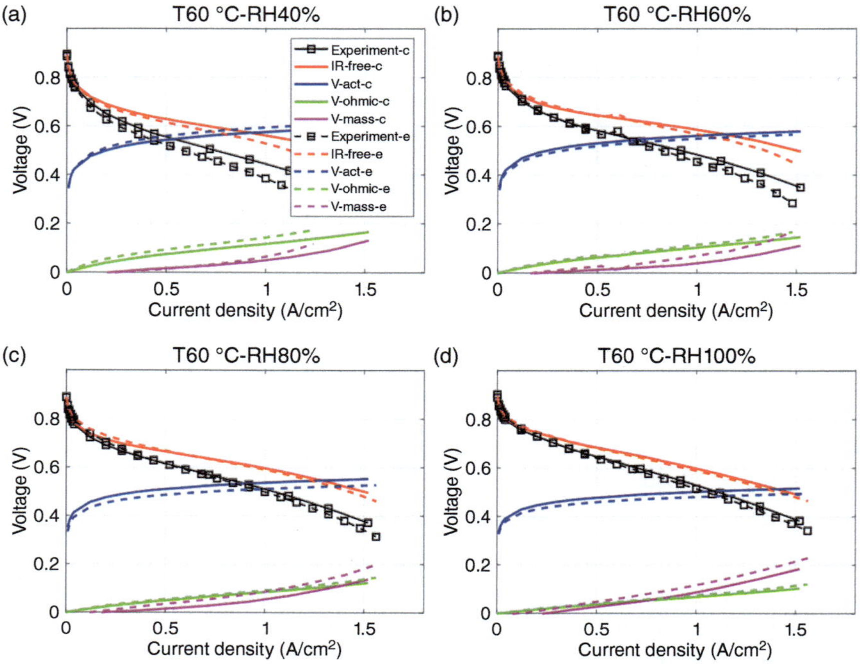

Figure 4.5 The polarization curves and voltage losses of two catalyst layer with a platinum loading of 0.1 mg/cm^2 under 60 °C and different humidity: (a) RH 40%, (b) RH 60%, (c) RH 80%, (d) RH 100%. (e: electrospinning nanofibers catalyst layer; C: conventional catalyst layer; IR-Free: polarization curves eliminating ohmic resistance; V-ohmic: ohmic losses; V-act: activation losses; V-mass: mass transport losses).

As shown in Figure 4.5, the polarization curves and loss separation data for nanofiber CLs and conventional CLs, both with a platinum loading of 0.1 mg/cm^2, were analyzed under varying humidity conditions at 60 °C. Across all humidity levels, the conventional CL consistently outperforms the nanofiber CL at this temperature. Loss separation reveals that ohmic losses remain nearly identical for both CLs, as both use the same commercial Nafion 211 PEMs. However, under medium- to high-humidity conditions (RH 60%–RH 100%), the nanofiber catalyst layers exhibit lower activation losses than their conventional counterparts, indicating that reactants more easily reach active sites within the nanofiber structure. Despite this advantage, the nanofiber CLs suffer from greater mass transport losses, which ultimately limit their overall performance relative to the conventional CLs at 60 °C.

In Figure 4.6, the polarization and loss separation data at 70 °C show different trends. Under RH 40% and RH 80% conditions, nanofiber CLs achieve better performance than do conventional layers. At RH 100%, both layers perform nearly equally. At RH 60%, the nanofiber CLs clearly surpass the performance of the traditional layers. Loss separation under these conditions highlights similar ohmic losses for both layers, while activation losses for the nanofiber CLs are slightly lower.

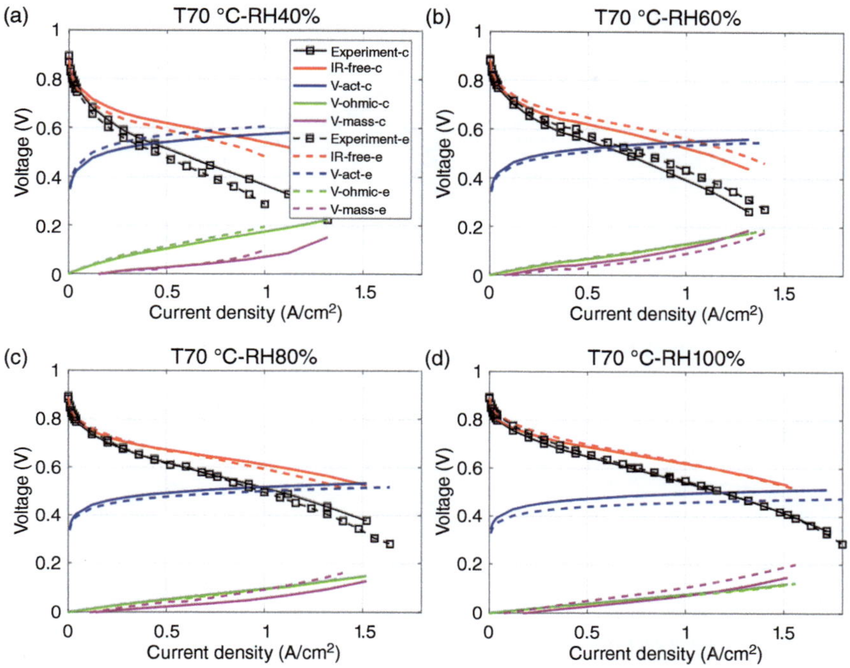

Figure 4.6 The polarization curves and voltage losses of two catalyst layer with a platinum loading of 0.1 mg/cm^2 under 70 °C and different humidity: (a) RH 40%, (b) RH 60%, (c) RH 80%, (d) RH 100%. (e: electrospinning nanofibers catalyst layer; C: conventional catalyst layer; IR-Free: polarization curves eliminating ohmic resistance; V-ohmic: ohmic losses; V-act: activation losses; V-mass: mass transport losses).

However, a more notable difference is observed in the mass transport losses, with the nanofiber CLs showing superior transport characteristics under certain humidity levels.

Figure 4.7 displays the polarization and loss separation results at 80 °C. Here, the nanofiber CLs outperform the conventional CLs at all humidity levels. The loss separation data indicate that while ohmic and mass transport losses are comparable between the two, the activation loss for nanofiber CLs is substantially lower, contributing significantly to their improved performance.

4.4 Conclusion

This chapter presents a detailed analysis of nanofiber and traditional CL performance under varying humidity and temperature conditions. It was found that increasing humidity generally enhances proton conductivity, lowering ohmic and transport losses, while higher current densities lead to reduced ohmic resistance, albeit at a decreasing rate. Under low to moderate humidity, performance drops

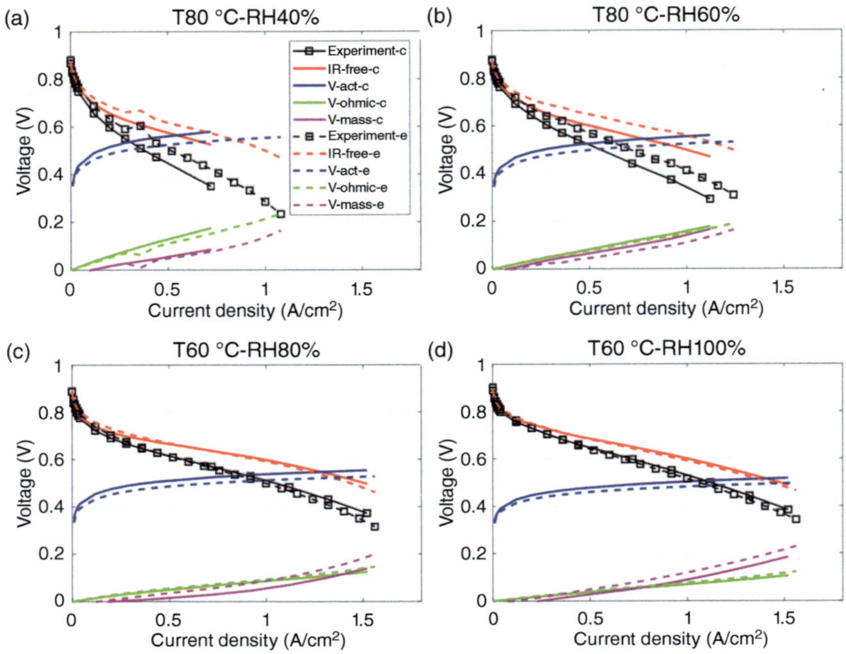

Figure 4.7 The polarization curves and voltage losses of two catalyst layer with a platinum loading of 0.1 mg/cm^2 under 80 °C and different humidity: (a) RH 40%, (b) RH 60%, (c) RH 80%, (d) RH 100%. (e: electrospinning nanofibers catalyst layer; C: conventional catalyst layer; IR-Free: polarization curves eliminating ohmic resistance; V-ohmic: ohmic losses; V-act: activation losses; V-mass: mass transport losses).

with rising temperature, but at high humidity levels, both catalyst types perform best at 70 °C. At 60 and 70 °C, traditional CLs show slightly better or equivalent performance, but at 80 °C, nanofiber CLs surpass them. Loss separation data reveal that both types have similar ohmic losses, while nanofiber layers demonstrate lower activation losses due to their larger specific surface area and increased active reaction sites. However, nanofiber layers also face higher mass transport losses, attributed to greater thickness and lower hydrophobicity, despite offering enhanced gas and proton transport properties.

References

1 Schmittinger, W. and Vahidi, A. (2008). A review of the main parameters influencing long-term performance and durability of Proton exchange membrane fuel cells. *Journal of Power Sources* 180 (1): 1–14.

2 Zawodzinski, T.A. (1993). Water uptake by and transport through Nafion® 117 membranes. *Journal of the Electrochemical Society* 140 (4): 1041–1047.

3 Mench, M.M. (2008). *Fuel Cell Engines*. New York: John Wiley & Sons Ltd.

4 Huang, J., Li, Z., and Zhang, J. (2017). Review of characterization and modeling of polymer electrolyte fuel cell catalyst layer: the blessing and curse of ionomer. *Frontiers in Energy* 11 (3): 334–364.

5 Issa, A.A., Al-Maadeed, M., Luyt, A.S. et al. (2016). Investigation of the physico-mechanical properties of electrospun PVDF/cellulose (nano)fibers. *Journal of Applied Polymer Science* 133 (26): 43594.

5

The Role and Performance of Gas Diffusion Layers in Proton Exchange Membrane Fuel Cell

5.1 Introduction

The gas diffusion layer (GDL) serves as a critical component in proton exchange membrane fuel cell (PEMFC), playing a vital role in maintaining the proper balance between mechanical strength, electrical conductivity, and porosity [1–3]. The GDL is generally composed of a support layer—most commonly made from carbon paper—and a microporous layer (MPL). The support layer must offer adequate mechanical properties, high electrical conductivity, and an optimized pore structure to facilitate efficient transport of reactants and by-products [4, 5].

Common materials used for the GDL support layer include:

- **Carbon Paper**: Produced by compressing carbon fibers, carbon paper features excellent electrical conductivity and a well-suited porosity, making it the most widely used material in PEMFC GDLs.
- **Carbon Cloth**: Though softer and more flexible than carbon paper, carbon cloth provides lower mechanical strength and is less commonly employed.
- **Metal Mesh**: While offering superior mechanical durability, metal mesh is susceptible to corrosion and typically requires anticorrosion coatings, which add complexity to its use as a GDL support layer.

The MPL is formed by combining carbon black with hydrophobic agents, most commonly polytetrafluoroethylene (PTFE) and applying the mixture to the carbon paper via spraying or rolling. The addition of the MPL significantly enhances GDL performance by redistributing water and gas, reducing interfacial resistance between the support layer and the catalyst layer (CL), and improving overall fuel cell efficiency. With increasing interest in PEMFCs, research on GDL materials, structures, and functionalities has grown rapidly [6]. Studies now address the intricate water and gas transport mechanisms within GDLs, explore various fabrication methods, and refine characterization techniques. Between 2000 and October 2020, a keyword search for "Proton Exchange Membrane Fuel Cell" in the Web of Science database yielded over 16,500 research papers, highlighting the substantial progress and growing knowledge base in this field.

Electrospinning for Proton Exchange Membrane Fuel Cells, First Edition.
Yong Liu and Mohideen Meerasahib Mohamedazeem.

5.2 Research on Gas Diffusion Layer and Water/Gas Transport Mechanisms

Recent research on GDLs has focused primarily on their dual-layer structure: the carbon paper support layer and the MPL. This section summarizes key findings and advancements concerning these two essential components. Studies have demonstrated that the performance of the carbon paper support layer is strongly influenced by its hydrophobic treatment and mechanical compression properties.

Lim et al. [7] investigated the use of fluorinated ethylene propylene (FEP) as a hydrophobic agent for carbon paper. Their findings indicated that a 10 wt% FEP content yielded the best fuel cell performance, as shown in their polarization curve tests (Figure 5.1). Higher FEP concentrations (up to 40 wt%) resulted in pore blockage, reducing the GDL's water management capability. Interestingly, the water contact angle remained stable across the tested FEP contents, suggesting that a moderate level of hydrophobicity—achieved with 10 wt% FEP—was sufficient to enhance water removal without compromising gas transport.

Ge et al. [8] examined the effect of compression on fuel cell performance, employing a specialized fixture to precisely control the degree of compression. The study compared the performance of GDLs made from carbon cloth and carbon paper under varying compression ratios. As illustrated in Figure 5.2, increasing compression negatively affected fuel cell performance at high current densities for both materials. At lower compression ratios, however, performance improved because the pore structure remained intact, promoting efficient water and gas transport. This finding highlights the importance of developing carbon paper support layers with higher compressive resistance to minimize deformation and maintain consistent voltage output at elevated current densities.

Park et al. [9] investigated the influence of long versus short carbon fibers on carbon paper used as GDL support layers in PEMFCs. Their research demonstrated that carbon paper made from long fibers had larger pores, which not only improved water permeability but also increased ohmic resistance due to a rougher surface.

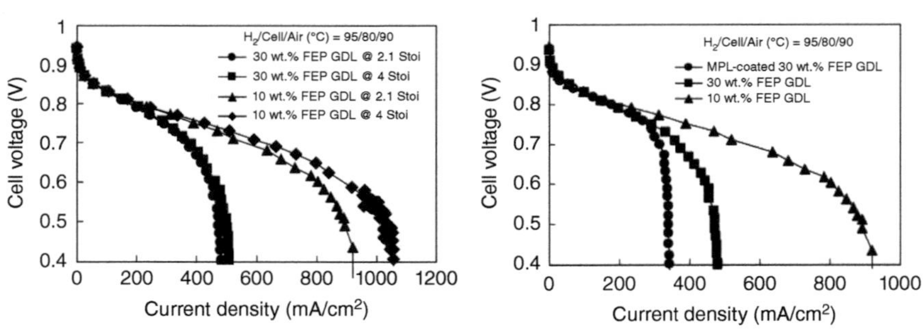

Figure 5.1 Schematic diagram of the influence of different content of fluorinated ethylene propylene in carbon paper on the test results of PEMFC polarization curve. Source: Lim and Wang [7]/with permission of Elsevier.

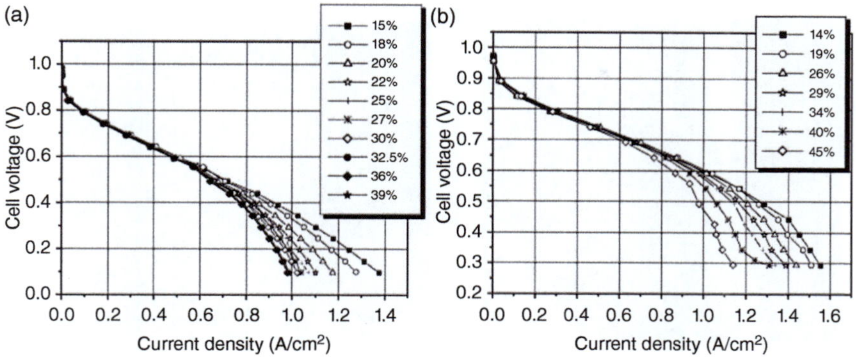

Figure 5.2 Polarization curves at different compression levels using (a) carbon cloth and (b) carbon fiber paper as gas diffusion layers (GDLs). Source: Ge et al. [8]/with permission of Elsevier.

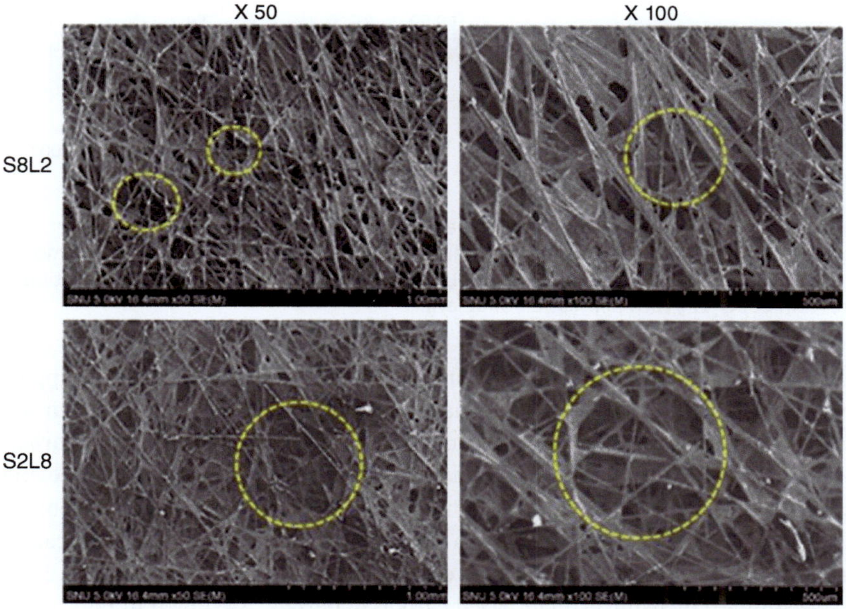

Figure 5.3 Scanning electron microscope images of carbon paper about short (S8L2) and long carbon fiber (S2L8). Source: Park et al. [9]/with permission of Elsevier.

Conversely, short-fiber carbon paper contained fewer large pores, negatively affecting water penetration into the MPL. Scanning electron microscope (SEM) images (Figure 5.3) of short (S8L2)- and long (S2L8)-fiber carbon papers supported these findings. Under 100% humidity, short-fiber carbon paper enhanced MPL water retention, while long-fiber paper showed 18.8% better water drainage. Electrochemical testing revealed that long-fiber carbon paper provided 10.3% higher voltage at 1.6 A/cm^2, attributed to larger pores facilitating water transport.

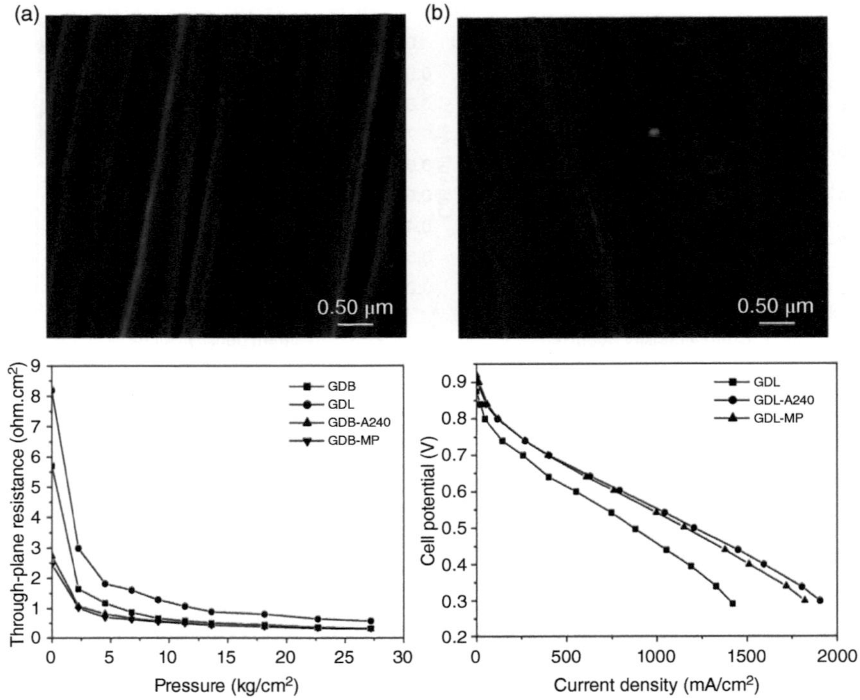

Figure 5.4 Comparison of physical properties and polarization curves of carbon fiber before and after modification. (a: before modification; b: after modification; GDB: commonly used carbon paper; GDL: commonly used GDL; GDB-A240 and GDB-MP: after modification.) Source: Lin et al. [12]/with permission of Elsevier.

Tang et al. [10] explored carbon nanotube–supported membranes as GDL support layers. With superior conductivity, mechanical strength, and hydrophilicity, carbon nanotubes emerged as promising GDL materials. Liu et al. [11] enhanced carbon paper performance by adding 10% phenolic resin and 2% carbon black, achieving a maximum power density of 0.613 W/cm^2 and a current density of 1.026 A/cm^2 at 40 °C (at 0.5 V). The resulting changes in pore size and hydrophobicity significantly boosted cell performance. Similarly, Lin et al. [12] improved conductivity and compressive strength by using asphalt-based carbon materials. As shown in Figure 5.4, their result indicated that, at 0.5 V, current density with asphalt-based materials was 37% higher than traditional carbon paper, driven by better fiber bonding and higher carbon content (over 90%), which improved conductivity and reduced voltage loss.

Yang et al. [13] developed ultrathin carbon paper by combining phenolic resin and graphene oxide. The addition of graphene oxide, which reduced porosity and water permeability, allowed the paper to achieve a tensile strength of 31 MPa and a resistivity of 2.1 mΩ·cm. This approach is also applicable to ultrathin proton exchange membrane (PEM), where graphene oxide improves water retention and proton conductivity under low humidity, while phenolic resin enhances membrane strength and durability.

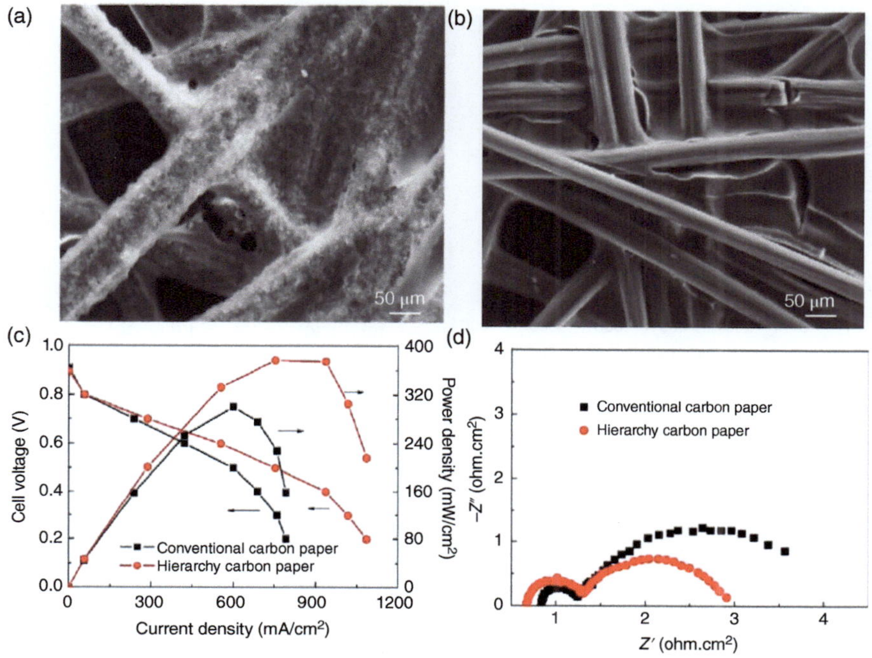

Figure 5.5 The effect of layered carbon paper and traditional carbon paper as the supporting layer of gas diffusion layer on proton exchange membrane fuel cell performance. (a: layered carbon paper; b: traditional carbon paper.) Source: Du et al. [14]/with permission of Elsevier.

Du et al. [14] designed layered carbon paper, which showed excellent hydrophobicity, low contact resistance, and high performance at both low and high current densities (Figure 5.5). SEM analysis revealed that larger pores in layered carbon paper helped expel water, reducing flooding and mass transport losses. Stuckey et al. grew carbon nanotubes (CNTs) in situ on carbon paper, achieving stable battery performance and a power density of 1 W/cm^2 under varied humidity conditions. Xie et al. enhanced carbon paper conductivity and gas permeability by chemically depositing acrylic acid and performing pyrolysis. Hung et al. created ultrathin carbon paper using 6-mm-long PAN carbon fibers, resulting in improved water/gas transmission and reduced mass transport losses.

Velan et al. [15] examined the effects of compression and high temperatures on GDL thickness, pore size, and conductivity. Their studies highlighted the need for high-hardness binders to reduce compression and suggested applying MPL before compression to enhance GDL density and minimize deformation. Delikaya et al. [16] used electrospinning to produce carbon fiber felts for GDLs as shown in Figure 5.6. Although the electrospun materials showed lower polarization and electrochemical performance compared to traditional carbon paper, their approach offers a novel fabrication method for future optimization.

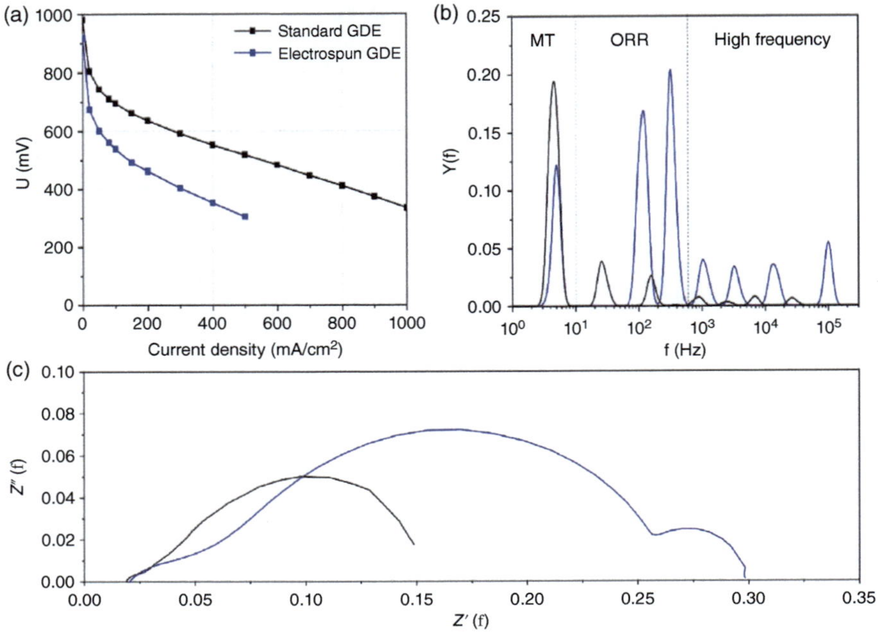

Figure 5.6 Electrospinning technology is used to prepare gas diffusion electrode and related performance test results. Source: Delikaya et al. [16]/with permission of Elsevier.

Overall, current research on GDL support layers focuses on several key areas: improving hydrophobicity through material selection, enhancing conductivity via carbon modifications, optimizing pore size by using fibers of varying lengths and multilayer structures, and strengthening mechanical properties by choosing high-strength fibers and effective binders.

5.2.1 Research on Microporous Layer

The MPL is a critical component of the GDL, typically constructed from carbon powder and PTFE, generally measuring tens to several tens of micrometers thick. Its primary functions include expelling excess water, delivering fuel, and reducing contact resistance between the GDL and CL. The structure of the MPL, which contains numerous micropores, facilitates effective water and gas transport within the fuel cell. From a keyword search on Web of Science using terms such as "Proton exchange membrane fuel cell," "GDL," and "MPL," 118 papers published between 2000 and October 2020 were identified. These studies largely focus on optimizing water management and gas diffusion [17–20]. Research highlights include investigations into the effects of MPL hydrophobicity, conductivity, pore size, porosity, and the benefits of employing double-layer or multilayer MPLs.

For example, Wang et al. [21] explored the impact of different carbon blacks on MPL performance. By mixing acetylene black (80 wt%) with Black Pearls 2000 (20 wt%), they observed significant improvements in the GDL's gas transmission

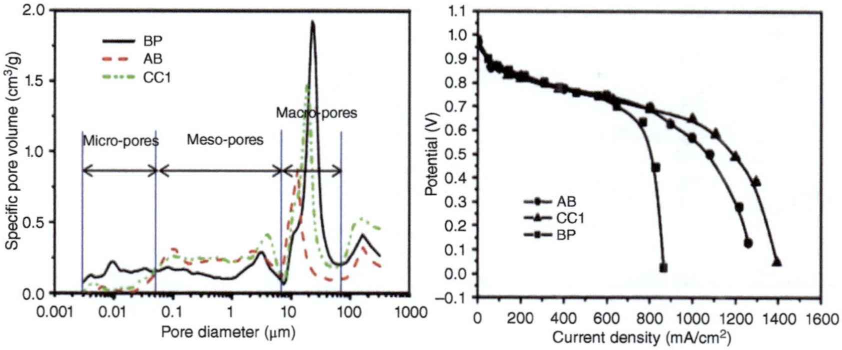

Figure 5.7 The influence of microporous layer prepared from different carbon blacks on proton exchange membrane fuel cell performance. (BP: black pearl powder 2000; AB: acetylene black; CC1: acetylene black is 80 wt%, Black Pearl powder 2000 is 20 wt%.) Source: Wang et al. [21]/with permission of Elsevier.

and water management capabilities (Figure 5.7). Experimental results demonstrated that the mixed carbon black MPL exhibited lower voltage loss at high current densities compared to single-type carbon black MPLs. This performance enhancement was linked to the increased proportion of mesopores and optimized pore size distribution resulting from the particle size differences—acetylene black measuring 35–45 nm and Black Pearls 2000 measuring 15 nm. The varied pore structures created at the contact points allowed for more controlled and favorable water and gas transmission properties in the GDL.

Zhan et al. [22] investigated the effects of uniform versus gradient porosity in GDLs. Their findings showed that gradient porosity facilitated more efficient transport of water and gases. However, when GDL porosity remained fixed, increasing MPL thickness reduced water transport efficiency. In related work, Chen-Yang et al. [23] demonstrated that 20 wt% PTFE content in the MPL provided the best fuel cell performance. Dai et al. [24] further examined PTFE levels, discovering that higher carbon loading in the MPL reduced the GDL's drainage capability. Additional studies by Chen et al. [25] on gradient porosity GDLs revealed that greater porosity differences improved fuel cell performance by enhancing water transport and reducing saturation. This created more pores for gas diffusion, leading to better efficiency.

Tseng et al. [26] conducted a comprehensive analysis of MPL's role in fuel cells under various conditions. They found that while MPL thickness had little effect at low current densities, excessively thin MPLs impeded gas diffusion and overly thick MPLs increased GDL resistance. Their research confirmed that a 20 wt% PTFE content yielded optimal performance. Prolonged sintering times enhanced PTFE distribution uniformity, which improved gas permeability. Tseng's findings point toward future work on optimizing MPL thickness and sintering processes.

Duan et al. [27] fabricated MPLs via electrospinning, achieving larger pores and improved performance, as shown in Figure 5.8. The electrospun MPLs demonstrated

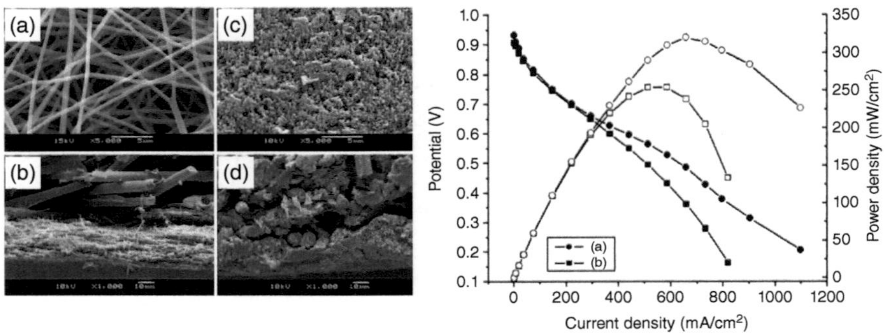

Figure 5.8 Surface morphology and corresponding fuel cell performance of MPL prepared by traditional technology. Source: Duan et al. [27]/with permission of Elsevier.

better limit current density and power peaks, attributed to their crack-free surfaces and three-dimensional porous structure. This method produced MPLs with higher specific surface area and porosity, improving overall fuel cell performance.

Sasabe et al. [28] used X-ray visualization to investigate water accumulation and removal in fuel cells. Their results confirmed that adding an MPL to the GDL improved water expulsion. However, when carbon cloth served as the support layer, water tended to collect in woven areas, hindering removal. FEP-based MPLs had smoother surfaces and fewer cracks, reducing contact resistance and enhancing water management.

Kitahara et al. [29] developed a dual-layer MPL, with the side closer to the carbon paper being hydrophobic and the side nearer the CL hydrophilic. This design improved moisture retention and CL humidification under nonhumidified conditions (Figure 5.9). They also found that reducing MPL thickness and increasing hydrophilicity enhanced cell performance. In addition, subsequent research on three-layer MPLs showed that varying hydrophobicity among layers could further improve performance. Chen et al. [31] examined different carbon blacks in cathode MPLs, showing that carbon black type significantly influenced performance under various humidity conditions.

Yu et al. [30] replaced anhydrous ethanol with a polyvinylpyrrolidone (PVP) solution to disperse PTFE, producing MPLs with higher power density and lower mass transport losses. PVP improved PTFE and carbon black distribution, creating more pores and enhancing water and gas transport. The hydrophilic properties of PVP also balanced the MPL surface's hydrophilic–hydrophobic character, boosting overall performance. Morgan and Kim et al. [32, 33] provided detailed analyses of how GDL and MPL parameters thickness, pore size, and porosity affect electrochemical performance, emphasizing tailored optimization. Leeuwne et al. [34] showed that graphene foam–based MPLs reduced ohmic and mass transport losses, crediting graphene's unique structure and high conductivity. Ebenezer et al. [35] introduced CNTs into MPLs, improving high-temperature resistance and thermal stability while outperforming conventional materials.

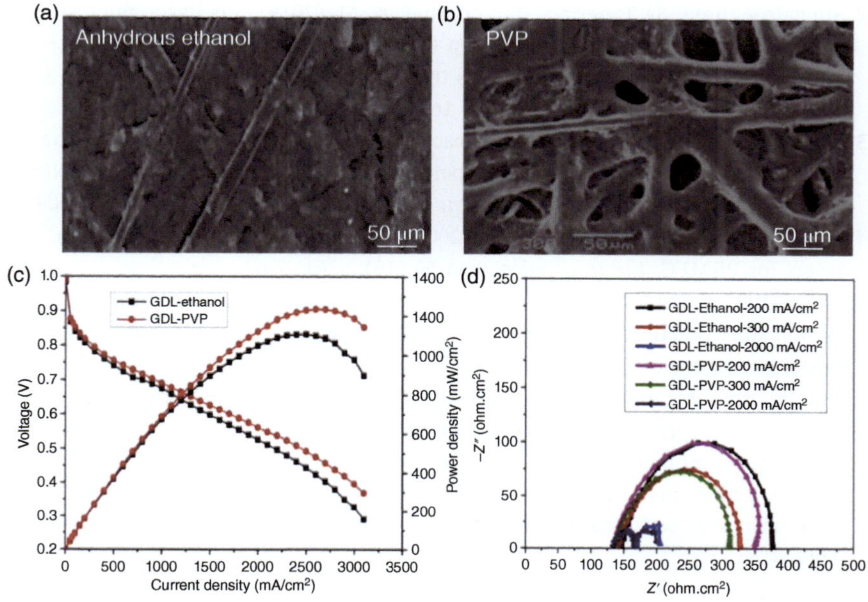

Figure 5.9 SEM images of the microporous layer (MPL) surface prepared with polyvinylpyrrolidone (PVP) aqueous solution at different magnifications: (a) 100× and (b) 300×. (c) Polarization curves of fuel cells using different cathode gas diffusion layers (GDLs). (d) Electrochemical impedance spectroscopy (EIS) spectra of GDLs measured at current densities of 200, 300, and 2000 mA/cm^2. Source: Yu et al. [30]/with permission of Elsevier.

5.3 Conclusion

This chapter has examined the critical role of GDLs and MPLs in fuel cells. Key findings include the advantages of gradient porosity for water and gas transport, the importance of optimizing PTFE content for improved performance, and the benefits of advanced MPL designs such as dual-layer structures and electrospun materials—for better conductivity and mass transport. Innovations in fabrication methods, material combinations, and structural modifications have consistently shown that fine-tuning the properties of GDLs and MPLs can significantly enhance fuel cell efficiency and durability. The research discussed here serves as a valuable guide for developing next-generation PEMFC components.

References

1 Taherian, R., Ghorbani, M.M., and Kiahosseini, S.R. (2018). A new method for optimal fabrication of carbon composite paper as gas diffusion layer used in proton exchange membrane of fuel cells. *Journal of Electroanalytical Chemistry* 815: 90–97.

2 Simaafrookhteh, S., Taherian, R., and Shakeri, M. (2019). Stochastic microstructure reconstruction of a binder/carbon fiber/expanded graphite carbon fiber paper for PEMFCs applications: mass transport and conductivity properties. *Journal of the Electrochemical Society* 166 (7): 3287–3299.

3 Chen, W. and Jiang, F.M. (2016). Impact of PTFE content and distribution on liquid-gas flow in PEMFC carbon paper gas distribution layer: 3D lattice Boltzmann simulations. *International Journal of Hydrogen Energy* 41 (20): 8550–8562.

4 Han, M., Xu, J.H., Chan, S.H. et al. (2008). Characterization of gas diffusion layers for PEMFC. *Electrochimica Acta* 53 (16): 5361–5367.

5 Yang, H.H., Tu, H.C., and Chiang, I.L. (2010). Carbon cloth based on PAN carbon fiber practicability for PEMFC applications. *International Journal of Hydrogen Energy* 35 (7): 2791–2795.

6 Joo, D., Jin, S.M., Jang, J.H. et al. (2018). Development of a self-supporting microporous layer on a metal mesh for carbon backing-free cathodes in proton exchange membrane fuel cells. *Fuel Cells* 18 (1): 57–62.

7 Lim, C. and Wang, C.Y. (2004). Effects of hydrophobic polymer content in GDL on power performance of a PEM fuel cell. *Electrochimica Acta* 49 (24): 4149–4156.

8 Ge, J.B., Higier, A., and Liu, H.T. (2006). Effect of gas diffusion layer compression on PEM fuel cell performance. *Journal of Power Sources* 159 (2): 922–927.

9 Park, J., Oh, H., Lee, Y.I. et al. (2016). Effect of the pore size variation in the substrate of the gas diffusion layer on water management and fuel cell performance. *Applied Energy* 171: 200–212.

10 Tang, J.M., Itkis, M.E., Wang, C. et al. (2006). Carbon nanotube free-standing membrane as gas diffusion layer in hydrogen fuel cells. *Micro & Nano Letters* 1 (1): 62–65.

11 Liu, C.H., Ko, T.H., Liao, Y.K. et al. (2008). Effect of carbon black concentration in carbon fiber paper on the performance of low-temperature proton exchange membrane fuel cells. *Journal of Power Sources* 178 (1): 80–85.

12 Lin, J.H., Chen, W.H., Su, S.H. et al. (2008). Carbon film coating on gas diffusion layer for proton exchange membrane fuel cells. *Journal of Power Sources* 184 (1): 38–43.

13 Yang, P.P., Xie, Z.Y., and Li, H.B. (2017). Graphene oxide reinforced ultra-thin carbon paper used for fuel cells and the mechanisms of reinforcement. *International Journal of Hydrogen Energy* 42 (16): 11699–11709.

14 Du, C.Y., Wang, B.R., and Cheng, X.Q. (2009). Hierarchy carbon paper for the gas diffusion layer of proton exchange membrane fuel cells. *Journal of Power Sources* 187 (2): 505–508.

15 Velan, V.S., Velayutham, G., Rajalakshmi, N. et al. (2014). Influence of compressive stress on the pore structure of carbon cloth based gas diffusion layer investigated by capillary flow porometry. *International Journal of Hydrogen Energy* 39 (4): 1752–1759.

16 Delikaya, O., Bevilacqua, N., Eifert, L. et al. (2020). Porous electrospun carbon nano fibers network as an integrated electrode@gas diffusion layer for high temperature polymer electrolyte membrane fuel cells. *Electrochimica Acta* 345: 1–6.

17 Wang, M., Medina, S., Pfeilsticker, J.R. et al. (2019). Impact of microporous layer roughness on gas-diffusion-electrode-based polymer electrolyte membrane fuel cell performance. *ACS Applied Energy Materials* 2 (11): 7757–7761.

18 Shan, J., Lin, R., Chen, X.D. et al. (2018). EIS and local resolved current density distribution analysis on effects of MPL on PEMFC performance at varied humidification. *International Journal of Heat and Mass Transfer* 12: 1076–1083.

19 Nanadegani, F.S., Lay, E.N., and Sunden, B. (2019). Effects of an MPL on water and thermal management in a PEMFC. *International Journal of Energy Research* 43 (1): 274–296.

20 Lu, Y.X., Steinberger-Wilckens, R., and Du, S.F. (2018). Evolution of gas diffusion layer structures for aligned Pt nanowire electrodes in PEMFC applications. *Electrochimica Acta* 279: 99–107.

21 Wang, X.L., Zhang, H.M., Zhang, J.L. et al. (2006). A bi-functional micro-porous layer with composite carbon black for PEM fuel cells. *Journal of Power Sources* 162 (1): 474–479.

22 Zhan, Z.G., Xiao, J.S., Zhang, Y.S. et al. (2007). Gas diffusion through differently structured gas diffusion layers of PEM fuel cells. *International Journal of Hydrogen Energy* 32 (17): 4443–4451.

23 Chen-Yang, Y.W., Hung, T.F., Huang, J. et al. (2007). Novel single-layer gas diffusion layer based on PTFE/carbon black composite for proton exchange membrane fuel cell. *Journal of Power Sources* 173 (1): 183–188.

24 Dai, W., Wang, H., Yuan, X.Z. et al. (2009). Measurement of water transport rates across the gas diffusion layer in a proton exchange membrane fuel cell, and the influence of polytetrafluoroethylene content and micro-porous layer. *Journal of Power Sources* 188 (1): 122–126.

25 Chen, F.L., Chang, M.H., and Hsieh, P.T. (2008). Two-phase transport in the cathode gas diffusion layer of PEM fuel cell with a gradient in porosity. *International Journal of Hydrogen Energy* 33 (10): 2525–2529.

26 Tseng, C.J. and Lo, S.K. (2010). Effects of microstructure characteristics of gas diffusion layer and microporous layer on the performance of PEMFC. *Energy Conversion and Management* 51 (4): 677–684.

27 Duan, Q.J., Wang, B.A., Wang, J.O. et al. (2010). Fabrication of a carbon nanofiber sheet as a micro-porous layer for proton exchange membrane fuel cells. *Journal of Power Sources* 195 (24): 8189–8193.

28 Sasabe, T., Deevanhxay, P., Tsushima, S. et al. (2011). Investigation on the effect of microstructure of proton exchange membrane fuel cell porous layers on liquid water behavior by soft X-ray radiography. *Journal of Power Sources* 196 (20): 8197–8206.

29 Kitahara, T., Nakajima, H., and Mori, K. (2012). Hydrophilic and hydrophobic double microporous layer coated gas diffusion layer for enhancing performance of polymer electrolyte fuel cells under no-humidification at the cathode. *Journal of Power Sources* 199: 29–36.

30 Yu, S.C., Li, X.J., Hao, J.K. et al. (2014). Preparation of microporous layer for proton exchange membrane fuel cell by using polyvinylpyrrolidone aqueous solution. *International Journal of Hydrogen Energy* 39 (28): 15681–15686.

31 Chen, H.H. and Chang, M.H. (2013). Effect of cathode microporous layer composition on proton exchange membrane fuel cell performance under different air inlet relative humidity. *Journal of Power Sources* 232: 306–309.

32 Morgan, J.M. and Datta, R. (2014). Understanding the gas diffusion layer in proton exchange membrane fuel cells. I. How its structural characteristics affect diffusion and performance. *Journal of Power Sources* 251: 269–278.

33 Kim, T., Lee, S., and Park, H. (2010). A study of water transport as a function of the micro-porous layer arrangement in PEMFCs. *International Journal of Hydrogen Energy* 35 (16): 8631–8643.

34 Leeuwner, M.J., Wilkinson, D.P., and Gyenge, E.L. (2015). Novel graphene foam microporous layers for PEM fuel cells: interfacial characteristics and comparative performance. *Fuel Cells* 15 (6): 790–801.

35 Ebenezer, D., Neelima, K., Jagannatham, M. et al. (2016). Carbon nanotubes and nanohorn hybrid composite buckypaper as microporous layer for proton exchange membrane fuel cell. *Fuel Cells* 16 (3): 349–355.

6

Preparation of Gradient Gas Diffusion Layer Process and Their Performance in Fuel Cell

6.1 Introduction

In a traditional carbon paper (gas diffusion layer [GDL]), the carbon paper and microporous layer (MPL) have significantly different pore sizes. Carbon paper contains larger macropores, while the MPL consists of smaller micropores. This abrupt pore size transition at their interface can hinder efficient water transport [1–4]. To address this issue, this chapter examines the incorporation of an intermediate transition layer between the carbon paper and the MPL. This layer, composed of carbon black, polytetrafluoroethylene (PTFE), and $CaCO_3$, is designed to provide a more gradual pore size transition. $CaCO_3$, acting as a pore-forming agent, is later removed through hydrochloric acid treatment, resulting in a smoother pore size gradient within the GDL.

6.2 Materials Preparation

The performance of a GDL is highly dependent on its preparation process. This section outlines the optimization steps for ultrasonic time, PTFE addition sequence, anhydrous ethanol content, and heating plate temperature—all factors critical to achieving well-dispersed GDLs with gradient pore sizes. The use of $CaCO_3$ as a pore-forming agent and the subsequent removal process are also detailed, culminating in the creation of a GDL with enhanced water management and gas transport properties.

6.2.1 Effect of Ultrasonic Time on Gas Diffusion Layer Preparation

Ultrasonic treatment plays a key role in dispersing carbon black within the ethanol solution. Continuous ultrasonication for extended periods, exceeding 60 minutes, resulted in ethanol evaporation and increased solution viscosity, making spraying impractical. In contrast, intermittent ultrasonication (3×30 minutes) provided better dispersion without nozzle clogging. By repeating these steps four times, a well-dispersed carbon black mixture was achieved, minimizing in-plane resistance and enhancing GDL performance.

Electrospinning for Proton Exchange Membrane Fuel Cells, First Edition.
Yong Liu and Mohideen Meerasahib Mohamedazeem.

6.2.2 Effect of Polytetrafluoroethylene Addition Sequence

Two methods were tested to integrate PTFE into the carbon black mixture. Simultaneously mixing PTFE and carbon black (Method 1) led to agglomeration due to heat from ultrasonication. Conversely, adding PTFE after four rounds of carbon black ultrasonication and stirring (Method 2) avoided agglomeration and produced a homogeneous mixture. Water contact angle tests confirmed uniform hydrophobicity (142°–145°) across the prepared GDL surface, validating Method 2 as the optimal approach.

6.2.3 Investigation of Ethanol Content and Heating Plate Temperature

Anhydrous ethanol concentration directly influenced material dispersion and spraying continuity. A balanced ratio of PTFE:carbon black:anhydrous ethanol (5:7:100 by weight) was identified. Additionally, heating plate temperature significantly affected PTFE's behavior. A temperature of 200 °C proved optimal, as lower temperatures slowed ethanol evaporation and higher temperatures caused PTFE flow, creating inconsistent hydrophobicity.

6.2.4 Preparation and Testing of Gradient Gas Diffusion Layer

Building upon the optimization of ultrasonication, PTFE addition timing, ethanol content, and heating parameters, a comprehensive process was established to produce a gradient GDL. The procedure began with dispersing 7 g of carbon black in anhydrous ethanol, followed by four cycles of 30-minute magnetic stirring and 30-minute ultrasonication. After achieving a uniform dispersion, 5 g of 60 wt% PTFE solution was introduced, subjected to 30 minutes of stirring, and another 30 minutes of ultrasonication. A predetermined amount of $CaCO_3$ was then added and ultrasonicated for 10–20 minutes, depending on the concentration ratio (1:4 or 1:2 relative to the total carbon black and PTFE mass), to introduce a gradient pore structure.

The resulting GDL was spray-coated and heat-treated in a tube furnace. After initial heating to 250 °C to remove activators from the PTFE surface, the temperature was raised to 350 °C for improved PTFE bonding with carbon black and carbon paper. Cooling followed, and the sample was immersed in concentrated hydrochloric acid for 24 hours, with periodic ultrasonication to remove the $CaCO_3$ pore-forming agent. Subsequent washing with deionized water and drying at 50 °C yielded the final gradient GDL with a refined pore-size distribution. The process is depicted in Figure 6.1.

The gradient GDL, consisting of a $CaCO_3$-modified microporous layer near the carbon paper and a PTFE-carbon black layer near the catalyst, effectively minimized abrupt pore-size transitions. This smooth gradient improved water management and gas transport, providing enhanced performance at high current densities. This innovative approach overcame the limitations of traditional single-layer GDLs, establishing a new benchmark for fuel cell efficiency.

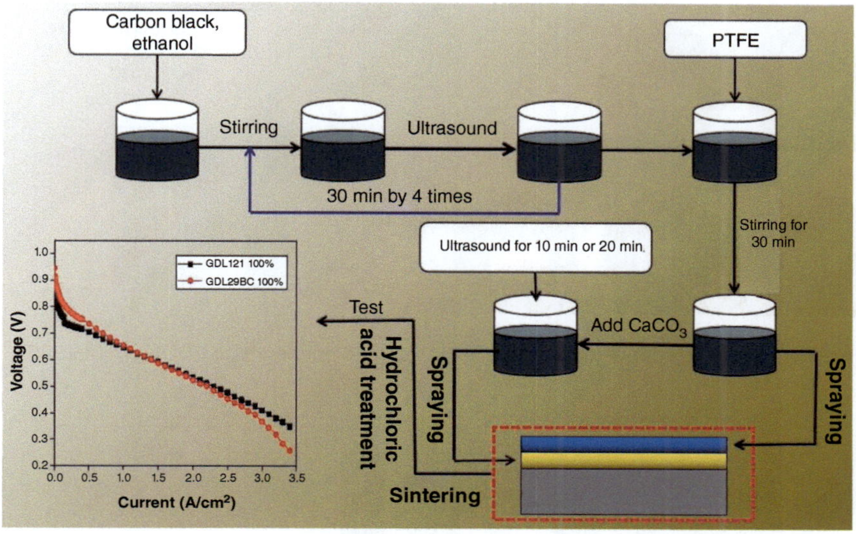

Figure 6.1 The preparation and testing process of gradient pore size GDL.

6.3 Results and Discussion

This section presents the characterization of the GDL's physical properties (hydrophobicity, conductivity, pore size, surface morphology) and electrochemical performance (polarization curves, power density, and electrochemical impedance).

6.3.1 Hydrophobicity

Hydrophobicity plays a crucial role in water transport within the GDL. When hydrophobicity is well balanced, water can be promptly expelled from the cathode catalyst layer, which reduces mass transfer resistance, improves gas transport efficiency, and lowers voltage losses [5, 6]. Conversely, inadequate hydrophobicity can lead to water flooding, hinder gas transport, and degrade the performance of platinum catalysts. As shown in Figure 6.2, the measured water contact angles for the self-prepared GDL samples were GDL141 (147°), GDL142 (148°), GDL121 (147°), and GDL122 (141°). In comparison, the commercial GDL29BC had a lower contact angle of 135°. This increased hydrophobicity in the prepared samples is attributed to the controlled use of $CaCO_3$ and optimized sonication processes, demonstrating the effectiveness of these modifications in enhancing water transport capabilities.

6.3.2 Water Permeability and Conductivity

The water permeability and conductivity test results for the GDL are presented in Figure 6.3. It can be observed that the commercial GDL29BC exhibits the lowest

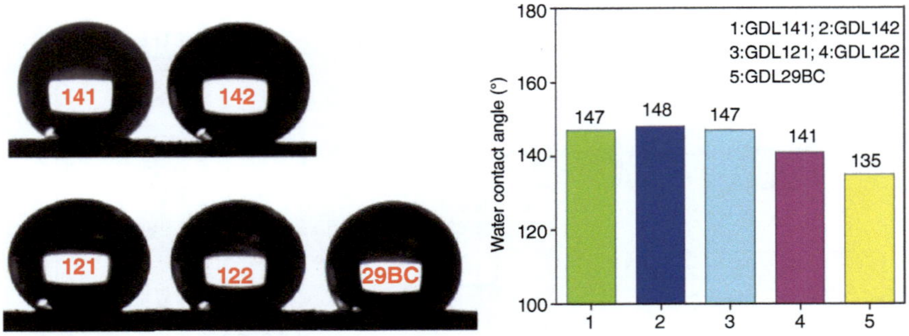

Figure 6.2 The water contact angle test results of gradient pore size GDL and GDL29BC.

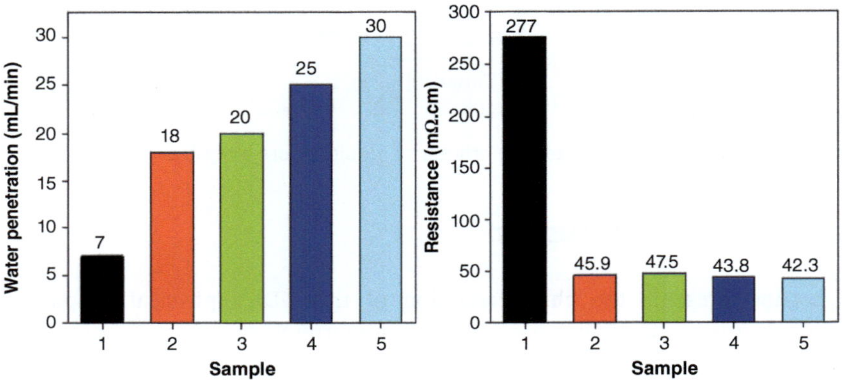

Figure 6.3 GDL's water permeability and electrical conductivity test results (1:GDL 29BC, 2:GDL 141, 3:GDL 142, 4:GDL121, 5:GDL122).

water permeability and the highest in-plane resistance. Low water permeability hampers efficient water transport, while elevated resistance values lead to increased Ohmic impedance. These factors cause higher voltage losses at the same current density, ultimately lowering the power density and efficiency of the fuel cell.

Furthermore, increasing the $CaCO_3$ content leads to a noticeable increase in water permeability. In the first case, repeated ultrasonic treatment generates more interconnected pores, enhancing the efficiency of water transport. In the second case, higher $CaCO_3$ content creates additional pores, further boosting water permeability. The in-plane resistance of the self-prepared GDL samples varies slightly, likely due to differences in $CaCO_3$ content and the duration of ultrasonic treatment. Additionally, the removal of $CaCO_3$ alters the effective contact area between particles, contributing to variations in in-plane resistance. These factors highlight the critical role of $CaCO_3$ and ultrasonic treatment in tuning the physical properties of the GDL.

6.3.3 Pore Size Distribution

The GDL pores comprise both nanopores and micropores, and their distribution plays a critical role in gas and water transport [7, 8]. Using a mercury intrusion porosimeter, we measured the GDL's pore size distribution, focusing on its influence on diffusion mechanisms [9]. During fuel cell operation, gas diffusion through the GDL is driven by concentration gradients: higher near the bipolar plates and lower near the cathode catalyst layer. This concentration differential drives gases toward the catalyst layer, with molecular diffusion as the dominant mechanism. To achieve molecular diffusion, pores need to be larger than 7 μm—about 100 times the free path of air molecules. Smaller pores, less than 7 μm, support Knudsen and mixed diffusion, which are less efficient.

Water generated at the cathode catalyst layer first occupies larger pores before filling smaller ones. At high current densities, the increased water production can lead to flooding if the pore size distribution is not optimized. Properly distributing pore sizes ensures both effective water removal and gas transport, thereby maintaining fuel cell performance and preventing operational failure [10, 11]. Figure 6.4 illustrates the pore size distribution of the gradient GDL, while Table 6.1 provides

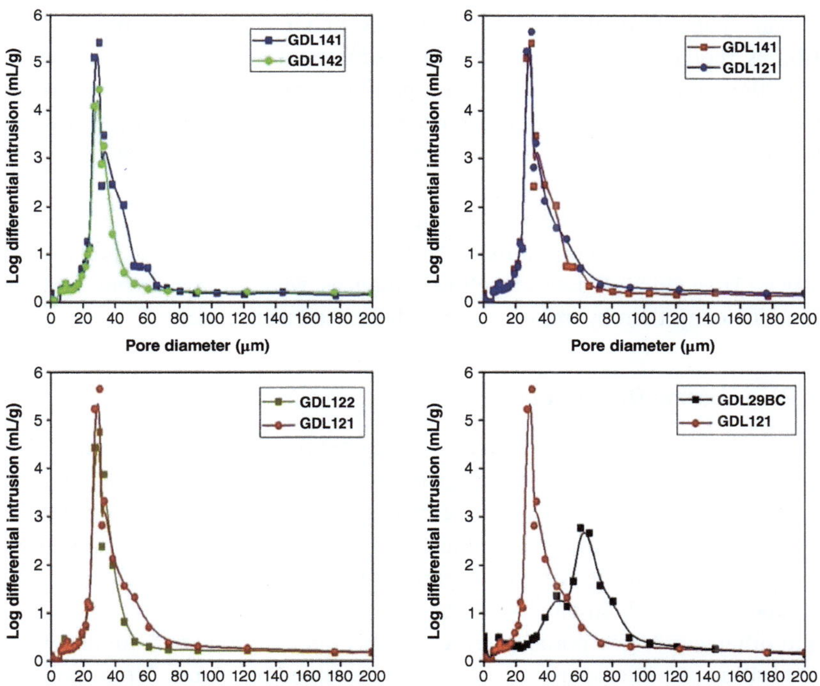

Figure 6.4 Comparison of gradient GDL pore size distribution interval.

Table 6.1 Pore content in different spatial intervals.

Sample	100 μm (mL/g)	20–100 μm (mL/g)	7–20 μm (mL/g)
GDL141	1.16	27.37	3.48
GDL142	0.95	20.68	3.26
GDL121	1.03	26.53	3.58
GDL122	0.93	22.33	3.48
GDL29BC	1.6	16.57	3.32

detailed pore content data. The results show that the commercial GDL29BC has a higher percentage of pores over 100 μm (1.6%) compared to the self-made gradient GDLs.

The spraying preparation method used for the self-made GDL likely caused some PTFE and carbon black to penetrate into the carbon paper, reducing the content of very large pores. In the 20–100 μm range, the self-made GDLs generally have a higher pore content than the commercial GDL, with GDL141 at 27.37% and GDL121 at 26.53%. This indicates that $CaCO_3$ addition effectively redistributes the pore sizes. Prolonged ultrasonic treatment further reduces the content of these larger pores, as the $CaCO_3$ particles break down into smaller sizes. In the 7-to-20-μm range, all samples showed minimal differences, suggesting that $CaCO_3$'s ability to form pores is limited in this range, resulting in a relatively stable distribution.

Overall, the self-made gradient GDL demonstrates a rational pore size distribution, achieved by incorporating $CaCO_3$ and using optimized preparation techniques. This tailored pore structure enhances water management and gas diffusion efficiency, thereby improving overall fuel cell performance.

6.3.4 Pore Volume Distribution

Mercury intrusion porosimetry was used to measure the overall porosity of the GDL, as the individual porosities of the carbon paper and microporous layer cannot be separately determined. In this study, commercial carbon paper was used, so variations in porosity are primarily attributed to changes in the microporous layer. High microporous layer porosity decreases the disparity between the microporous layer and carbon paper, potentially impairing water management.

On the other hand, low porosity increases this disparity but may hinder water transport within the microporous layer due to increased resistance. Thus, porosity alone cannot definitively indicate the GDL's suitability for water transport. As shown in Figure 6.5, the porosity of the GDL decreases with prolonged ultrasonic treatment when the $CaCO_3$ content remains constant. This decrease occurs because extended ultrasonication breaks down $CaCO_3$ particles into smaller fragments, thereby reducing the overall GDL porosity.

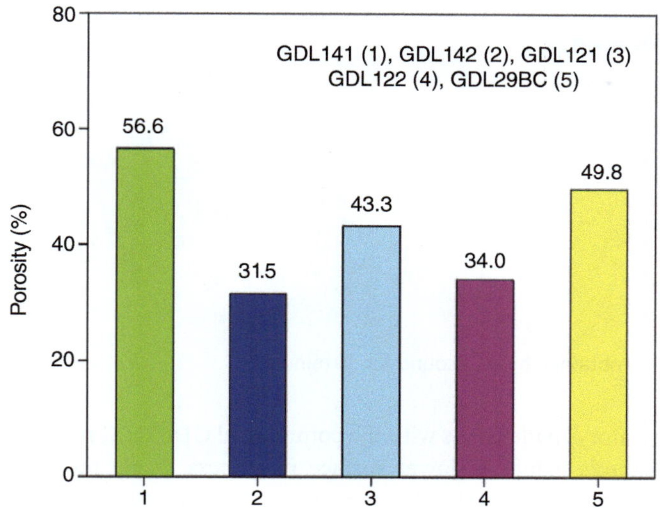

Figure 6.5 GDL porosity distribution.

6.3.5 Surface Roughness

The surface roughness of the GDL is a key factor influencing the effective contact area with the catalyst layer. Measurements were taken from GDL samples prepared by subjecting the carbon black mixture to ultrasonic treatment for varying durations (30, 60, 90, and 120 minutes). As ultrasonic time increased, surface roughness decreased, reflecting improved dispersion of carbon black in anhydrous ethanol and reduced particle aggregation. As shown in Figure 6.6, the surface roughness of the GDL ranged from 2.2 to 3.1 μm after 120 minutes of ultrasonic treatment. Variations in roughness across samples were largely due to differences in the amount of pore-forming agent and the ultrasonic treatment conditions applied to the pore-forming-agent-containing solution.

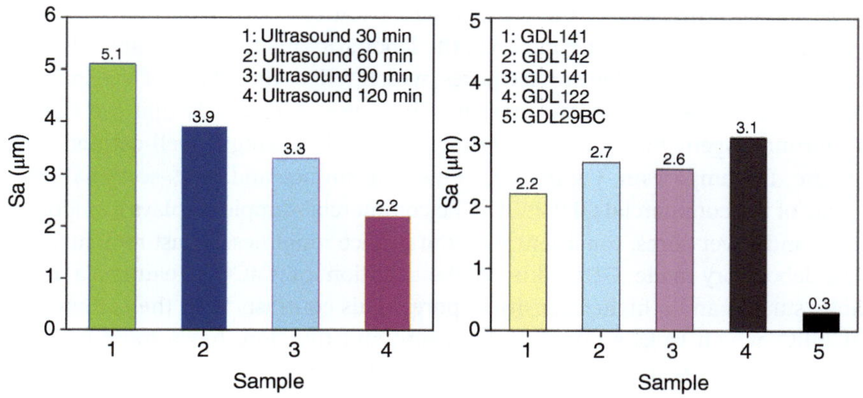

Figure 6.6 GDL surface roughness test results under different ultrasonic time.

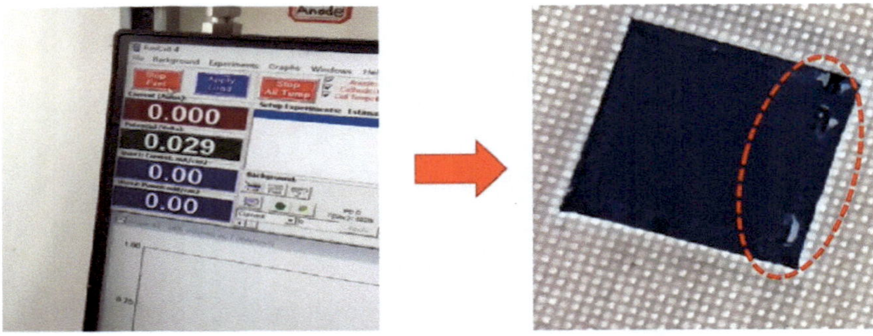

Figure 6.7 GDL activation obtained by ultrasound for 30 minutes.

A comparison of laboratory-made GDLs with the commercial GDL29BC revealed that the commercial sample exhibited lower surface roughness, likely resulting from differences in preparation methods. For instance, GDL produced with only 30 minutes of ultrasonic treatment had a coarse surface, and during cell testing, it caused visible damage to the proton exchange membrane (PEM) due to thermal expansion. This damage led to a short circuit, indicated by a sudden voltage drop to 0.029 V and perforations at the edge of the catalyst-coated membrane (CCM). Hot pressing the GDL after 30 minutes of ultrasonic treatment was used to mitigate surface coarseness, resulting in a roughness value (Sa) of 4.32 μm and improved cell performance during subsequent tests (Figure 6.7).

In contrast, GDL prepared using four cycles of 30-minute ultrasonic treatments exhibited stable performance during activation and polarization curve testing. This refined preparation method yielded a smoother surface and ensured consistent, reliable results in studying the effects of gradient-porosity GDLs on cell performance.

6.3.6 Surface Morphology

The surface morphology of the GDL provides critical insight into its structural characteristics. As shown in Figure 6.8, the laboratory-made GDL exhibits distinct particle stacking and visible surface pores, which reflect the effects of the spraying process. A cross-sectional view, presented in Figure 6.9, reveals that the $CaCO_3$ pore-forming agent has been successfully removed, leaving a well-defined pore structure. In comparison, Figure 6.10 shows the surface and cross-sectional morphology of the commercial GDL29BC. The commercial sample displays a smoother surface and fewer pores, consistent with the surface roughness measurements.

The laboratory-made GDL, due to the addition of $CaCO_3$, features a more chaotic surface and a higher density of pores. This contrasts with the commercial GDL29BC, which lacks a pore-forming agent and therefore has a more uniform surface and cross-section. These morphological differences highlight the impact of using $CaCO_3$ in the preparation process and the resulting enhancement of the GDL's porosity.

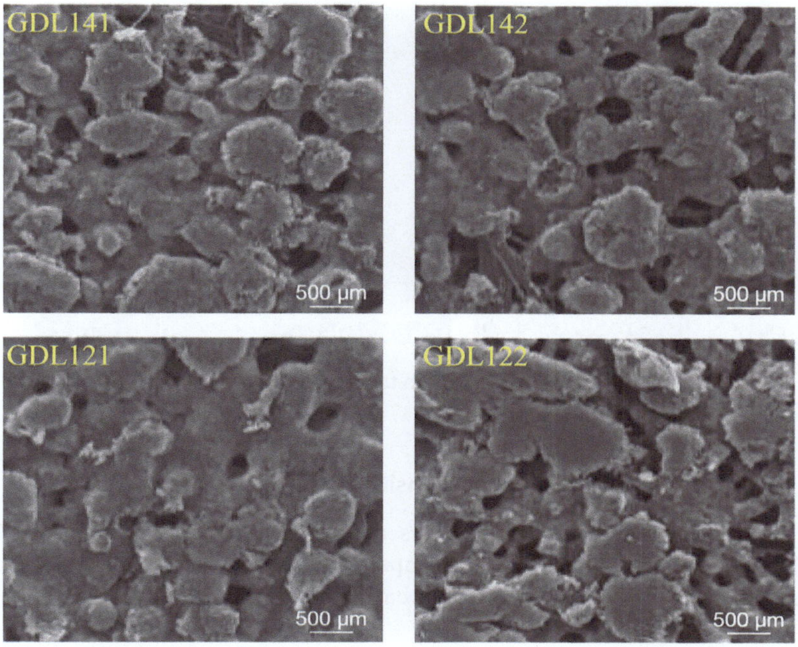

Figure 6.8 The surface topography of self-made GDL. Source: Li et al. [4]/ Springer Nature.

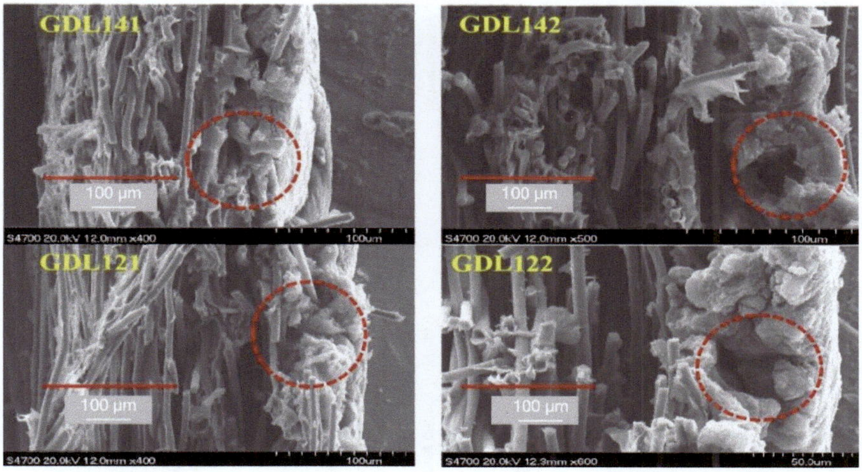

Figure 6.9 The cross-sectional morphology of self-made GDL. Source: Li et al. [4]/ Springer Nature.

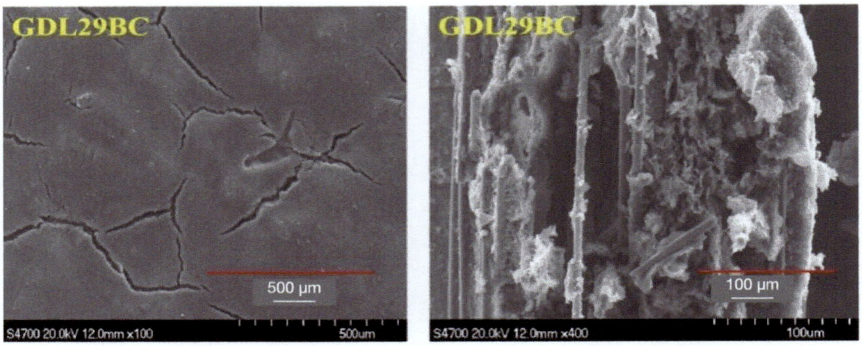

Figure 6.10 GDL29BC surface and cross-section topography. Source: Li et al. [4]/ Springer Nature.

6.3.7 Polarization Curve and Power Density

Figure 6.11 shows the polarization curves and power density comparisons of GDL141 and GDL142 under a battery temperature of 80 °C and humidification conditions of 60% and 100%. In the low current density region (0–0.5 A/cm^2),

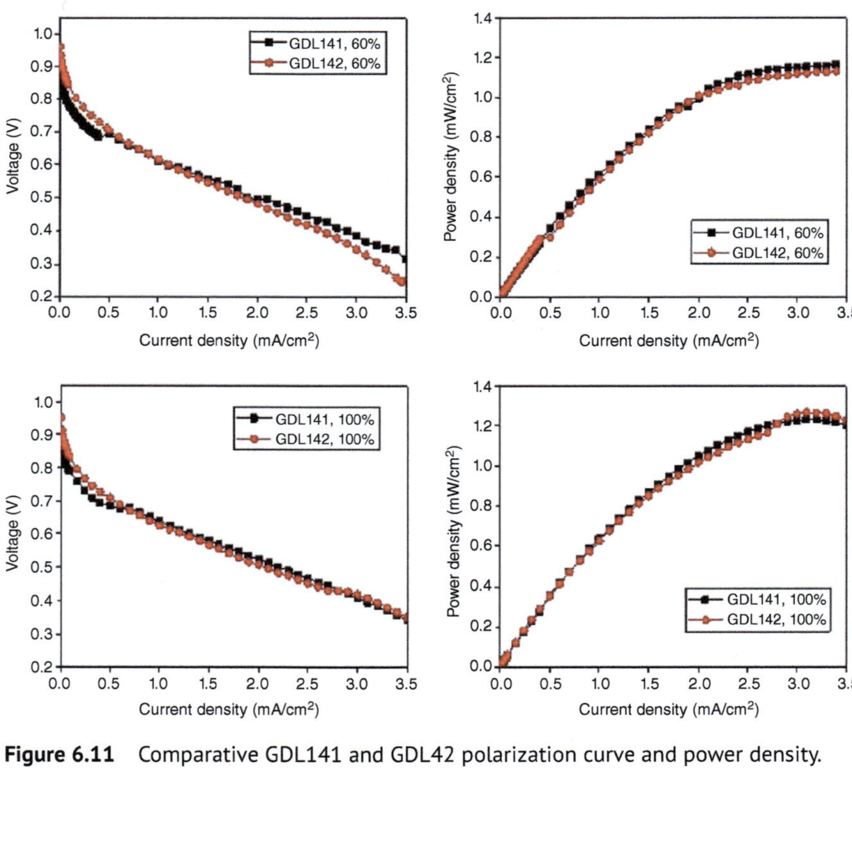

Figure 6.11 Comparative GDL141 and GDL42 polarization curve and power density.

GDL141 underperforms, which may be attributed to slight inconsistencies in the actual membranes despite theoretical uniformity. As current density increases, GDL141's performance improves due to its lower resistivity (45.9 mΩ·cm compared to GDL142's 47.5 mΩ·cm) and smaller surface roughness, which enhances the contact area with the catalyst layer. This improved contact area, coupled with increased water generation during the reaction, helps reduce contact resistance and improve proton conductivity, resulting in better performance at higher current densities.

Under high humidity (100%), GDL141 initially shows slightly worse performance in the low current density region due to a lower open circuit voltage. However, its higher pore content in the 7–20, 20–100, and >100 μm ranges ensures more efficient gas diffusion and reduced material transport losses at higher current densities. Overall, at 60% humidity, GDL141 demonstrates good polarization curve performance, while at 100% humidity, its performance is comparable to GDL142.

Figure 6.12 compares the performance of GDL141 with GDL121. GDL121 performs well under both high and low humidity conditions. At 60% humidity, the power densities of GDL121 and GDL141 are 1.22 and 1.12 W/cm^2, respectively,

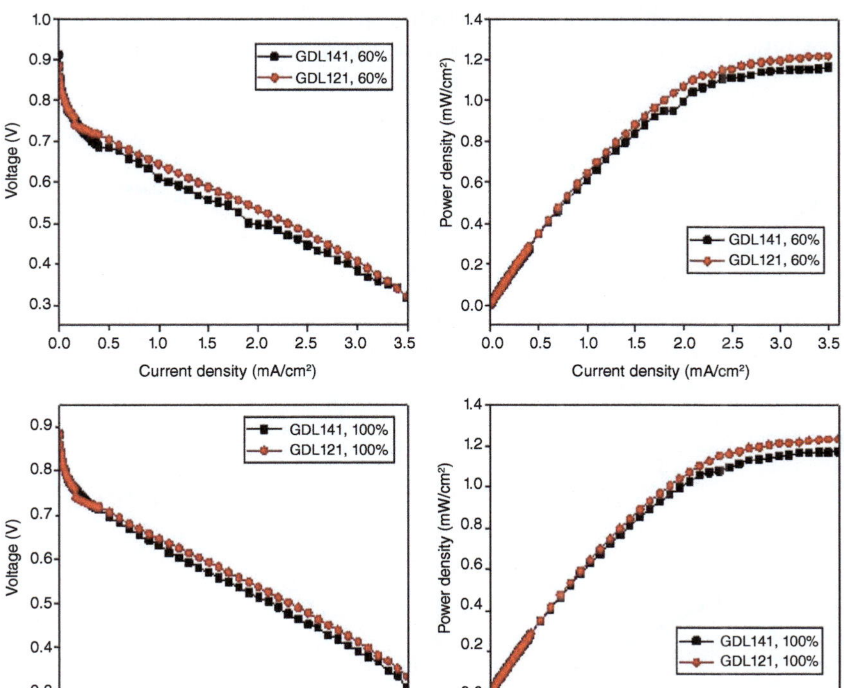

Figure 6.12 Polarization curve and power density in contrast about GDL141 and GDL121.

with corresponding maximum current densities of 3.499 and 3.399 A/cm^2. At 100% humidity, the power densities are 1.24 and 1.17 W/cm^2, with maximum current densities of 3.500 and 3.499 A/cm^2. In the low current density region, the polarization curves overlap under high and low humidity, reflecting stable membrane electrodes. At high current densities, however, GDL121 outperforms GDL141 under both humidity levels due to its higher pore content in the 7-to-20-μm range. These pores facilitate air and oxygen transport, reducing material transport losses and resulting in better performance.

Figure 6.13 further compares GDL121 and GDL122. Regardless of humidity levels (100% or 60%), GDL121 consistently delivers superior performance. This is attributed to GDL121's excellent hydrophobicity, low resistivity, and low surface roughness, combined with a beneficial pore distribution in the 7-to-20-μm range for gas transport and in the 20-to-100-μm and >100-μm ranges for water transport. Based on the comparison of GDL141, GDL142, GDL121, and GDL122, GDL121 emerges as the best-performing sample.

The final comparison between GDL121 and the commercial GDL29BC, shown in Figure 6.14, reveals that GDL121 outperforms GDL29BC under both 100% and 60% humidification. GDL121's higher pore content in the 7-to-100-μm range contributes to more efficient water and gas transport. While larger pores (>100 μm) in GDL29BC may reduce differentiation between the microporous layer and

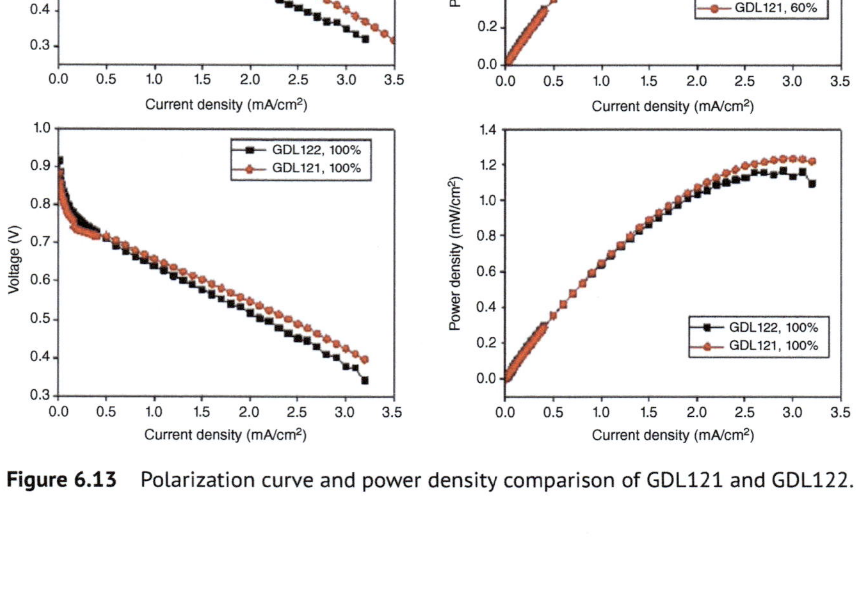

Figure 6.13 Polarization curve and power density comparison of GDL121 and GDL122.

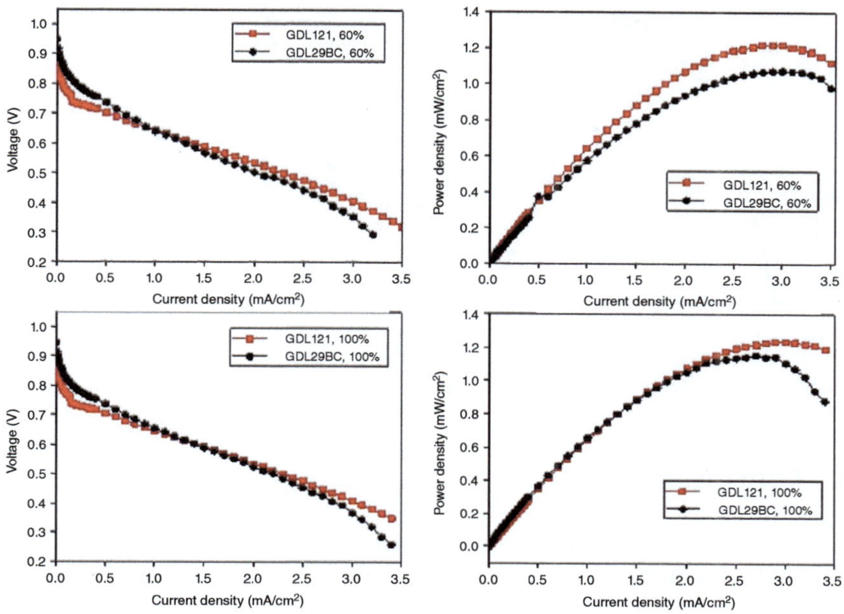

Figure 6.14 Comparison with polarization curves and power density about GDL121 and GDL29BC.

carbon paper, potentially hindering water and gas transport, the optimized pore sizes in GDL121 provide a more favorable structure, resulting in superior battery performance.

6.3.8 Electrochemical Impedance Spectroscopy Test

Following the results from the polarization curve and power density tests, which identified GDL121 as the top performer, electrochemical impedance spectroscopy (EIS) was used to compare the impedance characteristics of GDL121 and GDL29BC. Different loss mechanisms dominate at various current densities: activation losses are prominent at low current densities, ohmic losses arise from resistance within the battery components and the PEM, and mass transport impedance is most evident at high current densities. This analysis focused on mass transport impedance at a high current density of $I = 1.2\ A/cm^2$.

As illustrated in Figure 6.15, GDL121 exhibited consistently lower impedance than did GDL29BC across both high and low humidity conditions. This reduction in mass transport impedance for GDL121 is attributed to its optimized pore distribution, particularly within the 7-to-20-μm and 20-to-100-μm ranges, which facilitates more efficient gas and water transport. Consequently, GDL121 demonstrates superior performance in mass transport at high current densities compared to the commercial GDL29BC.

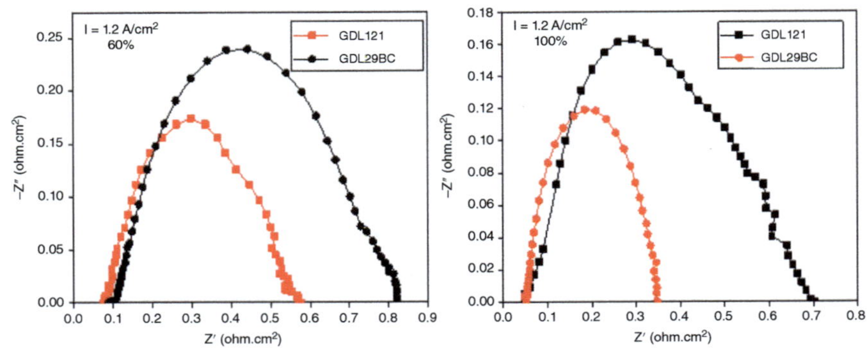

Figure 6.15 Electrochemical impedance spectroscopy test results of GDL121 and GDL29BC.

6.4 Conclusion

This chapter explored the effect of gradient pore size GDLs on water management and gas diffusion in PEM fuel cells, systematically refining the preparation process by studying ultrasonic time, PTFE addition sequence, and anhydrous ethanol content. The optimal approach involved four cycles of alternating magnetic stirring and ultrasonication, followed by adding PTFE and continuing the process, which produced well-dispersed mixtures with water contact angles between 142° and 145°, indicating excellent hydrophobicity and uniform PTFE distribution. $CaCO_3$ was used as a pore-forming agent to modify the pore size distribution. Adjusting the $CaCO_3$ content and ultrasonic treatment times successfully influenced pore size, leading to enhanced water and gas transport properties. The resulting gradient pore GDLs exhibited improved electrochemical performance, with GDL121 showing superior polarization curves, power densities, and lower mass transport impedance compared with the commercial GDL29BC. By redistributing pore sizes, the optimized GDL preparation enhanced both water removal and gas diffusion, ultimately improving fuel cell efficiency.

References

1 Ko, D., Do, S., Park, H.S. et al. (2018). the effect of through plane pore gradient GDL on the water distribution of PEMFC. *International Journal of Hydrogen Energy* 43 (4): 2369–2380.

2 Lee, H.F., Wang, P.C., and Chen-Yang, Y.W. (2019). An electrospun hygroscopic and electron-conductive core-shell silica@carbon nanofiber for microporous layer in proton-exchange membrane fuel cell. *Journal of Solid State Electrochemistry* 23 (3): 971–984.

3 Hou, S.Y., Ye, Y.K., Liao, S.J. et al. (2020). Enhanced low-humidity performance in a proton exchange membrane fuel cell by developing a novel hydrophilic gas diffusion layer. *International Journal of Hydrogen Energy* 45 (1): 937–944.

4 Li, T.Y., Wang, K.J., Wang, J.H. et al. (2020). Preparation of hierarchical-pore gas diffusion layer for fuel cell. *Journal of Materials Science* 55 (11): 4558–4569.

5 Chen, T., Liu, S.H., Zhang, J.W. et al. (2019). Study on the characteristics of GDL with different PTFE content and its effect on the performance of PEMFC. *International Journal of Heat and Mass Transfer* 128: 1168–1174.

6 Lin, G.Y. and Van Nguyen, T. (2005). Effect of thickness and hydrophobic polymer content of the gas diffusion layer on electrode flooding level in a PEMFC. *Journal of the Electrochemical Society* 152 (10): 1942–1948.

7 Pauchet, J., Prat, M., Schott, P. et al. (2012). Performance loss of proton exchange membrane fuel cell due to hydrophobicity loss in gas diffusion layer: Analysis by multiscale approach combining pore network and performance modelling. *International Journal of Hydrogen Energy* 37 (2): 1628–1641.

8 Gandomi, Y.A., Edmundson, M.D., Busby, F.C. et al. (2016). Water management in polymer electrolyte fuel cells through asymmetric thermal and mass transport engineering of the micro-porous layers. *Journal of the Electrochemical Society* 163 (8): 933–944.

9 Huang, B., Jian, Q.F., Luo, L.Z. et al. (2019). Cell and stack-level study of steady-state and transient behaviour of temperature uniformity of open-cathode proton exchange membrane fuel cells. *International Journal of Energy Research* 43 (14): 8148–8163.

10 Anyanwu, I.S., Hou, Y.Z., Xi, F.Q. et al. (2019). Comparative analysis of two-phase flow in sinusoidal channel of different geometric configurations with application to PEMFC. *International Journal of Hydrogen Energy* 44 (26): 13807–13819.

11 Zhou, X., Niu, Z.Q., Bao, Z.M. et al. (2019). Two-phase flow in compressed gas diffusion layer: Finite element and volume of fluid modeling. *Journal of Power Sources* 437: 1–10.

7

Impact of Optimizing the Thickness Direction of Gradient Gas Diffusion Layer on Water-Gas Management Capability of the Fuel Cell

7.1 Introduction

In Chapter 6, the preparation of gradient pore size gas diffusion layer (GDL) and the role of $CaCO_3$ as a pore-forming agent were examined. It was shown that using 5 g polytetrafluoroethylene (PTFE) (60 wt%), 7 g carbon black, 100 g anhydrous ethanol, and 5 g $CaCO_3$ with 10 minutes of ultrasonic treatment yielded a gradient GDL—designated as GDL121—that performed well under both 60% and 100% humidity at a battery temperature of 80 °C. GDL121 exhibited relatively low voltage loss at equivalent current densities, demonstrating its effectiveness in water–gas management.

Capillary pressure and its gradients play a crucial role in water transport efficiency within GDLs. When the GDL thickness is large, water transport efficiency can decrease due to longer paths. Conversely, when the thickness is small, the reduced differentiation between the microporous layer (MPL) and the carbon paper can also impair water management [1–4]. Based on these observations, this chapter investigates how varying GDL121 thickness affects water–gas management in the fuel cell. By roll pressing GDL121 from its original 270 μm thickness to 260, 240, and 220 μm, the study evaluates the physical and electrochemical properties of each variant. The objective is to identify the optimal thickness for performance and to understand the underlying reasons for these performance differences.

7.2 Materials Preparation

7.2.1 Preparation Process of Gradient Gas Diffusion Layers

The preparation of the gradient pore size GDLs followed the same methodology as outlined in Chapter 6, with one critical modification: the GDL was roll-pressed to achieve the desired thickness prior to sintering. This was followed by the usual sintering step, immersion in concentrated hydrochloric acid to remove $CaCO_3$, thorough washing, and drying. The adjusted preparation process for the gradient pore size GDLs of varying thicknesses is illustrated in Figure 7.1.

Electrospinning for Proton Exchange Membrane Fuel Cells, First Edition.
Yong Liu and Mohideen Meerasahib Mohamedazeem.

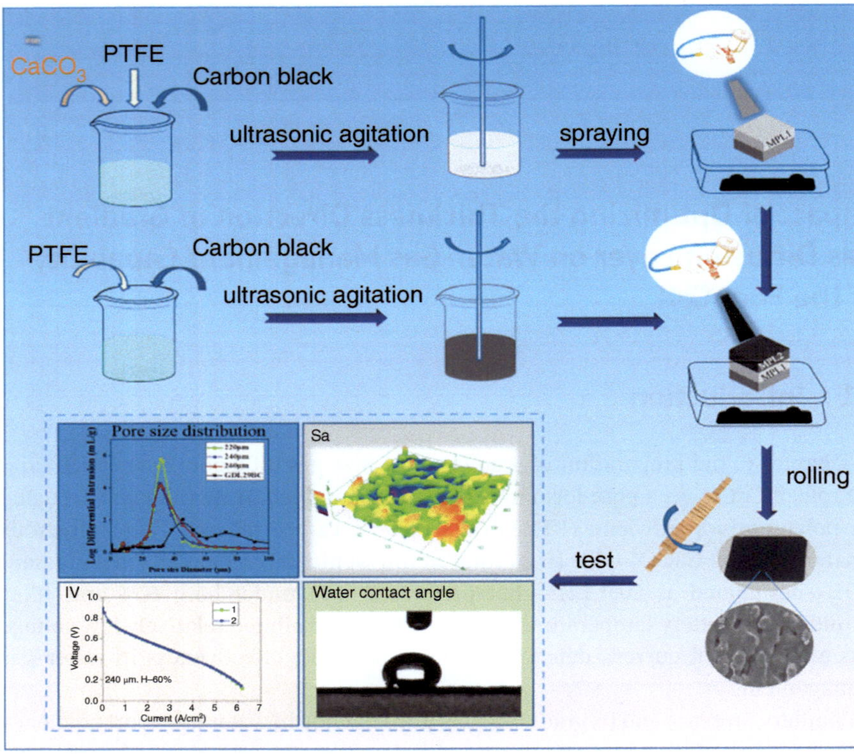

Figure 7.1 Schematic diagram of preparation process of GDL with different thicknesses about gradient pore size. Source: Li et al. [5]/with permission of Elsevier.

7.3 Results and Discussion

The in-plane resistance of the GDL strongly influences the fuel cell's ohmic impedance. Lower in-plane resistance reduces ohmic losses, decreasing voltage drops during operation [6, 7]. The hydrophobicity of the GDL samples, indicated by their water contact angles, reflects their ability to manage water transport. Surface morphology analysis provides insights into the material distribution on the GDL surface. Mercury intrusion porosimetry offers detailed statistics on pore size distribution, which, when combined with water and gas transport measurements, reveals the GDL's overall impact on water and gas transmission. Polarization curves and electrochemical impedance spectroscopy (EIS) are direct evaluation methods for battery performance. Polarization curves highlight voltage losses at a given current density and enable calculation of power density. EIS focuses on material transport impedance, helping to understand how GDL thickness affects transport efficiency [8, 9].

7.3.1 Physical Characterizations

Physical performance results are shown in Figure 7.2, where samples 1, 2, 3, and 4 correspond to GDLs of 220, 240, 260 μm, and GDL29BC, respectively. The in-plane resistance and water contact angle test results are depicted in Figure 7.2a. Among the samples, the 220-μm GDL has the highest in-plane resistance, measuring 44.67 mΩ·cm. GDLs of 240 and 260 μm show comparable in-plane resistance values, potentially due to PTFE redistribution, which increases resistance. The water contact angle results differ slightly among the samples, varying by only 2–5°.

Figure 7.2b shows the pore size and porosity results for the GDL samples of varying thicknesses and GDL29BC. Differences in pore sizes among the 220, 240, and 260 μm GDLs are primarily concentrated in the 20-to-60-μm range. As the GDL thickness decreases, the pore size distribution becomes more uniform, suggesting that thickness adjustments can help refine pore size distribution. Porosity differences among the three GDLs are minimal, ranging only from 1.7% to 3.1%, indicating that thickness variations have a limited impact on overall porosity. Table 7.1 provides a more

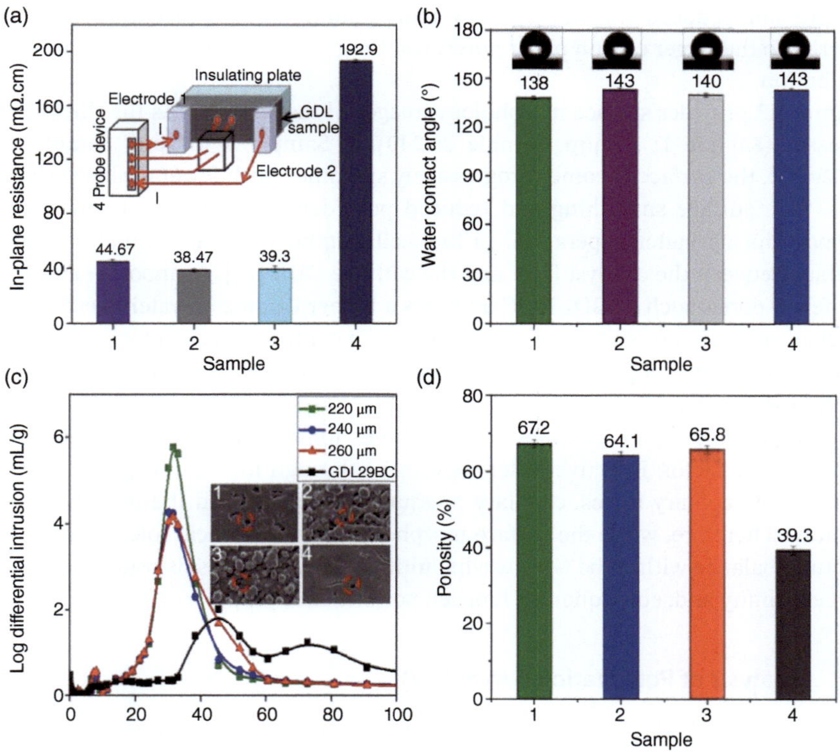

Figure 7.2 (a) GDL's in-plane resistance and water contact angle. (b) GDL's pore size distribution and porosity. Sample 1: GDL with thickness of 220 μm. Sample 2: GDL with thickness of 240 μm. Sample 3: GDL with thickness of 260 μm. Sample 4: GDL29BC. Source: Li et al. [5]/with permission of Elsevier.

Table 7.1 GDL pore size distribution statistics in different intervals.

Sample	100 μm	20–100 μm	7–20 μm
220	1	16.2	1.4
240	1	20.0	2.2
260	1	17.3	1.9
GDL29BC	1	5.2	0.8

detailed breakdown of the pore size distribution in the ranges of 7–20, 20–100 μm, and greater than 100 μm. For ease of comparison, pores larger than 100 μm were set as a reference value of 1, and the other ranges were scaled proportionally.

From Table 7.1, we see that the 240-μm-thick GDL has the highest pore sizes in both the 7-to-20-μm and 20-to-100-μm ranges, with values of 2.2 and 20.0, respectively, compared to the 220-μm, 260-μm, and GDL29BC samples. This increase in pore size at 240 μm can be attributed to the preparation process. The original GDL structure consisted of 200-μm carbon paper and two 35-μm MPLs, totaling 270 μm. Roll pressing reduced the thickness by redistributing carbon black, PTFE, and $CaCO_3$ into the larger carbon paper pores, resulting in noticeable shifts in pore size distribution.

Figure 7.3 provides surface morphology images of the samples. As the thickness decreases (Sample 1: 220 μm, Sample 2: 240 μm, Sample 3: 260 μm, Sample 4: GDL29BC), the surface becomes progressively smoother, with fewer visible surface pores. This surface smoothing and reduced pore density influence liquid water transport during battery operation. In fuel cells, liquid water accumulates at the interface between the catalyst layer and the cathode GDL [10]. A smoother surface with fewer pores, such as GDL29BC, requires a higher volume of water to establish sufficient capillary pressure for water transport. In contrast, the GDLs studied here exhibit some degree of pore distribution on their surfaces, facilitating liquid water transport at lower capillary pressures.

Although surface pore distribution contributes to water transport efficiency, it is not the sole factor. Effective water movement through the GDL is governed by a balance of capillary forces, capillary pressure differences, and the internal pore structure. Therefore, while the surface morphology plays a critical role, the overall structural balance within the GDL is what ultimately determines its water management capability and, consequently, fuel cell performance [11].

7.3.2 Analysis of Polarization Curve (I–V)

Figure 7.4 presents the polarization curve test results for GDLs of different thicknesses under 60% and 100% humidification at 80 °C. At 60% humidification, the 220-μm GDL exhibited the highest voltage losses and the lowest maximum

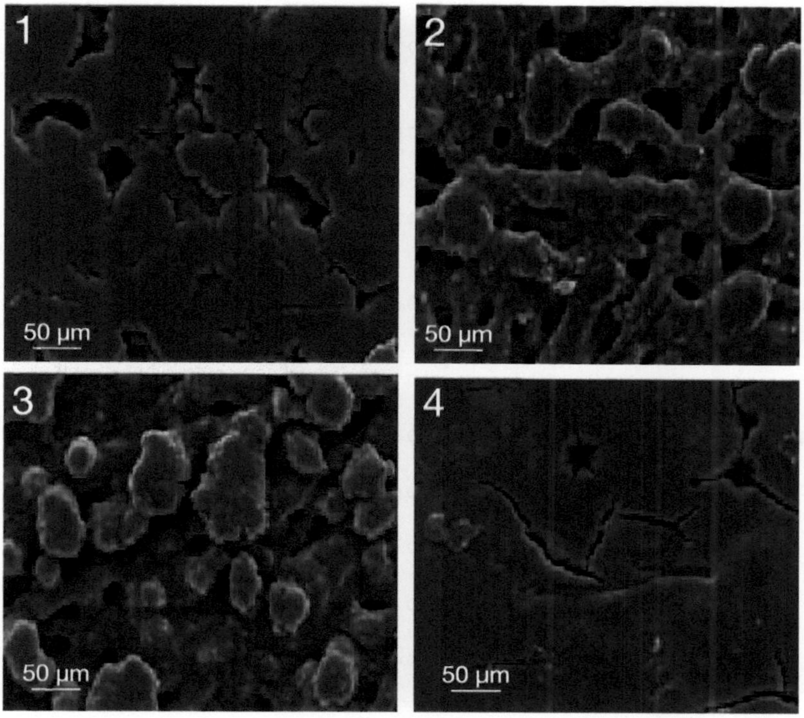

Figure 7.3 SEM test results of GDL and GDL29BC of different thickness (1 is GDL with the thickness of 220 μm, 2 is GDL with the thickness of 240 μm, 3 is GDL with the thickness of 260 μm, and 4 is GDL29BC). Source: Li et al. [5]/with permission of Elsevier.

current density. In contrast, the 240 and 260-μm GDLs showed similar performance at current densities below 3 A/cm^2. Above 3 A/cm^2, the 240-μm GDL outperformed the 260-μm GDL, displaying lower voltage losses and a higher maximum current density, suggesting superior water management capabilities under these conditions.

At 100% humidification, the 220-μm GDL again demonstrated higher voltage losses at lower current densities. Differences between the 240-μm and 260-μm GDLs emerged at current densities above 1.5 A/cm^2, where the 240-μm GDL consistently performed better. This performance gap widened with increasing current density. Compared to GDL29BC, the 240-μm GDL achieved superior performance in both low and high humidity conditions. At 60% humidification, the 240-μm GDL reached a maximum current density of 6.127 A/cm^2—22.6% higher than GDL29BC's 4.999 A/cm^2. At 100% humidification, the 240-μm GDL reached 6.844 A/cm^2, a 30% improvement over GDL29BC's 5.25 A/cm^2. These findings indicate that the 240-μm GDL provides the best combination of water management and electrochemical performance, particularly when compared to GDL29BC.

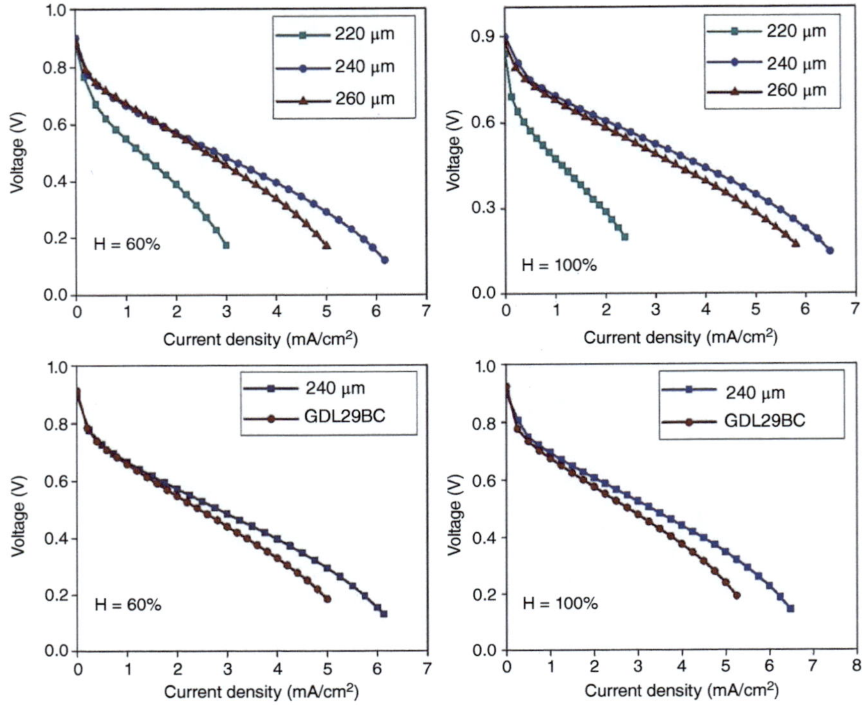

Figure 7.4 Polarization curve test results.

7.3.3 Analysis of Power Density

Power density test results are shown in Figure 7.5, under 60% and 100% humidification at 80°C. At 60% humidification, the 240-μm GDL achieved the highest maximum power density of 1.583 W/cm^2, outperforming the other three GDLs. Under 100% humidification, the 240-μm GDL also demonstrated the highest

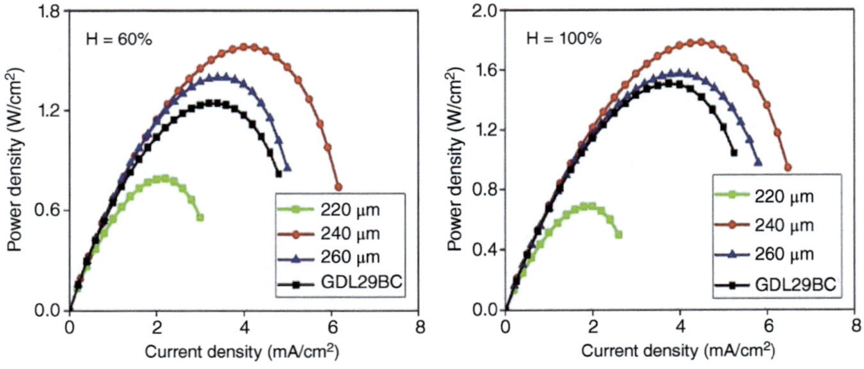

Figure 7.5 The power density test results when the cell temperature is 80 °C and the humidity is 60% and 100%.

maximum power density at 1.78 W/cm^2. These results clearly show that GDL thickness strongly influences power density, and optimizing thickness leads to the 240-μm GDL having the best performance.

7.3.4 Water Transport Mechanism

To further understand the impact of GDL thickness on water transport performance, we consider the water transport mechanism in terms of capillary pressure and pore characteristics. The driving force for water transport, as described by Eq. (7.1), is determined by the difference between the pressure of liquid water and the pressure of gas within the GDL pores. Specifically, the capillary pressure depends on factors such as the critical pore radius ($R_{critical}$), surface tension of water (γ), and the water contact angle (θ). A smaller capillary pressure may indicate reduced water transport efficiency, potentially caused by either a lower liquid water pressure or higher gas pressure.

Under identical conditions—constant battery temperature, humidification levels, gas flow rates, and membrane electrode properties—the amount of water generated per unit time should remain consistent. However, if the generated water fills most of the available hydrophilic and hydrophobic pores, it can obstruct gas flow to the catalyst layer. This obstruction increases gas pressure within the MPL, causing water retention in the GDL and hindering the delivery of reaction gases to the catalyst. Consequently, material transport resistance rises, adversely affecting the overall performance of the fuel cell.

$$P_C = P_l - P_g \tag{7.1}$$

$$r_c = \frac{-2\sigma \cos\theta}{P_C} \tag{7.2}$$

7.3.5 Comparison of Capillary Pressure Differences in Gas Diffusion Layers with Varying Thicknesses

Figure 7.6 illustrates the distribution and transport of liquid water in the MPL. Figure 7.6a shows the schematic of water transport within a single-layer microporous structure, while Figures 7.6b and c depict water transport in MPLs of varying thicknesses. Similarly, in Figure 7.6b, we select the midpoint of the MPL. Equation (7.3) represents the capillary pressure difference per unit thickness, which determines the efficiency of water transport. By calculating Eq. (7.3), we arrive at Eq. (7.4). The results show that under identical pore size distribution, hydrophobicity, and thickness, both samples exhibit the same capillary pressure difference.

We also compared the capillary pressure differences between the remaining half-thickness MPL shown in Figure 7.6a and MPL1 in Figure 7.6b. As porogens were introduced into MPL1, a noticeable increase in capillary pressure was observed. Based on Eq. (7.2), the calculated results are presented in Eq. (7.5). These findings confirm that incorporating porogens into the MPL enhances the

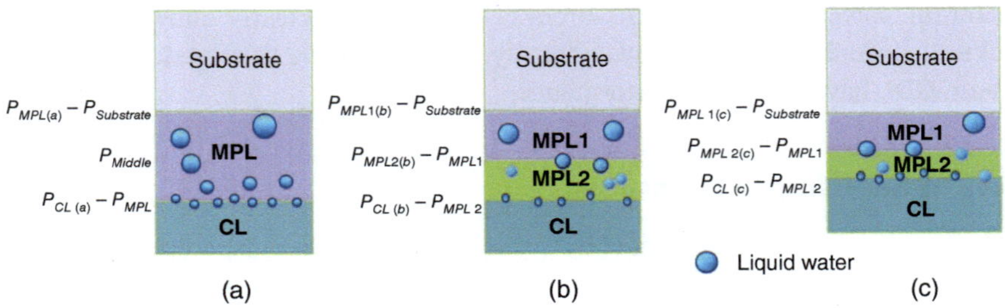

Figure 7.6 Schematic diagram of water transport in the microporous layer. (a) Conventional gas diffusion layer (GDL), (b) two-layered microporous GDL, and (c) GDL of different thickness.

capillary pressure difference, which improves water transport efficiency and thereby contributes to better water management within the fuel cell.

$$\Delta P = \frac{P_1 - P_2}{\delta} \tag{7.3}$$

$$\frac{P_{\mathrm{CL(a)}} - P_{\mathrm{Middle}}}{\delta} = \frac{P_{\mathrm{CL(b)}} - P_{\mathrm{MPL2(b)}}}{\delta} \tag{7.4}$$

$$\frac{P_{\mathrm{Middle}} - P_{\mathrm{MPL(a)}}}{\delta} < \frac{P_{\mathrm{MPL2(b)}} - P_{\mathrm{MPL1(b)}}}{\delta} \tag{7.5}$$

Figures 7.6b and c illustrate the capillary pressure distribution of gradient pore-size GDLs with different thicknesses at various positions. From the perspective of thickness, we compared the capillary pressure differences at MPL2 per unit thickness in Figures 7.6b and c. The calculated results from Eq. (7.6) show that as the thickness of the MPL decreases, the capillary pressure difference tends to increase.

In addition, we compared the capillary pressure differences at MPL1 in Figures 7.6b and c. The comparison indicates that reducing the MPL's thickness can increase the capillary pressure difference, facilitating faster water removal. However, while thinner GDLs or MPLs can achieve higher capillary pressure differences, a reduced GDL thickness diminishes the distinct characteristics between the MPL and the carbon paper. This leads to a more uniform pore-size distribution, which can impair both water and gas transport, ultimately lowering water management efficiency.

Table 7.1 presents the pore-size distribution of GDLs with thicknesses of 220, 240, and 260 μm, along with the polarization curve results in Figure 7.4. Combining the pore-size distribution ranges, capillary pressure differences, and the polarization curve data, we conclude that pore size largely determines whether water and gas can be transported effectively, while capillary pressure differences influence the efficiency of water transport within the MPL. The polarization curve results in Figure 7.4 indicate that the 240-μm-thick GDL allows for rapid water and gas transport.

$$\frac{P_{\mathrm{CL(c)}} - P_{\mathrm{MPL2(c)}}}{\delta_c} > \frac{P_{\mathrm{CL(b)}} - P_{\mathrm{MPL2(b)}}}{\delta_b} (\delta_c < \delta_b) \tag{7.6}$$

$$\frac{P_{\mathrm{MPL2(c)}} - P_{\mathrm{MPL1(c)}}}{\delta_{c2}} > \frac{P_{\mathrm{MPL2(b)}} - P_{\mathrm{MPL1(b)}}}{\delta_{b2}} (\delta_{c2} < \delta_{b2}) \tag{7.7}$$

7.3.6 Electrochemical Impedance Spectroscopy

EIS tests were performed under conditions of 80 °C cell temperature, humidification levels of 60% and 100%, and a current density of 2.5 A/cm^2 as shown in Figure 7.7. At 60% humidification, the 240-μm GDL exhibited the lowest mass transport impedance, followed by the 260-μm GDL, with the 220-μm GDL showing the highest impedance. As shown in the polarization curves in Figure 7.4, the 240-μm GDL demonstrated the highest limiting current density and the smallest voltage loss at a given current density. These results indicate that the 240-μm GDL excels in water management and gas transport.

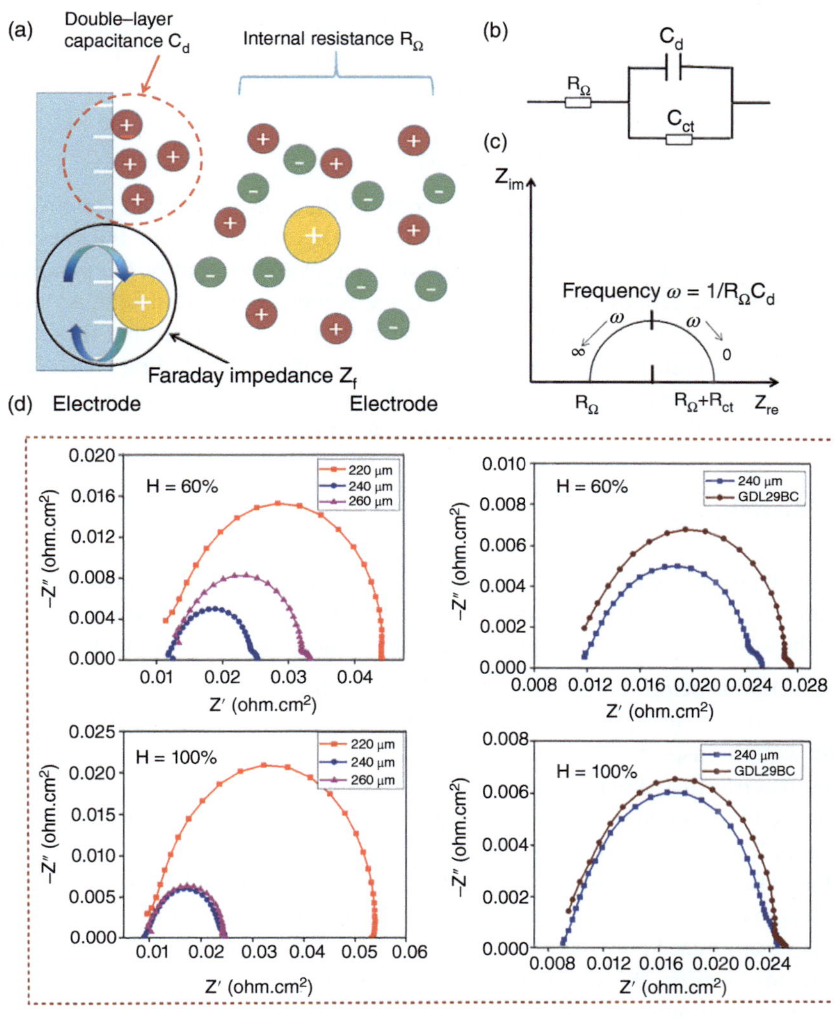

Figure 7.7 (a) Schematic representation of the electrochemical interface, illustrating the double layer capacitance, internal resistance and Faradaic impedance, (b) Equivalent circuit model used for EIS fitting, (c) Nyquist plot and (d) (EIS) analysis of gas diffusion layers (GDLs) under different humidification conditions at a cell temperature of 80 °C and a current density of 2.5 A/cm^2.

The improved performance of the 240-μm GDL can be explained by two key factors:

1) A broader pore size distribution, particularly in the 7-to-20-μm and 20-to-100-μm ranges, which enhances both water management and gas transport efficiency.
2) A larger capillary pressure difference, which accelerates water transport through the GDL.

At 100% humidification, the 220-μm GDL again showed the highest mass transport impedance. While the 240-μm GDL had slightly lower mass transport impedance than the 260-μm GDL, the difference was small. The 240-μm GDL's advantage is still attributed to its broader pore size distribution, which supports more efficient water and gas transport. Compared to GDL29BC, the 240-μm GDL consistently exhibited lower mass transport losses, underscoring the benefits of its optimized pore structure.

7.3.7 Polarization Curve Reproducibility

Reproducibility tests were carried out to confirm the consistency and stability of the polarization curves for the three self-made GDLs with different thicknesses, along with GDL29BC. Figure 7.8 demonstrates that at 60% humidity, the polarization curves for the 220, 240, and 260-μm GDLs remained highly consistent. At 100% humidity, slight discrepancies were observed between repeated tests, but these differences were quantified as being under 5% for all samples, including GDL29BC. These results confirm that the self-made GDLs, along with the commercial GDL, exhibit good stability and reproducibility in performance.

7.4 Conclusion

This chapter optimized the thickness of GDLs with gradient pore structures, starting from an initial thickness of 270 μm and reducing it to 220, 240, and 260 μm. Among these, the 240-μm GDL exhibited the best performance. Key findings include:

1) **Pore Distribution:** The 240-μm GDL displayed the widest pore size distribution in the critical 7-to-20-μm and 20-to-100-μm ranges, contributing to enhanced gas and water transport.
2) **Electrochemical Performance:** In polarization curve tests, the 240-μm GDL achieved the lowest voltage losses and highest limiting current density at both 60% and 100% humidity.
3) **Mass Transport Resistance:** EIS tests confirmed that the 240-μm GDL had the lowest mass transfer resistance under both humidification conditions, consistent with its optimized pore distribution.
4) **Reproducibility:** Polarization curve reproducibility tests showed less than 5% variation between repeated measurements, demonstrating excellent stability.

Overall, the 240-μm GDL proved superior in water and gas management, making it an optimal thickness for improved proton exchange membrane fuel cell performance.

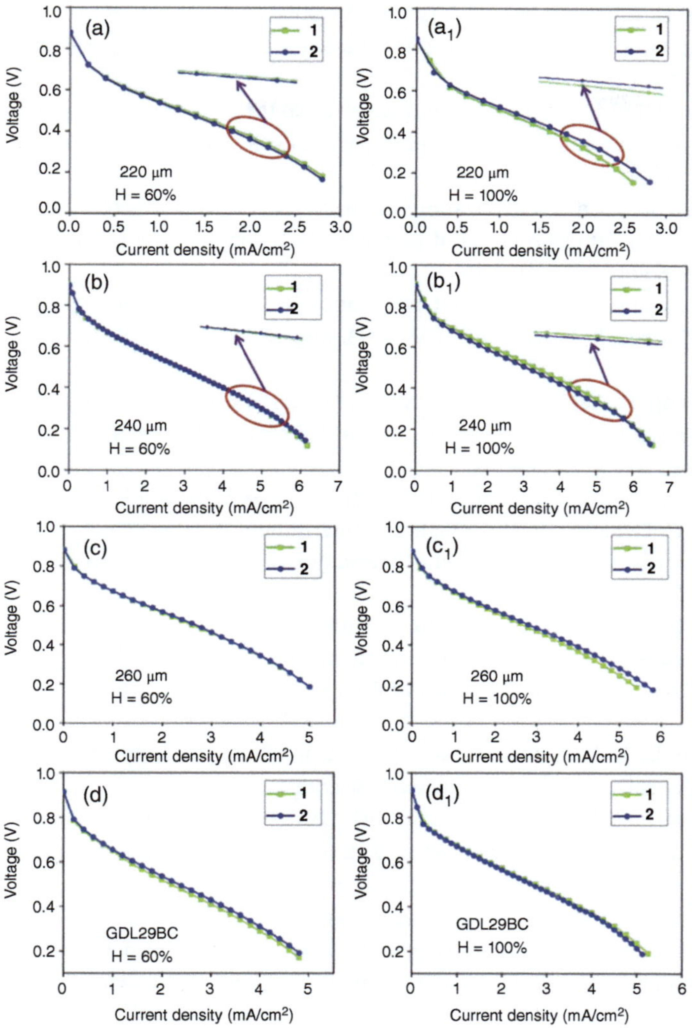

Figure 7.8 Repeated polarization test results of fuel cells using four different gas diffusion layers (GDLs). The anode and cathode platinum loadings were 0.1 mg/cm^2 and 0.2 mg/cm^2, respectively. Curves labeled "1" represent the first test, and "2" represent the second test. (a, a1) Polarization curves for GDL with 220 μm thickness at 60% and 100% relative humidity (RH); (b, b1) for 240 μm GDL at 60% and 100% RH; (c, c1) for 260 μm GDL at 60% and 100% RH; (d, d1) for commercial GDL29BC at 60% and 100% RH.

References

1 Holzer, L., Pecho, O., Schumacher, J. et al. (2017). Microstructure-property relationships in a gas diffusion layer (GDL) for polymer electrolyte fuel cells, Part II: Pressure-induced water injection and liquid permeability. *Electrochimica Acta* 241: 414–432.

2 Flipo, G., Josset, C., Giacoppo, G. et al. (2015). Eruptive water transport in PEMFC: A single-drop capillary model. *International Journal of Hydrogen Energy* 40 (45): 14667–14675.

3 Mortazaui, M. and Tajiri, K. (2014). Liquid water breakthrough pressure through gas diffusion layer of proton exchange membrane fuel cell. *International Journal of Hydrogen Energy* 39 (17): 9409–9419.

4 Liu, T.L. and Pan, C. (2012). Visualization and back pressure analysis of water transport through gas diffusion layers of proton exchange membrane fuel cell. *Journal of Power Sources* 207: 60–69.

5 Li, T., Wang, K., Wang, J. et al. (2021). Optimization of GDL to improve water transferability. *Renewable Energy* 179: 2086–2093. https://doi.org/10.1016/j.renene.2021.08.026.

6 Hao, L. and Cheng, P. (2012). Capillary pressures in carbon paper gas diffusion layers having hydrophilic and hydrophobic pores. *International Journal of Heat and Mass Transfer* 55 (1-3): 133–139.

7 Lu, Z.J., Daino, M.M., Rath, C. et al. (2010). Water management studies in PEM fuel cells, part III: Dynamic breakthrough and intermittent drainage characteristics from GDLs with and without MPLs. *International Journal of Hydrogen Energy* 35 (9): 4222–4233.

8 Chen, L., Feng, Y.L., Song, C.X. et al. (2013). Multi-scale modeling of proton exchange membrane fuel cell by coupling finite volume method and lattice Boltzmann method. *International Journal of Heat and Mass Transfer* 63: 268–283.

9 Cho, J.I.S., Marquis, J., Trogadas, P. et al. (2020). Optimizing the architecture of lung-inspired fuel cells. *Chemical Engineering Science* 215: 1–40.

10 Nazari, M., Shakerinejad, E., and Kayhani, M.H. (2018). Tailored surface wettability of gas diffusion layer in polymer electrolyte membrane fuel cells: Proposing a pore scale-two phase design. *Fuel Cells* 18 (6): 698–710.

11 Zhang, D., Cai, Q., and Gu, S. (2018). Three-dimensional lattice-Boltzmann model for liquid water transport and oxygen diffusion in cathode of polymer electrolyte membrane fuel cell with electrochemical reaction. *Electrochimica Acta* 262: 282–296.

8

Study on the Effect of Gradient Hydrophobic Treatment of Gas Diffusion Layer on Water Management Performance in Fuel Cells

8.1 Introduction

In Chapter 7, the effect of gas diffusion layers (GDL) thickness on proton exchange membrane fuel cell (PEMFC) performance was explored, with a 240-μm gradient GDL showing superior results due to its optimal pore size distribution. Building on that foundation, this chapter shifts focus to the influence of GDL hydrophobicity on water management and overall cell performance. Capillary pressure is a key factor in water transport within GDLs. It guides liquid water, formed on the catalyst layer surface, into hydrophobic or hydrophilic pores [1–3]. At low current densities, where water production is minimal, hydrophilic pores dominate as the primary water transport pathway. Regardless of the operating current density, the hydrophilic or hydrophobic nature of the GDL directly affects the fuel cell's water management efficiency [4, 5].

This chapter examines dual-microporous-layer GDLs with varying hydrophobicity levels. Initially, polytetrafluoroethylene (PTFE) and carbon black were combined in different ratios to prepare the microporous layers (MPLs) via spray-coating. The layer adjacent to the carbon paper was kept constant at a PTFE-to-carbon black ratio of 3:7, while the layer near the cathode catalyst was produced with ratios of 3:7, 2:8, and 1:9. After fabrication, posttreatment processes were applied to enhance the distinct hydrophobic properties of each layer. The physical and electrochemical characteristics of these GDLs were then evaluated to understand their impact on cell performance.

Additionally, a second set of GDLs was prepared with the layer near the carbon paper set to a ratio of 2:8 and the layer near the cathode catalyst layer set to 2:8 and 1:9. This control group confirmed the consistency of the findings. Through these analyses, the performance and water management capabilities of the gradient hydrophobic GDLs were thoroughly assessed.

8.2 Materials Preparation

The gradient hydrophobic GDLs were fabricated by preparing PTFE and carbon black mixtures at mass ratios of 3:7, 2:8, and 1:9. These mixtures were spray-coated to

Electrospinning for Proton Exchange Membrane Fuel Cells, First Edition.
Yong Liu and Mohideen Meerasahib Mohamedazeem.

form dual-microporous-layer structures with different hydrophobicity levels. Afterward, the samples underwent rolling, sintering, concentrated hydrochloric acid immersion, deionized water rinsing, and drying, yielding gradient hydrophobic GDLs. The pore-forming agent content and ultrasonic treatment time followed the procedures outlined in Chapter 7. Carbon paper served as the substrate, with anhydrous ethanol as the solvent. Carbon black (Vulcan XC-72) was chosen for its small particle size (30 nm) and high conductivity. PTFE was used to provide hydrophobicity, and hydrochloric acid was used to remove $CaCO_3$ from the samples.

8.3 Results and Discussion

Herein, we evaluate gradient hydrophobic GDLs using electrochemical and physical performance tests. Polarization curves reveal voltage behavior at different current densities, while power density metrics highlight the cell's output and peak performance. Electrochemical impedance spectroscopy (EIS) focuses on mass transport impedance, which reflects how efficiently gases and water move through the GDL under high-current, low-frequency conditions.

Physical characterizations assess surface hydrophobicity via water contact angle measurements, while surface morphology and roughness provide insights into GDL preparation quality. In-plane resistance, a key factor influencing ohmic impedance, is also measured to ensure minimal voltage losses.

8.3.1 Analysis of Polarization Curve (I–V)

The polarization curves in Figure 8.1 show the performance of three GDL samples: (1) 3:7, 3:7: uniform PTFE-to-carbon black ratio; (2) 3:7, 2:8: gradient hydrophobic structure with a more hydrophobic MPL near the catalyst; and (3) 3:7, 1:9: higher hydrophilicity near the catalyst layer.

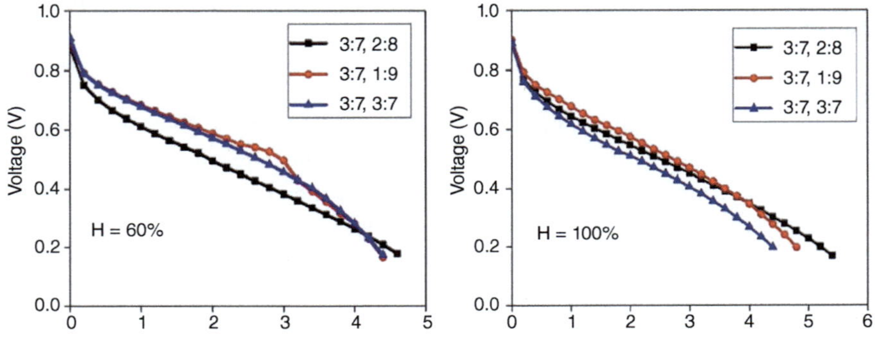

Figure 8.1 Polarization curve test results of three samples (The temperature of fuel cell is 80 °C; the test in hydrogen and oxygen environment; humidity is 60% and 100%, respectively; and the Pt loading capacity of anode and cathode are both 0.1 mg/cm^2).

At 60% humidification and current densities below 3 A/cm^2, the 3:7, 1:9 sample displayed the lowest voltage loss due to its greater hydrophilicity, which facilitated water removal under low water production conditions. At higher current densities, the 3:7, 2:8 sample outperformed the others by maintaining higher limiting current densities. Its more hydrophobic pores prevented water blockage and sustained effective gas transport. In contrast, the 3:7, 1:9 sample experienced voltage drops at high current densities, likely due to water accumulation obstructing gas flow.

At 100% humidification, the 3:7, 1:9 sample performed well across current ranges, but its limiting current density was lower than that of the 3:7, 2:8 sample. The 3:7, 2:8 configuration consistently achieved the highest limiting current density at both humidification levels, confirming the advantage of a balanced hydrophobic pore structure in handling higher water production rates and improving overall gas transport efficiency.

8.3.2 Analysis of Power Density

Figure 8.2 presents the power density data for the three GDL samples at humidification levels of 60% and 100%. Among the tested samples, the one with a PTFE-to-carbon black ratio of 1:9 (on the catalyst side) achieved the highest maximum power densities, reaching 1.53 W/cm^2 at 60% humidification and 1.44 W /cm^2 at 100% humidification.

At 60% humidification, the superior performance of the PTFE:carbon black = 1:9 sample is attributed to two key factors [6, 7]:

1) Water produced at the catalyst layer efficiently enters the hydrophilic pores, humidifying the incoming gas. This effect enhances the sample's performance at low current densities, as corroborated by the polarization curve results.
2) The balanced mix of hydrophilic and hydrophobic pores supports simultaneous water transport through both pore types, enabling the sample to sustain high power and current densities.

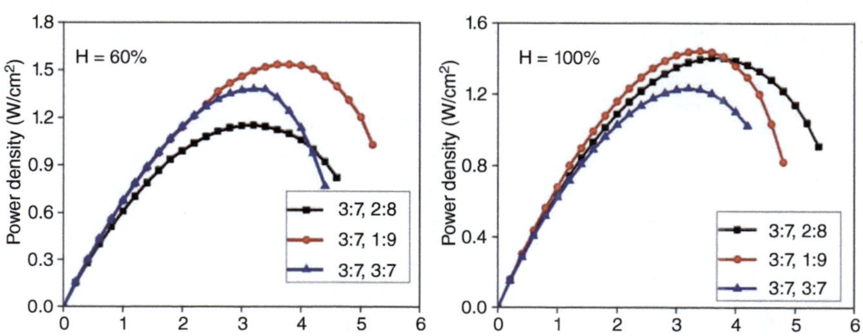

Figure 8.2 The power density test results of three samples under different humidity (The temperature of fuel cell is 80 °C; the test in hydrogen and oxygen environment; humidity is 60% and 100%, respectively; and Pt loading anode of the catalyst layer is 0.1 mg/cm^2 and cathode is 0.1 mg/cm^2).

At 100% humidification, the PTFE:carbon black = 1:9 sample maintained its high power density thanks to its optimal pore distribution. This configuration allowed efficient water removal at medium current densities without compromising Pt catalytic activity or gas transport efficiency.

8.3.3 Analysis of Electrochemical Impedance Spectroscopy

Figure 8.3 depicts the EIS results for the three GDL samples at 2.5 A/cm^2 under humidification conditions of 60% and 100%. When the PTFE:carbon black ratio is fixed at 3:7 in the MPL adjacent to the carbon paper, a lower PTFE content in the layer near the catalyst side correlates with a smaller intersection point on the left side of the real axis. This observation suggests that a reduced PTFE content lowers the ohmic resistance of the GDL. In the low-frequency region, the PTFE:carbon black = 2:8 sample shows the smallest intersection point on the right side of the real axis under both humidification conditions. This indicates that the 2:8 sample exhibits the lowest mass transport impedance among the tested configurations.

By contrast, GDLs lacking a gradient hydrophobic structure have higher mass transport impedance under both 60% and 100% humidity, emphasizing the effectiveness of gradient hydrophobic treatment. This treatment optimizes the balance of hydrophobic and hydrophilic properties within the GDL, ultimately reducing mass transport losses and improving overall fuel cell performance.

8.3.4 Analysis of Physical Performance

As shown in Figure 8.4, increasing PTFE content leads to higher water contact angles, indicating a shift in the hydrophilicity and hydrophobicity balance within the GDL. The surface roughness measurements—6.2, 4.1, and 4.6 μm—reveal that the roughness variations are relatively small and unlikely to significantly affect overall cell performance.

Figure 8.5 highlights the surface morphology and in-plane resistance of the samples. The spray-coating process produced a visible pore distribution and agglomerated particles, enabling effective water transport at low and high

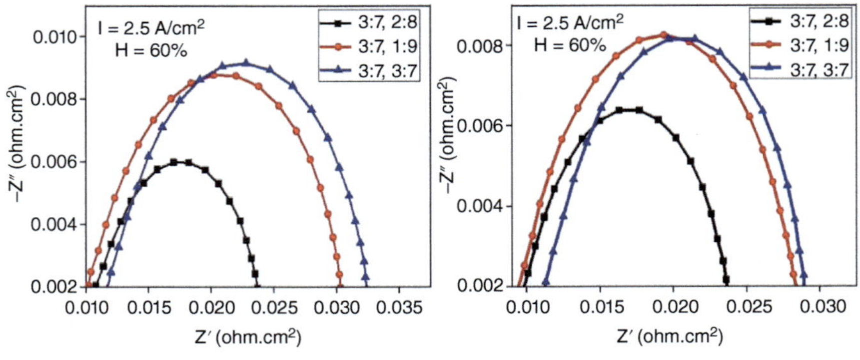

Figure 8.3 EIS test results of three samples under different humidity (current density $I = 2.5$ A/cm^2).

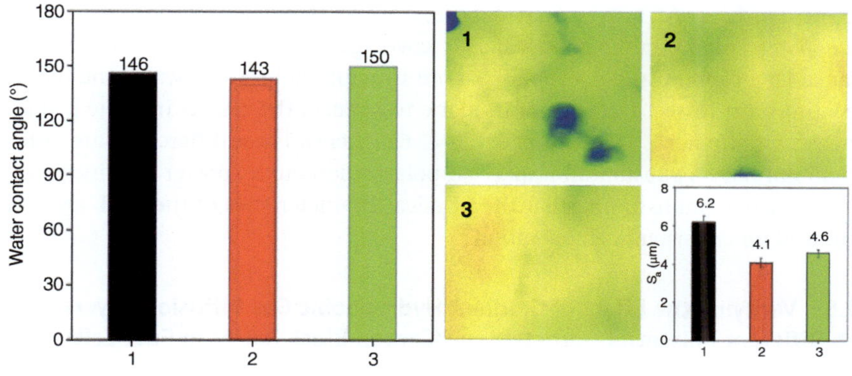

Figure 8.4 Hydrophobicity and coarseness test results. (Samples 1, 2, and 3 represent near the catalyst layers side PTFE:carbon black ratios of 2:8, 1:9, and 3:7, respectively.)

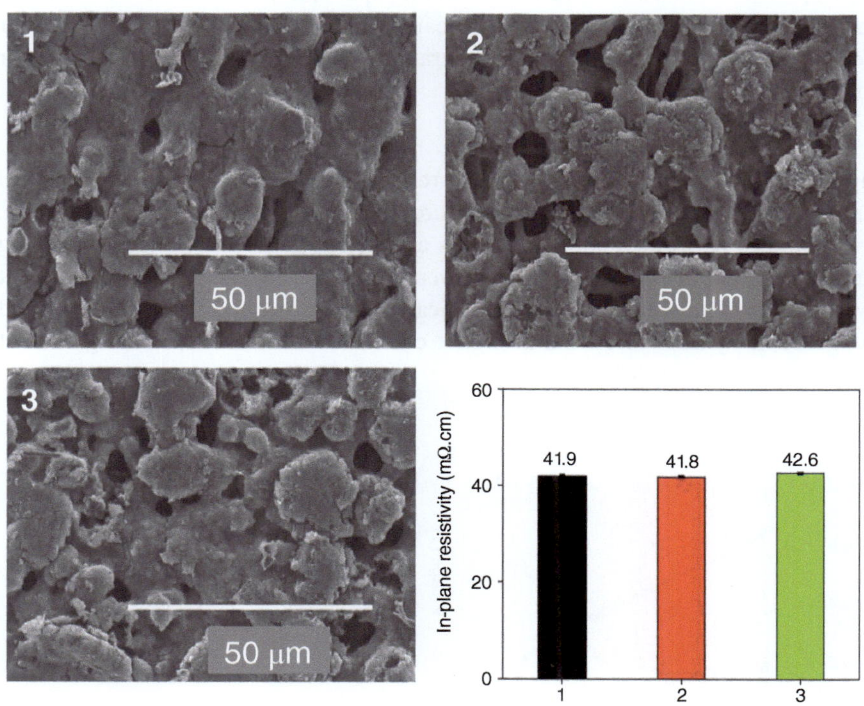

Figure 8.5 Surface topography and in-plane resistance test results (Samples 1, 2, and 3 represent PTFE:carbon black ratios near the side of catalyst layer of 2:8, 1:9, 3:7, respectively.)

current densities. At low current densities, water is efficiently managed within these pores, maintaining good catalyst coverage. At high current densities, a balanced hydrophilic and hydrophobic pore structure helps expel water and reduce mass transport losses. In terms of in-plane resistance, the results indicate a slight increase with higher PTFE content (41.9–42.6 mΩ·cm). These differences are minor and do not have a substantial impact on polarization curve, power density, or EIS performance, demonstrating that the physical characteristics of the GDL are well optimized for consistent cell operation.

8.3.5 Verifying the Effect of Gradient Hydrophobic Gas Diffusion Layers with Different Polytetrafluoroethylene/Carbon Black Ratios in Fuel Cells

Chapter 7 evaluated the impact of gradient hydrophobic treatment in the MPL near the catalyst side, keeping the carbon paper side constant. The tests showed that hydrophobic treatment improved limiting current density, power density, and reduced mass transport impedance. However, it remained unclear how altering the PTFE:carbon black ratio near the carbon paper side would affect performance. To address this, GDLs with a 2:8 PTFE:carbon black ratio near the carbon paper were prepared and tested in two configurations: one with a 2:8 ratio and another with a 1:9 ratio near the catalyst side.

8.3.5.1 Analysis of Polarization Curve

Figure 8.6 presents polarization curve results for these GDLs under 60% and 100% humidification at a cell temperature of 80 °C. Under 60% humidification, both samples performed similarly across all current densities, achieving limiting current densities of 5.2 A/cm^2 for the untreated GDL and 5.0 A/cm^2 for the gradient hydrophobic GDL. Under 100% humidification, the untreated GDL outperformed the gradient hydrophobic GDL at higher current densities, with limiting current densities of 5.4 versus 5.0 A/cm^2.

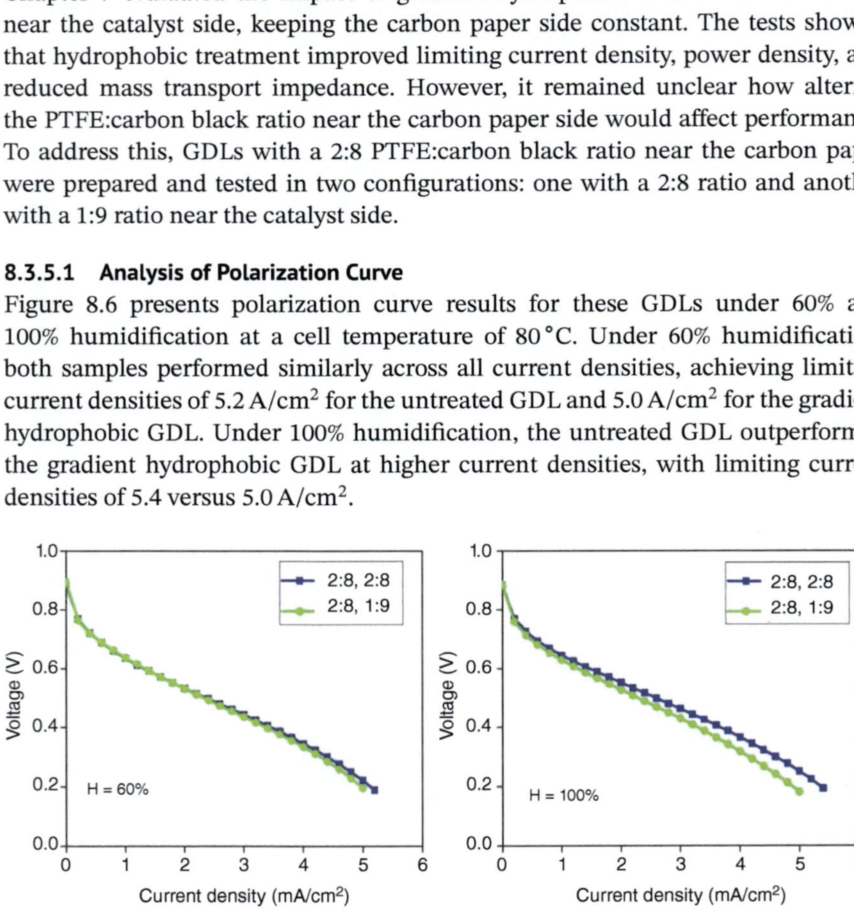

Figure 8.6 Polarization curve test results of two samples. Among them, 2:8, 2:8 represent that the mass ratio of PTFE:carbon black in the two microporous layers is both 2:8; samples 2:8, 1:9 represent PTFE:carbon black ratio of 2:8 in the microporous layer near the carbon paper side, and PTFE:carbon black ratio of 1:9 near the cathode catalyst layer side. Pt loading of the catalyst layer for both anode and cathode is 0.1 mg/cm^2.

At 60% humidification, the small amount of water produced at low current densities was effectively managed by the existing hydrophilic pores, even in the more hydrophobic samples. At higher current densities, water production increased, but both configurations handled water transport adequately under moderate humidification conditions. Under 100% humidification, however, the untreated GDL excelled at high current densities due to its ability to better manage the larger water volumes generated. The gradient hydrophobic GDL with a 1:9 ratio near the catalyst side had more hydrophilic pores, which retained water longer and relied on hydrophobic pores for transport. When water moved from the first MPL to the second, the higher hydrophilic pore content near the carbon paper side caused water accumulation, reducing overall transport efficiency.

Figure 8.6 reveals that gradient hydrophobic treatment does not consistently enhance cell performance, especially under high humidification. Further analyses of power density and electrochemical impedance are needed to better understand how different PTFE:carbon black ratios near both the carbon paper and catalyst sides affect the GDL's water management and overall performance.

8.3.5.2 Analysis of Power Density

Figure 8.7 illustrates the power density outcomes under 60% and 100% humidity at a cell temperature of 80 °C for the two GDLs. At 60% humidity, the untreated GDL achieved a power density of 1.40 W/cm^2, while the gradient hydrophobic GDL reached 1.36 W/cm^2. At 100% humidity, the untreated GDL demonstrated a power density of 1.47 W/cm^2 compared to 1.32 W/cm^2 for the gradient hydrophobic GDL. These results indicate that under 60% humidification, gradient hydrophobic treatment provided no significant enhancement in power density, consistent with the polarization curve findings. At 100% humidification, the untreated GDL's power density was markedly higher. This disparity can be attributed to the substantial water generation at the cathode catalyst side under fully humidified conditions and high current densities. The 1:9 PTFE:carbon black GDL near the catalyst

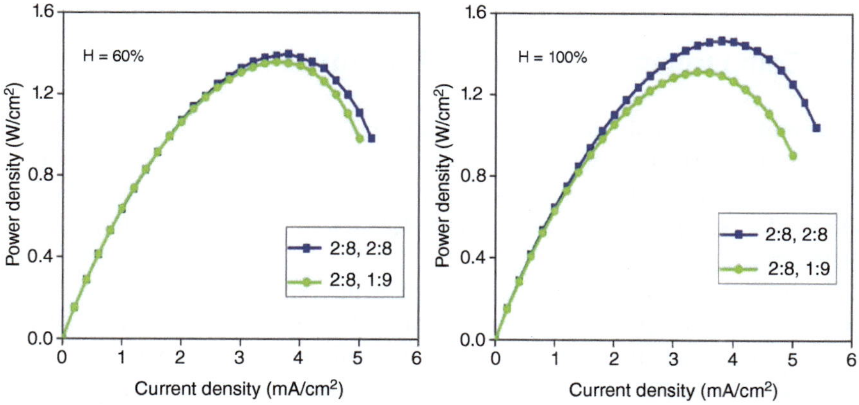

Figure 8.7 Power density test results of the two GDLs.

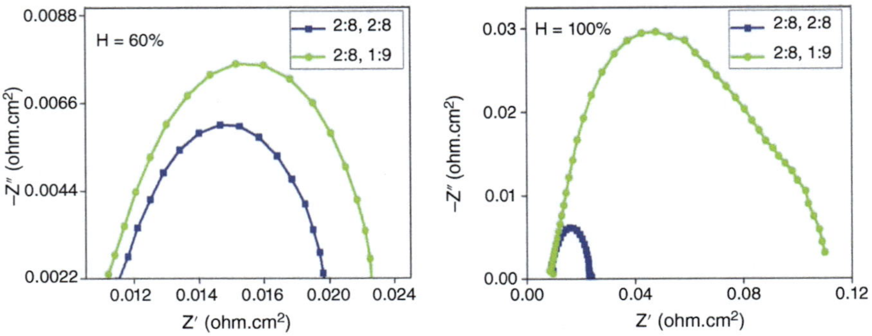

Figure 8.8 The EIS test result of the sample when $I = 2.5\,A/cm^2$.

side exhibited a pore distribution imbalance that impeded water removal, leading to increased mass transport impedance and diminished power density.

8.3.5.3 Analysis of Electrochemical Impedance Test Results

Figure 8.8 presents EIS results for the untreated and gradient-treated GDLs at a current density of $2.5\,A/cm^2$. At both 60% and 100% humidity, the untreated GDL consistently exhibited lower mass transport impedance. This suggests that gradient hydrophobic treatment did not necessarily decrease mass transport impedance. Instead, the overall cell performance improvements observed with gradient hydrophobic GDLs appeared to rely on achieving a balanced distribution of hydrophilic and hydrophobic pores. This balance enabled water absorption through hydrophilic pores and transport through hydrophobic pores, enhancing water management. Conversely, if the gradient hydrophobic treatment led to reduced performance, it was likely due to an unfavorable hydrophilic-to-hydrophobic pore ratio, resulting in poor water removal and elevated mass transport impedance at higher current densities [8, 9].

8.4 Comparison of Membrane Electrodes

The membrane electrodes used in Chapters 7 and 8 demonstrate notable differences in performance. In Chapter 7, a 240-μm GDL achieved a peak power density of $1.78\,W/cm^2$ under 100% humidification. By contrast, in Chapter 4, the same 240-μm GDL (sample 3:7, 3:7) reached only $1.23\,W/cm^2$, resulting in a difference of $0.55\,W/cm^2$ (Figure 8.9b).

Herein, the performance of the as-prepared membrane electrode (CCM-4) is compared with the previously reported membrane electrode (CCM-3 from Chapter 7) and the commercial GDL29BC. As shown in Figure 8.9a, CCM-3 achieved a peak power density of $1.506\,W/cm^2$, while CCM-4 reached only $0.734\,W/cm^2$, resulting in a significant performance difference of $0.87\,W/cm^2$. This disparity is primarily attributed to differences in proton exchange membrane (PEM) properties and Pt loading between the two membrane electrodes.

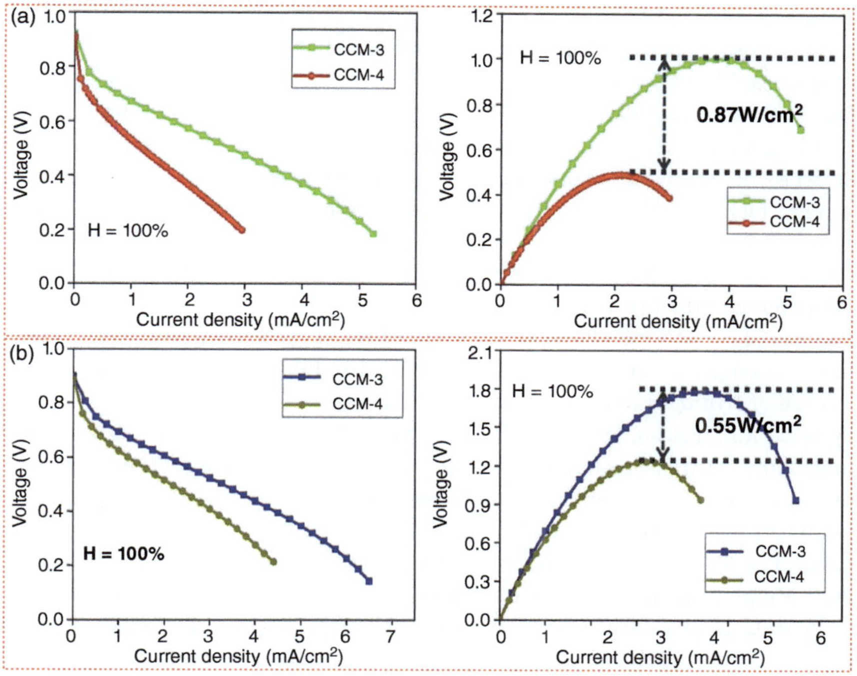

Figure 8.9 (a) Test results of GDL29BC when using membrane electrodes in Chapters 7 and 8, respectively. (b) Test results of 240-μm GDL when using membrane electrodes in Chapters 7 and 8.

8.5 Conclusion

In this chapter, we explored the influence of hydrophilic and hydrophobic properties in GDLs. Initial tests, using a PTFE:carbon black ratio of 3:7 near the carbon paper, showed no significant performance differences at lower current densities under 60% humidification. However, at higher current densities, GDLs with a 2:8 ratio near the catalyst layer achieved better water and gas management. Under 100% humidification, gradient hydrophobic GDLs demonstrated clear performance advantages across various current densities, supported by polarization curve, power density, and EIS data. Further testing with a 2:8 ratio near the carbon paper yielded inconsistent results. While untreated GDLs performed slightly better at 100% humidification, the study highlighted that precise adjustment of PTFE:carbon black ratios in both MPLs is essential. Optimizing this balance can enhance water management, reduce mass transport losses, and ultimately improve peak power density.

References

1 Wu, R., Zhu, X., Liao, Q. et al. (2013). Liquid and oxygen transport in defective bilayer gas diffusion material of proton exchange membrane fuel cell. *International Journal of Hydrogen Energy* 38 (10): 4067–4078.

2 LaManna, J.M. and Kandlikar, S.G. (2011). Determination of effective water vapor diffusion coefficient in PEMFC gas diffusion layers. *International Journal of Hydrogen Energy* 36 (8): 5021–5029.

3 Quick, C., Ritzinger, D., Lehnert, W. et al. (2009). Characterization of water transport in gas diffusion media. *Journal of Power Sources* 190 (1): 110–120.

4 Schweiss, R., Steeb, M., and Wilde, P.M. (2010). Mitigation of water management in PEM fuel cell cathodes by hydrophilic wicking microporous layers. *Fuel Cells* 10 (6): 1176–1180.

5 Chang, S.M. and Chu, H.S. (2006). Transient behavior of a PEMFC. *Journal of Power Sources* 161 (2): 1161–1168.

6 Radhakrishnan, V. and Haridoss, P. (2010). Effect of cyclic compression on structure and properties of a gas diffusion layer used in PEM fuel cells. *International Journal of Hydrogen Energy* 35 (20): 11107–11118.

7 Ferreira, R.B., Falcao, D.S., Oliveira, V.B. et al. (2017). Experimental study on the membrane electrode assembly of a proton exchange membrane fuel cell: Effects of microporous layer, membrane thickness and gas diffusion layer hydrophobic treatment. *Electrochimica Acta* 224: 337–345.

8 Chun, J.H., Park, K.T., Jo, D.H. et al. (2011). Development of a novel hydrophobic/hydrophilic double micro porous layer for use in a cathode gas diffusion layer in PEMFC. *International Journal of Hydrogen Energy* 36 (14): 8422–8428.

9 Qin, Y.Z., Yin, Y., Jiao, K. et al. (2019). Effects of needle orientation and gas velocity on water transport and removal in a modified PEMFC gas flow channel having a hydrophilic needle. *International Journal of Energy Research* 43 (7): 2538–2549.

9

Fundamentals of Proton Exchange Membranes

9.1 Introduction

Proton exchange membranes (PEMs), also known as polymer electrolyte membranes, are specialized cation exchange materials containing strong acidic functional groups (such as sulfonic acid groups). These membranes facilitate the exchange of protons with their surrounding medium without undergoing dissolution. Typically, PEMs are ionic polymers large molecules with ionizable sites covalently bonded to their polymer backbone or sidechains [1–4].

A critical feature of PEMs is their reliance on water molecules to enable proton dissociation and transport, classifying them as hydrated acidic polymers. To achieve optimal proton conductivity, it is essential to maximize the degree of charge dissociation and the concentration of ionic charge [5]. However, the presence of sulfonic acid groups also increases the polymer's affinity for water (hygroscopicity), potentially causing highly sulfonated polymers to dissolve. To address this challenge, careful selection and molecular design of polymer materials are required. Strategies include optimizing the polymer structure to enhance the acidity of the sulfonic acid groups while maintaining polymer stability and exploiting inherent phase separation properties to develop high-performance membrane materials [6–9].

Another category of PEMs is acid-doped polymer membranes. These membranes consist of a polymer matrix doped with strong acids possessing high boiling points (e.g., H_2SO_4, H_3PO_4), enabling proton conduction under anhydrous conditions [10–12]. Among these, polybenzimidazole-phosphoric acid (PBI-PA) membranes stand out for their exceptional proton conductivity and physicochemical stability [13–16]. Unlike hydrated acid polymer membranes, the ion exchange mechanism in acid-doped membranes takes place within phosphoric acid rather than water. Because of the high boiling point of the acid, these membranes offer anhydrous proton conductivity, making them suitable for high-temperature polymer electrolyte membrane fuel cells [17, 18].

The primary role of PEMs is to provide proton transport from the anode to the cathode in a fuel cell, where protons combine with reduced oxygen atoms (O_2^-) to form water. Proper operation requires that oxygen atoms at the cathode be fully

Electrospinning for Proton Exchange Membrane Fuel Cells, First Edition.
Yong Liu and Mohideen Meerasahib Mohamedazeem.

reduced; otherwise, they may react with hydrogen diffusing from the anode to produce hydrogen peroxide (H_2O_2). This compound can decompose into highly reactive hydroxyl peroxide (–OOH) and hydroxyl radicals (–OH), accelerating polymer degradation under elevated temperatures and pressures within the fuel cell [19–21]. Consequently, PEMs must meet rigorous performance standards to ensure high efficiency and durability. An ideal PEM exhibits high proton conductivity, robust mechanical strength, superior thermal and chemical stability, and cost-effectiveness. To meet these stringent demands, PEMs must satisfy the following requirements [22–24]:

1) High proton conductivity, no less than 100 mS/cm.
2) Excellent electrical insulation under all operating conditions.
3) Low hydrogen and oxygen permeability, with hydrogen leakage current not exceeding 2 mA/cm^2.
4) High stability against hydrolysis, thermal stress, and oxidative conditions.
5) Adequate water absorption and retention to maintain proton conductivity at low relative humidity (RH).
6) Strong mechanical properties when hydrated, ensuring reliable membrane electrode assembly (MEA) operation.
7) A low production cost, not exceeding \$20/m^2.

By adhering to these criteria, PEMs can effectively support the efficient and durable operation of polymer electrolyte membrane fuel cells.

9.1.1 Partially Fluorinated Proton Exchange Membranes

Cost has long been a critical factor limiting the widespread use of perfluorosulfonic acid (PFSA) PEMs. To reduce costs and simplify the synthetic process, researchers have increasingly focused on non-PFSA PEMs [25]. One key approach involves creating partially fluorinated PEMs by blending fluorinated and nonfluorinated compounds. This involves replacing some fluorinated groups with nonfluorinated ones, modifying perfluorocarbon structures to lower fluorine content without entirely losing their beneficial properties.

The first partially fluorinated PEM was made from poly(trifluorobenzene sulfonate). However, its durability proved insufficient for the prolonged operation required in fuel cells. To address these limitations, Ballard Power Systems in Canada developed an improved sulfonated film known as BAM3G. The chemical composition of this membrane, as illustrated in Figure 9.1a, is based on poly(trifluorobenzene). By maintaining the main perfluorinated backbone and introducing nonfluorinated side chains, researchers reduced the fluorine content and enhanced the membrane's stability [26].

Wei et al. [27] evaluated the performance of BAM3G membranes through a series of rigorous tests. The results, shown in Figure 9.1b, demonstrate that these membranes outperform conventional PFSA membranes such as Nafion and Dow, while also being more cost-effective. Furthermore, the lifespan of BAM3G membranes has significantly increased, exceeding 10,000 hours.

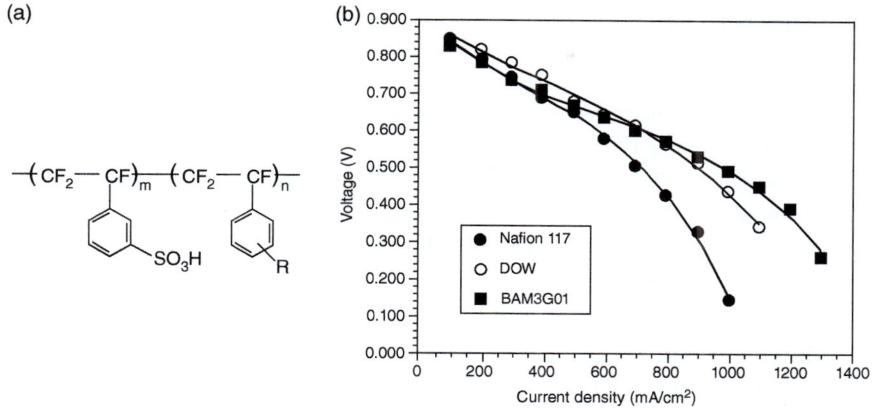

Figure 9.1 (a) Chemical structural formula of BAM3G membrane, (b) Performance comparison of BAM3G membrane with Nafion 117 membrane and Dow membrane.

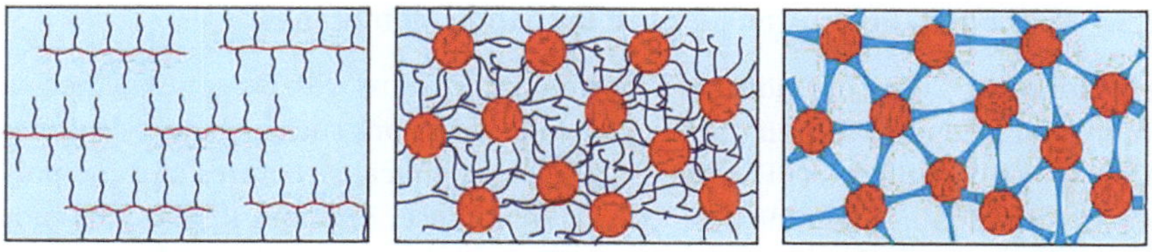

Figure 9.2 Water channel formed by phase separation of PVDF-g-PSSA in DMSO solvent.

The chemical grafting method is a well-established approach for preparing partially fluorinated PEMs. Sadeghi et al. [28] successfully created a sulfonated polyphenylene film (PVDF-g-PSSA) by irradiating polydifluoroethylene materials such as polyvinylidene fluoride (PVDF) and then grafting sodium styrene sulfonate (SSS) onto the PVDF powder. This process formed proton-conducting channels, as illustrated in Figure 9.2. After casting the solute and steaming it to form a film, the PVDF was melted and filled into the resultant pores, producing a composite membrane. Initial experiments revealed that the PVDF-g-PSSA membrane retained the intrinsic physical and chemical stability of the original PVDF material. Qualitatively, the proton conductivity showed a positive correlation with the grafting degree. When the grafting level reached approximately 35%, the proton conductivity improved significantly, reaching about 70 mS/cm^2.

9.1.2 Fluorine-Free Proton Exchange Membranes

While numerous studies have explored the application of fluorinated PEMs, many of these materials exhibit certain performance limitations and environmental concerns. The presence of C—F bonds in fluorinated polymers can lead to issues such as pollution and complex disposal. As a result, researchers have sought high-quality alternatives that use nonfluorinated (C–H) main chain polymers, which not only are more cost-effective but also pose fewer environmental challenges.

Aromatic polymers, such as polysulfone (PSU), polyaryletherketone (PAEK), and polyimides (PIs), as well as their sulfonated derivatives, have become a focal point in fluorine-free PEM research. These polymers are known for their thermal stability and chemical robustness. A common approach to endow these materials with proton conductivity is the incorporation of sulfonic acid groups into the polymer's side chains. This sulfonation process typically involves reacting the polymer with sulfonating agents such as concentrated sulfuric acid, sulfur trioxide, or chlorosulfonic acid. C–H main chain polymers offer several distinct advantages over their fluorinated counterparts [29, 30]:

1) Lower cost and widespread material availability.
2) Reduced environmental impact due to the absence of fluorine, making waste products easier to degrade.
3) Improved water retention due to polar groups, helping the membrane maintain hydration at elevated temperatures.

9.1.3 Sulfonated Polysulfone Proton Exchange Membranes

Polysulfones are a class of amorphous thermoplastic resins with exceptional thermal stability and mechanical strength. They include variations such as polyarylsulfones (PASs), polyethersulfones (PESs), and PSU. The chemical structures of these polymers are depicted in Figure 9.3 [31, 32]. Among these, PSU and PES are the most extensively studied sulfonated polysulfones. By introducing sulfonic acid groups into their molecular structures, these polymers become capable of proton conduction, making them well-suited for use in high-quality PEMs.

Incorporating inorganic materials into sulfonated polymers is a significant area of focus in PEM research. For example, Baradie et al. [33] demonstrated that combining phosphoric acid–based fillers with sulfonated polyurethane resulted in a composite PEM that performed well under fuel cell operating conditions (80 °C and 96% RH). Comparisons with Nafion 117 revealed that the composite membrane achieved 80% of Nafion 117's overall performance, attributable to the presence of unused phosphorus. This composite also exhibited improved mechanical strength and reduced gas permeability, enhancing the membrane's durability and functionality.

To further leverage the intrinsic advantages of polysulfone polymers, Ureña et al. [34] developed a novel approach by synthesizing a multiblock copolymer of sulfonated polysulfone (SPSU) and sulfonated polyphenylsulfone (SPPSU). The resulting composite membrane, incorporating polyether chain segments from polyphenylsulfone (PPSU), was formed through a synthesis and sulfonation process illustrated in Figure 9.4. The SPSU/SPPSU composite membrane displayed excellent mechanical properties, thermal stability, and reliable performance under operating conditions at 70 °C. Comprehensive tests revealed a proton conductivity of 34.1 mS/cm, a peak power density of 400 mW/cm^2, and a current density of 1100 mA/cm^2, all of which highlight the membrane's robust performance.

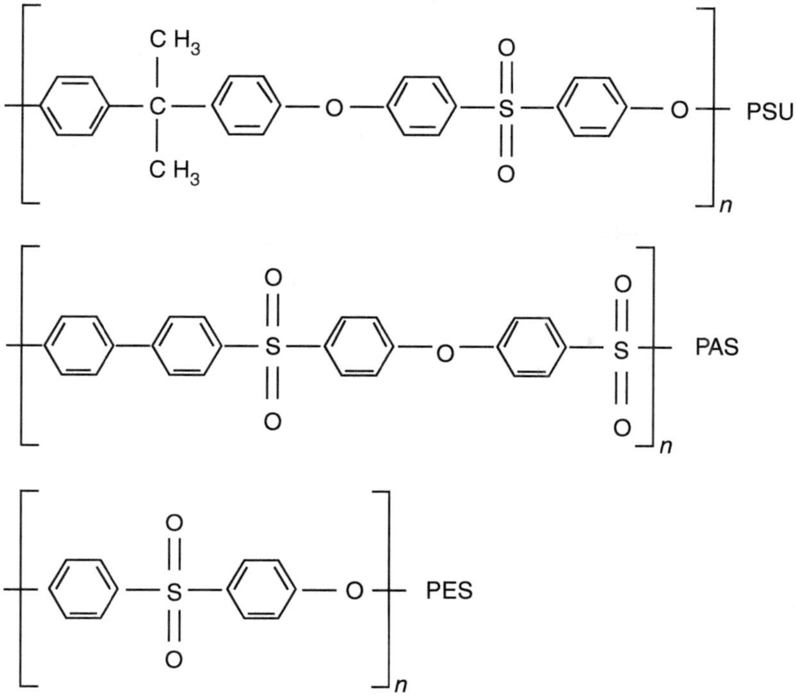

Figure 9.3 Chemical structure of polysulfone polymers.

Divya et al. [35] investigated a composite membrane consisting of a polydopamine-coated disulfide molybdenum (PDM) nano-layer and sulfonated polyethers (SPESs). This innovative membrane demonstrated enhanced thermal, mechanical, and oxidative stability, along with lower methanol permeability compared to traditional SPES and Nafion membranes. Notably, its proton conductivity reached up to 5.64×10^{-3} S/cm, surpassing both pure SPES and Nafion 212 membranes. This improved conductivity is attributed to the strong acid–base interactions formed between amine groups ($-NH_2$, –NH–) in PDM and sulfonate groups ($-SO_3^-$) in SPES. These interactions facilitate proton transport by creating a more robust proton conduction pathway within the nanocomposite membrane. Furthermore, the PDM structure contributes to reducing methanol permeability by distorting the transport paths, which in turn hinders methanol crossover.

9.1.4 Sulfonated Polyaryletherketone Proton Exchange Membranes

PAEKs are cost-effective engineering thermoplastics that possess excellent thermal and chemical stability. When sulfonated, these materials retain the inherent advantages of PAEKs while also exhibiting high proton conductivity. Research on

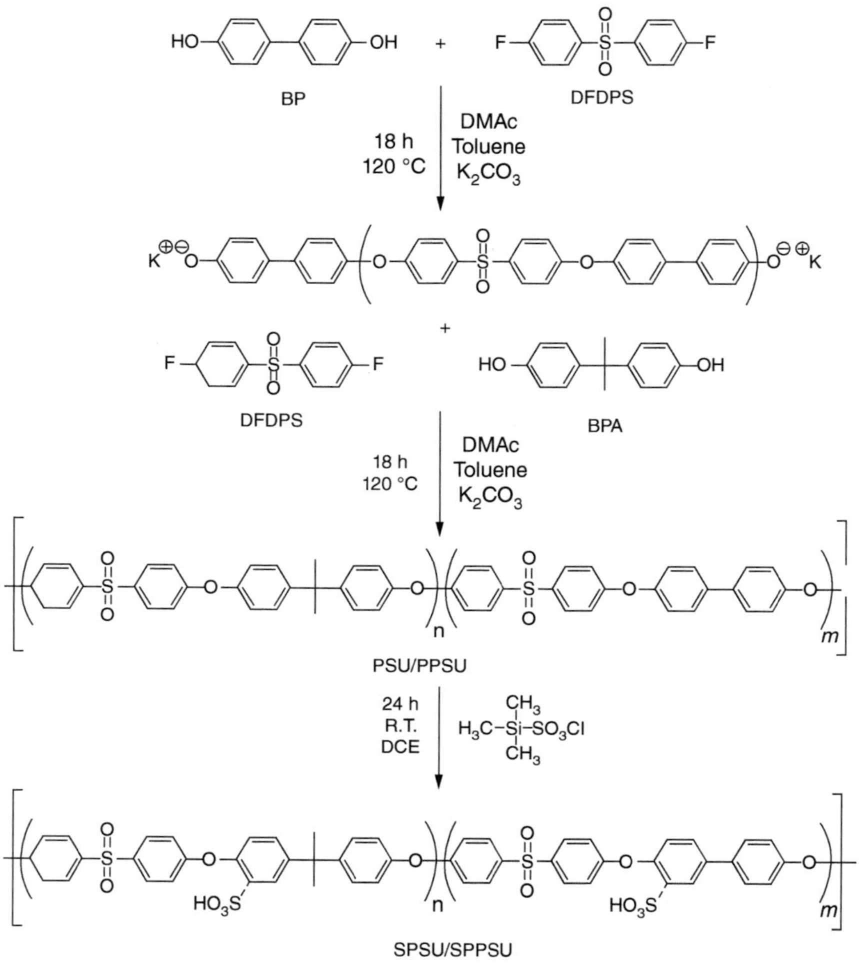

Figure 9.4 PSU/PPSU synthesis and sulfonation process.

sulfonated PAEKs has been ongoing for decades, focusing on various derivatives such as polyetherketone (PEK), polyetheretherketone (PEEK), and polyetherketoneketone (PEKK). The chemical structures of these materials are depicted in Figure 9.5 [36–38].

Sulfonation plays a critical role in the performance of sulfonated PAEK membranes. For example, Dai Xianzhi et al. [39] used concentrated sulfuric acid as a sulfonating agent and systematically studied the effects of different sulfonation degrees by varying reaction times. Their findings revealed that prolonged reaction times resulted in higher sulfonation levels. However, the sulfonation rate tended to increase more rapidly with higher sulfuric acid concentrations, eventually stabilizing. The sulfonated poly(ether ether ketone) (SPEEK) membranes with an optimal,

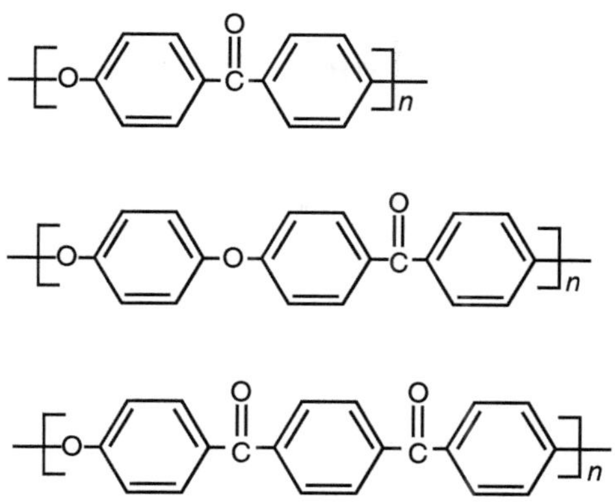

Figure 9.5 Chemical structure of polyaryletherketone polymers.

moderate sulfonation degree exhibited the most uniform microstructure and demonstrated superior thermal stability.

The proton conduction mechanism in PEMs strongly depends on the water content of the membrane. Veldre et al. [40] explored the use of ionic liquid (IL) additives in SPEEK membranes, demonstrating that when the membrane remains well hydrated, water molecules can replace the IL at higher humidity levels. This ensures that the membrane retains high moisture content even at elevated temperatures, maintaining stable performance.

Hydrophilic inorganic particles have also been investigated as a means of enhancing both the physical properties and water content of membranes. Nunes et al. [41] prepared SPEEK-based inorganic composite membranes using SiO_2, TiO_2, and ZrO_2 as modifiers. Among these, ZrO_2 had the greatest impact, significantly improving the membrane's mechanical strength and resistance to alcohol permeability. However, the addition of inorganic particles was observed to reduce proton conductivity. By combining ZrO_2 with phosphate, a balance was achieved, maintaining mechanical integrity while limiting the reduction in proton conductivity to between 10% and 30%.

In another study, Martina et al. [42] investigated sulfonated polyetherketone membranes by incorporating sulfonated silica (S-SiO_2) into a blend of SPEEK and sulfonated poly(vinylidene fluoride-co-hexafluoropropylene) (SPVdF-HFP). X-ray diffraction (XRD) and FTIR analyses confirmed the phase structure of the composite membrane. The optimized blend, consisting of 80 wt% SPEEK and 20 wt% SPVdF-HFP, with 2 wt% S-SiO_2, demonstrated high water absorption (36.5%), low swelling (15.9%), and a proton conductivity of 7.9×10^{-2} S/cm.

Organic clay is another modifier frequently used to enhance polymer properties. Hande et al. [43] investigated the cross-linking of SPEEK with cloisite-30b (a type of organically modified clay) using 1,4-bis(hydroxymethyl)benzene (BHMB) as a cross-linker. As shown in Figure 9.6, varying the clay content affected the physical and thermal properties of the membrane. While the addition of clay had minimal impact on proton conductivity, it significantly improved the mechanical strength, oxidative stability, and thermal stability of the membrane.

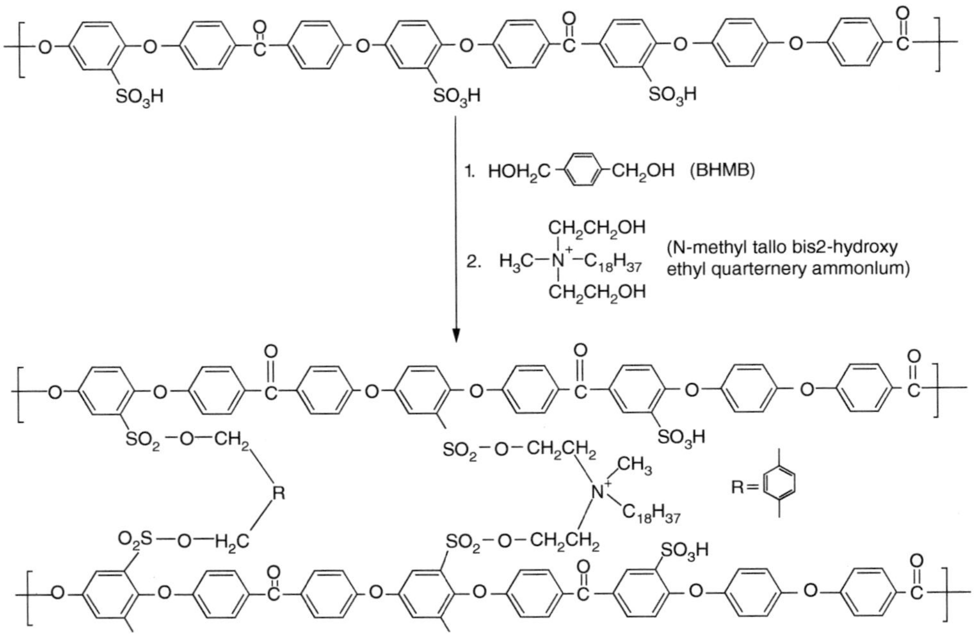

Figure 9.6 Schematic of cross-linking of N-methyl tallow bis2-hydroxyethyl quaternary ammonium with BHMB cross-linked SPEEK.

9.1.5 Sulfonated Polyimide Proton Exchange Membranes

PI is a high-performance polymer that stands out due to its exceptional thermal resistance—up to 400 °C—along with excellent film-forming capabilities and chemical stability. When sulfonated, polyimide is converted into sulfonated polyimide (SPI), enabling proton conduction. The chemical structure of PI is shown in Figure 9.7 [44].

The majority of SPI membrane studies have involved modifying SPIs to enhance their properties. Ali et al. [45], for example, reinforced SPI membranes by incorporating graphene-based ultrasonic clay (GSC) through a solution casting method. The resulting composite membrane, depicted schematically in Figure 9.8, exhibited a significant reduction in thickness—48.6% lower than the original SPI film—while improving proton conduction pathways. Impedance spectroscopy measurements showed that the SPI-GSC composite's proton conductivity increased from 6.02×10^{-10} S/cm in the pure SPI membrane to 6.61×10^{-6} S/cm, demonstrating its potential as a fuel cell PEM.

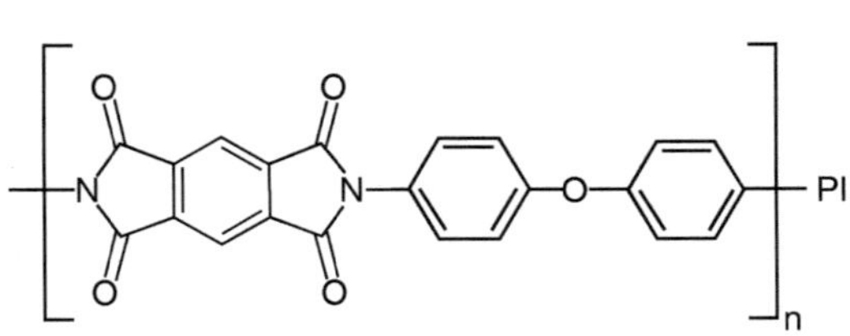

Figure 9.7 Chemical structure of polymerize polymer.

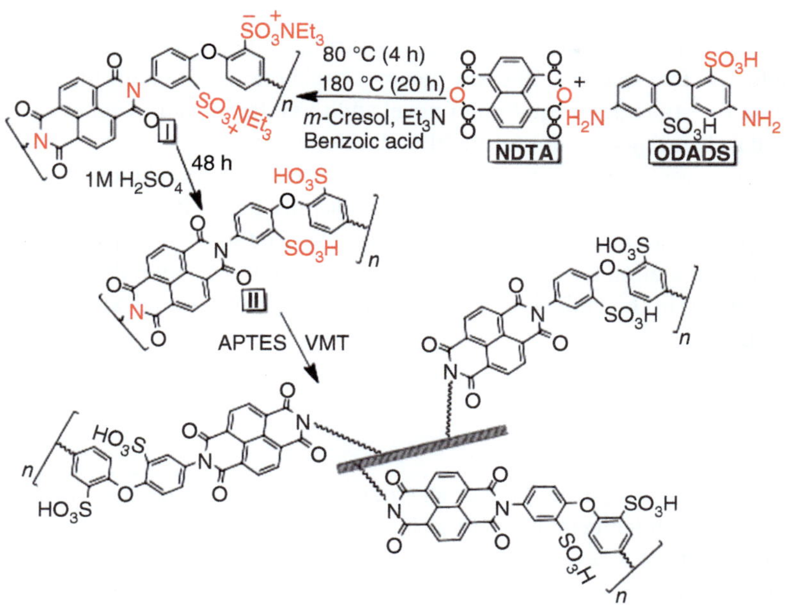

Figure 9.8 Schematic diagram of the reaction between GSC and SPI.

Polymerization techniques have also been explored as a means to achieve uniform SPI membranes with desirable properties. Asano et al. [46] synthesized a range of SPIs containing both main-chain and side-chain aliphatic groups. The resulting membranes achieved ion exchange capacities (IEC) between 1.78 and 2.33 meq/g, with reduced fuel gas permeability compared to Nafion membranes. At an IEC of 1.95 meq/g, the SPI membranes exhibited proton conductivity slightly lower than Nafion under temperatures below 100 °C. However, at temperatures above 100 °C, the proton conductivity of SPI matched or exceeded that of Nafion, reaching a peak of 0.18 S/cm at 140 °C. In addition, these membranes demonstrated a lifespan of over 5000 hours in single-cell fuel cell environments.

9.2 Conclusion

In this chapter, we provided a comprehensive review of PEM materials, highlighting the significant progress made in developing high-performance membranes suitable for various fuel cell applications. The discussion has covered both partially fluorinated and fluorine-free PEMs, examining their synthesis methods, material compositions, and modifications to enhance conductivity, stability, and durability. Furthermore, the incorporation of inorganic fillers, composite structures, and advanced sulfonation techniques has been shown to significantly improve key performance parameters, including thermal stability, mechanical strength, and proton conductivity, while reducing costs and environmental impacts.

In particular, advancements in SPSUs, PAEKs, and PIs demonstrate the diverse approaches researchers have adopted to meet the stringent demands of modern fuel cell technologies. By exploring different polymer backbones, side-chain functionalization, and the strategic integration of novel additives, these studies collectively point toward a clear trajectory for future PEM development.

As the field moves forward, the ongoing challenge remains to balance conductivity, stability, and cost in a way that ensures not only high efficiency but also practical implementation at a commercial scale. The findings summarized here provide a solid foundation for the continued refinement of PEM materials and offer valuable insights into the pathways that hold the most promise for advancing fuel cell technology.

References

1 Park, C.H., Lee, C.H., Guiver, M.D. et al. (2011). Sulfonated hydrocarbon membranes for medium-temperature and low-humidity proton exchange membrane fuel cells (PEMFCs). *Progress in Polymer Science* 36 (11): 1443–1498.

2 Chandan, A., Hattenberger, M., El-Kharouf, A. et al. (2013). High temperature (HT) polymer electrolyte membrane fuel cells (PEMFC) – A review. *Journal of Power Sources* 231: 264–278.

3 Zhu, L.Y., Li, Y.C., Liu, J. et al. (2022). Recent developments in high-performance Nafion membranes for hydrogen fuel cells applications. *Petroleum Science* 19 (3): 1371–1381.

4 Li, G., Kujawski, W., and Rynkowska, E. (2022). Advancements in proton exchange membranes for high-performance high-temperature proton exchange membrane fuel cells (HT-PEMFC). *Reviews in Chemical Engineering* 38 (3): 327–346.

5 Neelakandan, S., Wang, L., Zhang, B. et al. (2022). Branched polymer materials as proton exchange membranes for fuel cell applications. *Polymer Reviews* 62 (2): 261–295.

6 Li, N. and Guiver, M.D. (2014). Ion transport by nanochannels in ion-containing aromatic copolymers. *Macromolecules* 47 (7): 2175–2198.

7 Yang, X., Kim, J.H., and Kim, Y.J. (2022). Enhanced proton conductivity of poly(ether sulfone) multi-block copolymers grafted with densely pendant sulfoalkoxyl side chains for proton exchange membranes. *Polymer* 242: 124604.

8 Kusoglu, A. and Weber, A.Z. (2017). New insights into perfluorinated sulfonic-acid ionomers. *Chemical Reviews* 117 (3): 987–1104.

9 Shin, D.W., Guiver, M.D., and Lee, Y.M. (2017). Hydrocarbon-based polymer electrolyte membranes: Importance of morphology on ion transport and membrane stability. *Chemical Reviews* 117 (6): 4759–4805.

10 Rao, Z., Lan, M., Zhu, D. et al. (2022). Synergistically promoted proton conduction of proton exchange membrane by phosphoric acid functionalized carbon nanotubes and graphene oxide. *Journal of Membrane Science* 659: 120810.

11 Peng, J., Fu, X., Liu, D. et al. (2022). An effective strategy to enhance dimensional-mechanical stability of phosphoric acid doped polybenzimidazole membranes by introducing in situ grown covalent organic frameworks. *Journal of Membrane Science* 655: 120603.

12 Tang, H., Geng, K., Wu, L. et al. (2022). Fuel cells with an operational range of –20 °C to 200 °C enabled by phosphoric acid-doped intrinsically ultramicroporous membranes. *Nature Energy* 7 (2): 153–162.

13 Li, T., Yang, J., Chen, Q. et al. (2023). Construction of highly conductive cross-linked polybenzimidazole-based networks for high-temperature proton exchange membrane fuel cells. *Materials* 16 (5): 1932.

14 Aili, D., Zhang, J., Jakobsen, M.T.D. et al. (2016). Exceptional durability enhancement of PA/PBI based polymer electrolyte membrane fuel cells for high temperature operation at 200 °C. *Journal of Materials Chemistry A* 4 (11): 4019–4024.

15 Asadi Tashvigh, A., Luo, L., Chung, T.S. et al. (2018). Performance enhancement in organic solvent nanofiltration by double crosslinking technique using sulfonated polyphenylsulfone (sPPSU) and polybenzimidazole (PBI). *Journal of Membrane Science* 551: 204–213.

16 Qu, E., Hao, X., Xiao, M. et al. (2022). Proton exchange membranes for high temperature proton exchange membrane fuel cells: Challenges and perspectives. *Journal of Power Sources* 533: 231386.

17 Wang, J., Dai, Y., Wan, R. et al. (2021). Grafting free radical scavengers onto polyarylethersulfone backbones for superior chemical stability of high temperature polymer membrane electrolytes. *Chemical Engineering Journal* 413: 127541.

18 Jin, Y., Liu, R., Che, X. et al. (2021). New high temperature polymer electrolyte membranes based on poly(ethylene imine) crosslinked poly(ether ketone cardo). *Journal of The Electrochemical Society* 168 (5): 054524.

19 Curtin, D.E., Lousenberg, R.D., Henry, T.J. et al. (2004). Advanced materials for improved PEMFC performance and life. *Journal of Power Sources* 131 (1): 41–48.

20 Xie, H., Xie, X., Hu, G. et al. (2022). Ta–TiOx nanoparticles as radical scavengers to improve the durability of Fe–N–C oxygen reduction catalysts. *Nature Energy* 7 (3): 281–289.

21 Rui, Z. and Liu, J. (2020). Understanding of free radical scavengers used in highly durable proton exchange membranes. *Progress in Natural Science: Materials International* 30 (6): 732–742.

22 Kraytsberg, A. and Ein-Eli, Y. (2014). Review of advanced materials for proton exchange membrane fuel cells. *Energy & Fuels* 28 (12): 7303–7330.

23 Chen, F., Dong, W., Lin, F. et al. (2021). Composite proton exchange membrane with balanced proton conductivity and swelling ratio improved by gradient-distributed POSS nanospheres. *Composites Communications* 24: 100676.

24 Energy.gov. (2024). DOE technical targets for polymer electrolyte membrane fuel cell components [EB/OL], February 25. https://www.energy.gov/eere/fuelcells/doe-technical-targets-polymer-electrolyte-membrane-fuel-cell-components.

25 Hickner, M.A., Ghassemi, H., Kim, Y.S. et al. (2004). Alternative polymer systems for proton exchange membranes (PEMs). *Chemical Reviews* 104 (10): 4587–4612.

26 Basura, V.I., Chuy, C., Beattie, P.D. et al. (2001). Effect of equivalent weight on electrochemical mass transport properties of oxygen in proton exchange membranes based on sulfonated α, β, β-trifluorostyrene (BAM®) and sulfonated styrene-(ethylene-butylene)-styrene triblock (DAIS-analytical) copolymers. *Journal of Electroanalytical Chemistry* 501 (1-2): 77–88.

27 Wei, J., Stone, C., and Steck, A.E.1995). Trifluorostyrene and substituted trifluorostyrene copolymeric compositions and ion-exchange membranes formed there from. US Patent 5,422,411, filed 6 June 1995.

28 Sadeghi, S., Şanlı, L.I., Güler, E. et al. (2018). Enhancing proton conductivity via sub-micron structures in proton conducting membranes originating from sulfonated PVDF powder by radiation-induced grafting. *Solid State Ionics* 314: 66–73.

29 Peng, G., Xiaoxia, G., Hongjie, X. et al. (2007). Research progress of non fluorinated sulfonated polymer proton exchange membrane materials (I). *Polymer Bulletin* 4: 1–12.

30 Peng, G., Xiaoxia, G., Hongjie, X. et al. (2007). Research progress of non fluorinated sulfonated polymer proton exchange membrane materials (Part 2). *Polymer Bulletin* 5: 12–22.

31 Lufrano, F., Gatto, I., and Staiti, P. (2001). Sulfonated polysulfone ionomer membranes for fuel cells. *Solid State Ionics* 145: 47–51.

32 Nolte, R., Ledjeff, K., and Bauer, M. (1993). Partially sulfonated poly(arylene ether sulfone) – A versatile proton conducting membrane material for modern energy conversion technologies. *Journal of Membrane Science* 83: 211–220.

33 Baradie, B., Poinsignon, C., Sanchez, J.Y. et al. (1998). Thermostable ionomeric filled membrane for H_2/O_2 fuel cell. *Journal of Power Sources* 74 (1): 8–16.

34 Ureña, N., Pérez-Prior, M.T., Del Río, C. et al. (2019). Multiblock copolymers of sulfonated PSU/PPSU poly(ether sulfone)s as solid electrolytes for proton exchange membrane fuel cells. *Electrochimica Acta* 302: 428–440.

35 Divya, K., Rana, D., Saraswathi, M.S.S.A. et al. (2019). Sulfonated poly(ether sulfone) composite membranes customized with polydopamine coated molybdenum disulfide nanosheets for renewable energy devices. *Polymer* 175: 255–264.

36 Silva, V.S., Ruffmann, B., Vetter, S. et al. (2005). Characterization and application of composite membranes in DMFC. *Catalysis Today* 104: 205–212.

37 Zaidi, S.M.J., Mikhailenko, S.D., Robertson, G.P. et al. (2000). Proton conducting composite membranes from polyether ether ketone and heteropolyacids for fuel cell applications. *Journal of Membrane Science* 173: 17–34.

38 Muthu Lakshmi, R.T.S., Jochen, M.H., Schlenstedt, K. et al. (2005). Sulphonated poly(ether ether ketone) copolymers: Synthesis, characterisation and membrane properties. *Journal of Membrane Science* 261: 27–35.

39 Xianzhi, D. and Chunxia, Z. (2016). Effect of sulfonation degree on Microstructure and thermal properties of sulfonated peek. *New Chemical Materials* 44 (12): 74–76.

40 Veldre, K., Sala, E., Āboltiņa, E. et al. (2018). Hydration behaviour of sufonated polyetheretherketone (SPEEK) membranes. *Key Engineering Materials* 762: 220–225.

41 Nunes, S.P., Ruffmann, B., Rikowski, E. et al. (2002). Inorganic modification of proton conductive polymer membranes for direct methanol fuel cells. *Journal of Membrane Science* 203 (1-2): 215–225.

42 Martina, P., Gayathri, R., and Pugalenthi, M.R. (2020). Nanosulfonated silica incorporated SPEEK/SPVdF-HFP polymer blend membrane for PEM fuel cell application. *Ionics* 26: 3447–3458.

43 Hande, V.R., Rath, S.K., Rao, S. et al. (2011). Cross-linked sulfonated poly(ether ether ketone)(SPEEK)/reactive organoclay nanocomposite proton exchange membranes (PEM). *Journal of Membrane Science* 372 (1-2): 40–48.

44 Yin, Y., Fang, J.H., Cui, Y.F. et al. (2003). Synthesis, proton conductivity and methanol permeability of a novel sulfonated polyimide from 3-(2′,4′∼diamiphenoxy) propane sulfonic acid. *Polymer* 44: 4509–4518.

45 Ali, N., Ali, F., Saeed, S. et al. (2019). Structural characteristics and electrochemical properties of sulfonated polyimide clay-based composite fabricated by a solution casting method. *Journal of Materials Science: Materials in Electronics* 30 (21): 19164–19172.

46 Asano, N., Aoki, M., Suzuki, S. et al. (2006). Aliphatic/aromatic polyimide ionomers as a proton conductive membrane for fuel cell applications. *Journal of the American Chemical Society* 128 (5): 1762–1769.

10

Design of Proton Exchange Membrane

10.1 Introduction

The design of new proton exchange membrane (PEM) largely revolves around several key strategies [1–5]:

1) **Hydrophilic/Hydrophobic Microphase Separation**: By carefully designing the molecular structure, membranes can achieve a microphase separation that mirrors the structure found in Nafion membranes. This involves creating a hydrophilic phase that forms interconnected proton transport channels while maintaining a hydrophobic phase for structural stability. Approaches include introducing long side-chain sulfonic acid groups, incorporating highly sulfonated main chains, or preparing hydrophilic/hydrophobic block copolymers. Such designs regulate ionic cluster size, connectivity, and phase separation morphology, thereby enhancing the membrane's microphase separation structure.
2) **Stability and Durability Enhancements**: Operating in harsh conditions—such as high temperature, high humidity, and exposure to oxidizing radicals—places stringent demands on membrane stability. To address this, physical or chemical cross-linking methods are used to stabilize ion transport channels, reduce swelling, improve mechanical strength, and enhance resistance to free radicals.
3) **Composite Membranes with Functionalized Nanomaterials**: Incorporating organic and inorganic nanomaterials—such as carbon dots, carbon nanotubes, graphene oxide (GO), and functionalized nanofibers—can improve the overall performance of PEMs. Functional groups such as sulfonic acid, phosphoric acid, or heteropolyacids are introduced onto these nanomaterials, enhancing proton conductivity when integrated into composite membranes.
4) **Microporous Materials**: Advanced microporous materials, including metal-organic frameworks (MOFs), covalent organic frameworks (COFs), and polymeric microporous materials (PIMs), provide unique pore structures that can be tailored to achieve high proton conductivity. By introducing plasmonic conductors into these pore channels and optimizing channel size and continuity, these materials offer new pathways for effective proton transport.

Electrospinning for Proton Exchange Membrane Fuel Cells, First Edition.
Yong Liu and Mohideen Meerasahib Mohamedazeem.

5) **Novel Proton Transport Channels**: Innovative methods such as applying external magnetic or electric fields, or other emerging techniques, are explored to construct new proton transport channels, further broadening the range of design possibilities for next-generation PEMs.

10.2 Design of Molecular Structure

Molecular structure design plays a critical role in developing advanced PEMs that exhibit clear microphase separation and effective proton transport. By synthesizing long side-chain, locally highly sulfonated, or block-type polymers, researchers have produced PEMs with distinct hydrophilic/hydrophobic phase structures, enabling interconnected proton-conductive channels while maintaining excellent stability.

Long side-chain sulfonated PEMs, such as those based on sulfonated polyimides (SPIs) and sulfonated polysulfones (SPSUs), have garnered significant attention. These materials resemble Nafion in their molecular design: a hydrophobic, chemically stable backbone combined with strongly hydrophilic sulfonic acid-functionalized side chains. This structure encourages microphase separation, creating effective proton conduction pathways while preserving the membrane's mechanical and chemical integrity. Kim et al. [6] synthesized a poly(ether sulfone) with a long side chain bearing sulfonic acid groups. The resulting structure (Figure 10.1a) provided proton transport properties comparable to Nafion while maintaining high mechanical strength.

Lafitte and Jannasch [7] introduced aromatic side chains connected by keto groups, each containing multiple sulfonic acid groups adjacent to the sulfone group. The structure (Figure 10.1b) yielded a highly stable backbone, an obvious hydrophilic/hydrophobic phase separation, and a well-formed proton transport network. However, excessive sulfonation led to membrane swelling. Asano et al. [8] developed SPIs by polymerizing diamine monomers with sulfonate-functionalized aliphatic hydrocarbon side chains and dianhydride. The resulting structure (Figure 10.1c) offered excellent hydrolysis resistance and stable operation at 140°C and 100% relative humidity (RH), achieving a proton conductivity of 0.18 S/cm and consistent performance over 5000 hours in fuel cell tests. Norsten et al. [9] prepared a highly fluorinated aromatic polymer backbone with aliphatic side chains containing multiple benzenesulfonic acid groups. This design (Figure 10.1d) resulted in a membrane with a hydrophilic/hydrophobic phase structure similar to Nafion and a proton conductivity of 0.58 S/cm at 80 °C, compared to 0.3 S/cm for Nafion under the same conditions. Through these molecular design approaches, significant progress has been made in tailoring PEMs to achieve high proton conductivity, thermal stability, and mechanical robustness, thereby advancing their application in fuel cell technology [10].

Long et al. [11] polymerized a highly fluorinated dichloromonomer with a polyphenylene dichloromonomer, yielding polymers with localized high degrees of sulfonation after sulfuric acid treatment. These materials, depicted in Figure 10.2, exhibited distinct phase-separated structures and proton conductivity levels comparable to those of Nafion.

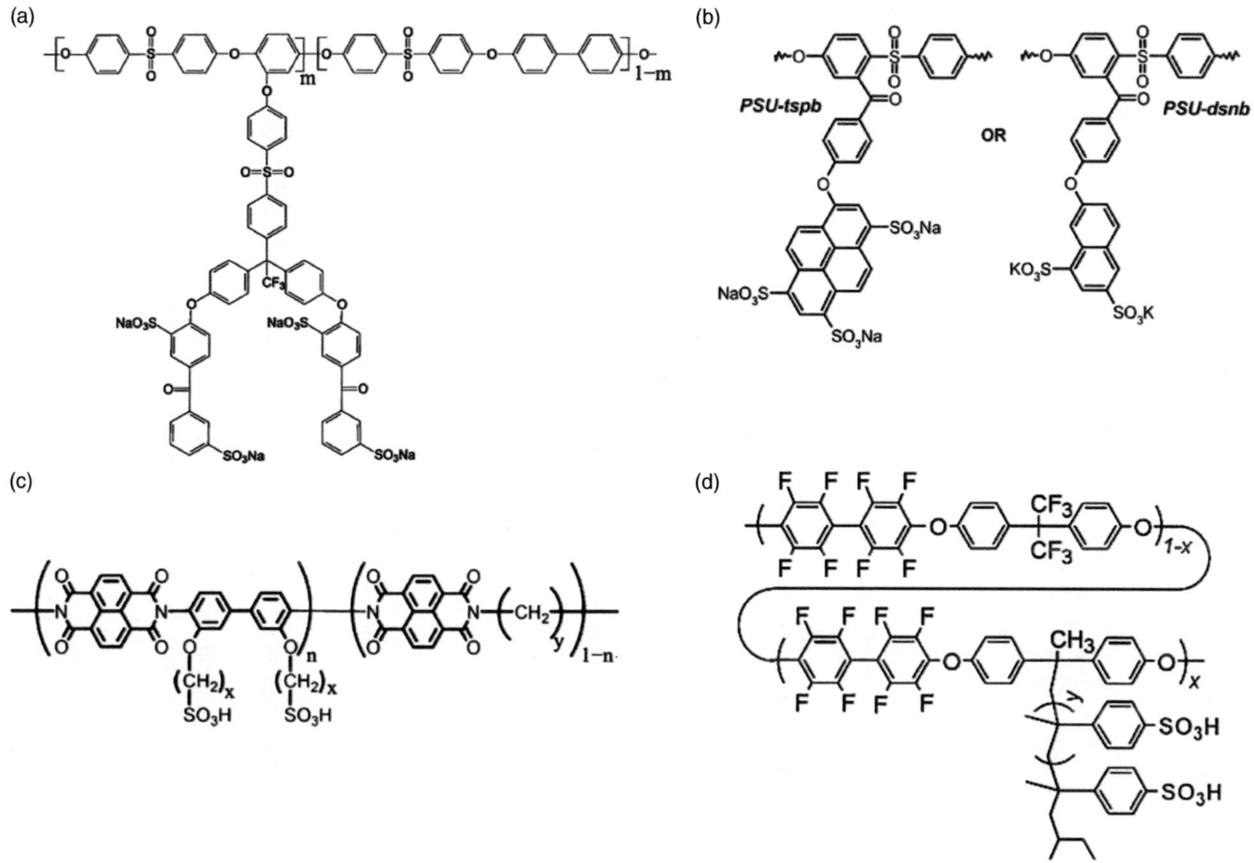

Figure 10.1 Sulfonated long side-chain polymer structures.

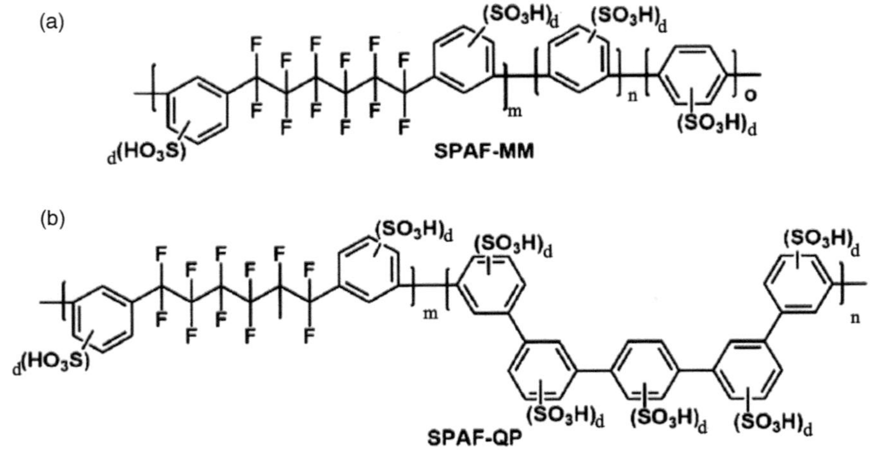

Figure 10.2 Structures of proton exchange membranes containing densely sulfonated quinquephenylene groups.

Block copolymers, composed of alternating hydrophilic and hydrophobic chain segments, provide another route to PEMs with pronounced microphase separation. For example, Kins et al. [12] synthesized sulfonated polyimide block copolymers (Figure 10.3a), while Takamuku and Jannasch [13] combined sulfonated hydrophilic chain segments with fluorinated hydrophobic segments to produce sulfonated polyethersulfone block copolymers (Figure 10.3b). Both approaches resulted in clear microphase-separated structures, with localized high sulfonation further enhancing their performance.

Ahn et al. [14] developed a multiblock copolymer incorporating highly sulfonated poly(phenylsulfone) (sPPSS) hydrophilic chains and hydrophobic perfluorocyclobutyl (PFCB) chains (Figure 10.3c). The sharp hydrophilic/hydrophobic contrast produced a pronounced phase separation. The resulting membrane achieved a proton conductivity of 41.9 mS/cm under low humidity conditions (50% RH), while demonstrating strong chemical and physical durability.

10.3 Physical and Chemical Cross-Linking Modification

Physical and chemical cross-linking are crucial for mitigating excessive swelling in polymer membranes, as well as enhancing their mechanical and chemical stability. Qian et al. [15] created sulfonated polyarylethersulfone (SPAES) polymers containing amphiphilic side chains (Figure 10.4a) and blended these with sulfonated poly(ether ether ketone) (SPEEK) in varying ratios. This strategy balanced proton conductivity and permeability while improving the membranes' structural stability and durability.

Kwon et al. [16] designed a physically cross-linked membrane by combining a highly sulfonated polyfluorenyl ether sulfone with an imidazole-functionalized polyethersulfone (Figure 10.4b). The multisulfonated fluorenyl groups provided

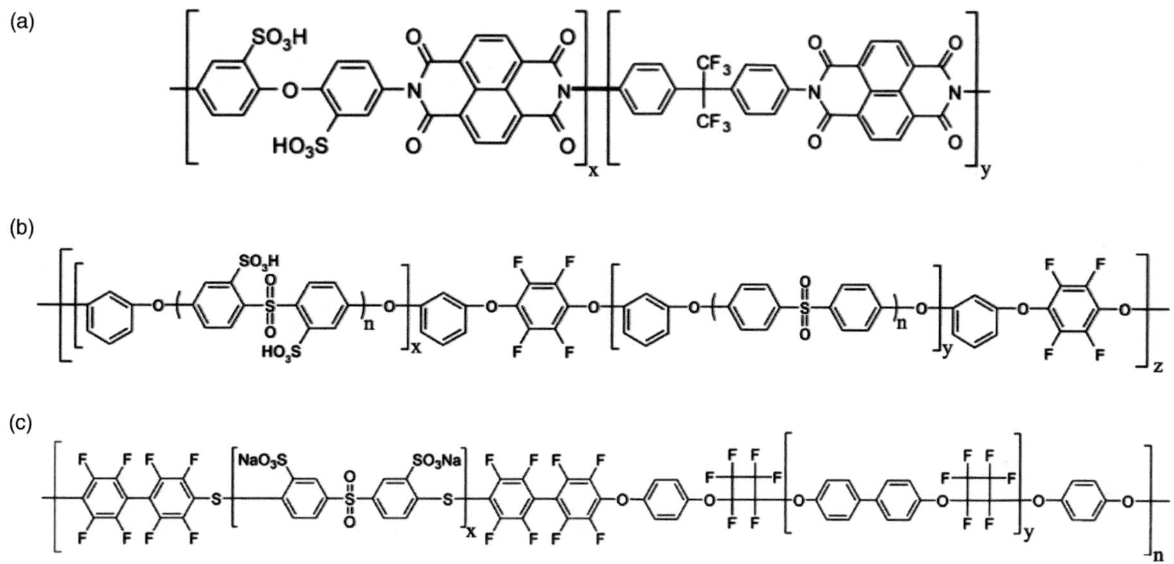

Figure 10.3 Structural formulas of hydrophilic/hydrophobic block copolymers.

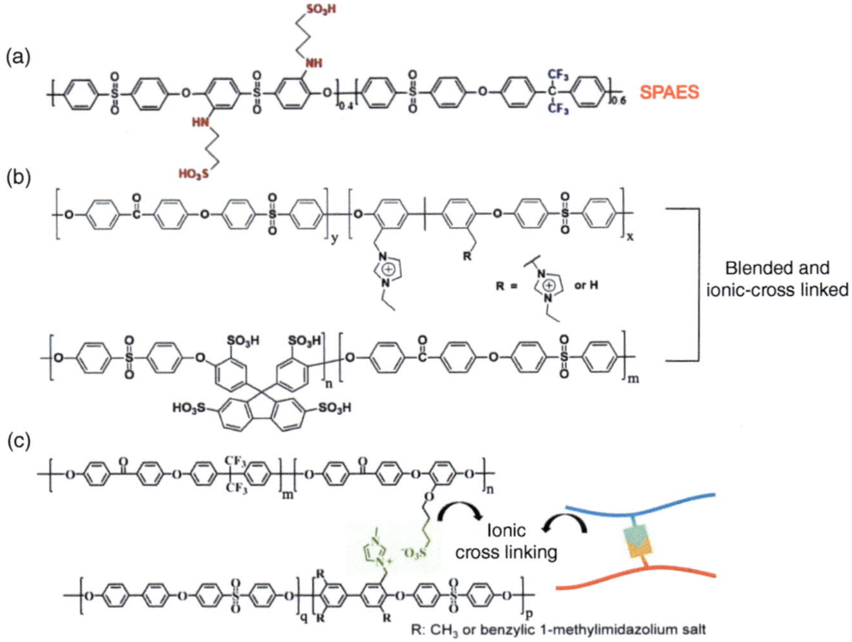

Figure 10.4 (a) Sulfonated polyarylethersulfone polymer with amphoteric side chains; (b) blended cross-linked membrane of sulfonated fluorene-based poly(ether sulfone) and imidazole-functionalized poly(ether sulfone); (c) ionic cross-linking membrane with nanophase separation.

abundant proton transport sites, while ionic interactions with the imidazole groups formed a stable cross-linked structure. The membrane demonstrated high physical and chemical stability, with a proton conductivity of over 0.24 S/cm at 80 °C. Chen et al. [17] fabricated a chemically cross-linked membrane using sulfonated polyaryl ether ketone and imidazole-functionalized polyaryl ether sulfone (Figure 10.4c). This ionic cross-linking approach achieved nanophase separation and constructed efficient ion transport channels, resulting in excellent ion selectivity and high ionic conductivity.

10.4 Enhancing Stability and Performance Through Chemical Cross-Linking

For polymer membranes that rely on sulfonate groups to facilitate proton conductivity, increasing the degree of sulfonation is a straightforward way to improve conductivity. However, excessive sulfonation often results in significant swelling and weakened mechanical strength. Chemical cross-linking offers a proven solution, strengthening the membrane structure and reducing fuel permeability while maintaining proton conductivity and overall cell performance. Matsushita and Kim [18] demonstrated that using ethylene glycol and propylene glycol as cross-linking agents for highly SPSUs could significantly enhance membrane stability. As shown in Figure 10.5, the cross-linked membranes were resistant to boiling water, achieved a tensile strength of 43.5 MPa, a modulus of 1.9 GPa, and maintained a proton conductivity of 0.23 S/cm at 120 °C and 90% RH. Furthermore, the antioxidant stability was noteworthy, with only 2% mass loss after one hour of testing, highlighting the membrane's excellent chemical stability.

Yaguchi et al. [19] used a P_2O_5/methanesulfonic acid solution to cross-link sulfonated polyimide membranes. By forming sulfone bonds between sulfonic acid groups and benzene rings (Figure 10.6a), the cross-linked membranes exhibited significantly reduced hydrogen permeability and improved fuel cell performance. Chen et al. [20] expanded on this approach, blending highly sulfonated polyethersulfone and low sulfonation polyimide, then cross-linking them using a similar P_2O_5/methanesulfonic acid process (Figure 10.6b). The resulting membranes showed enhanced mechanical strength and controlled swelling while retaining high proton conductivity. In fuel cell performance tests, these cross-linked membranes maintained stable operation for over 200 hours without significant performance degradation.

Functionalized nanomaterials have been widely explored for their potential to enhance proton conductivity and stability in PEMs. Both organic and inorganic nanocomposites have been integrated into polymer matrices, incorporating materials such as carbon quantum dots, nanotubes, GO, nanofibers, heteropolyacids, and zeolites. By forming composite membranes, significant improvements have been achieved in properties such as proton conductivity, fuel permeability, mechanical strength, and water retention. The relatively simple preparation of composite membranes has made them a focal point of research in this field.

- **Carbon Quantum Dots**: Hydrophilic carbon quantum dots are particularly effective at increasing proton conductivity. Wu et al. [21] prepared both hydrophilic and hydrophobic carbon quantum dots and blended them with Nafion. During the membrane's hydrophilic and hydrophobic phase separation, the hydrophilic dots aggregated in hydrophilic regions while the hydrophobic dots aggregated in hydrophobic regions. This phase separation enhanced proton transport, significantly improving the conductivity of the resulting blended membrane.
- **Nanotube Doping**: Nanotube doping is another well-studied approach. He et al. [22] doped SPEEK with phosphorylated silica nanotubes, while another study incorporated modified multiwalled carbon nanotubes into Nafion [23]. The high surface area of these nanotubes allowed for the integration of many acidic functional groups, and their one-dimensional structure connected previously isolated hydrophilic channels, resulting in substantial increases in conductivity. Furthermore, nanotube doping improved the mechanical strength and reduced the fuel permeability of the membrane.

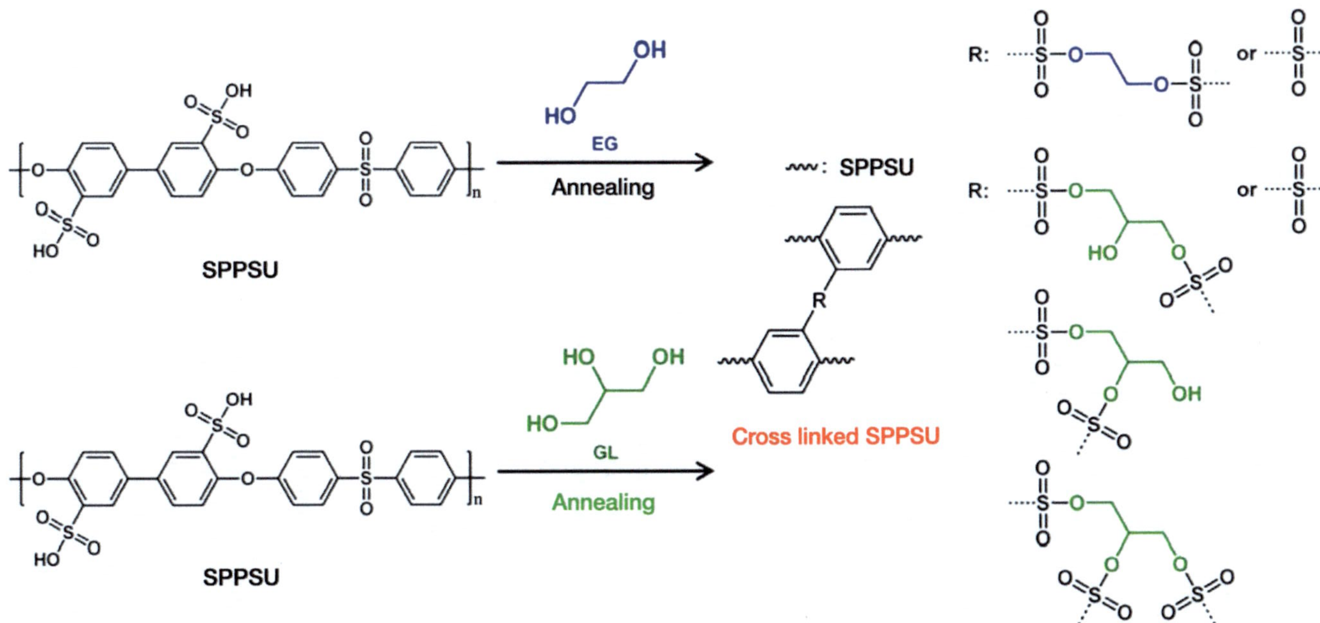

Figure 10.5 Sulfonated polyethersulfone crosslinking via glycol/propanetriol.

(a)

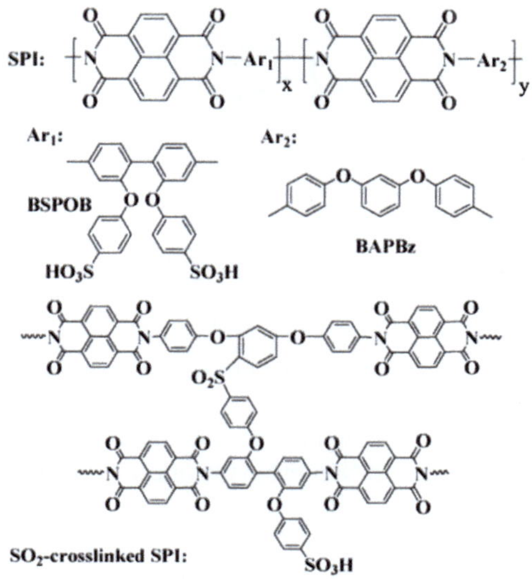

(b)

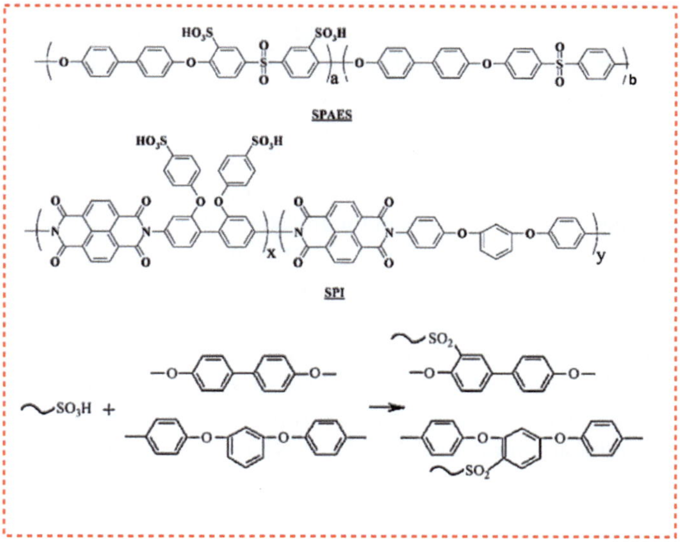

Figure 10.6 Sulfonated polyimide cross-linking through the sulfone group.

- **Water Retention and Low-Humidity Conductivity**: Adding nanomaterials that retain water can further enhance membrane performance, particularly at low humidity. For instance, composite membranes made by blending phosphotungstic acid-filled carbon nanotubes with Nafion showed improved ion transport channels and reduced phosphotungstic acid loss. Due to the water retention properties of phosphotungstic acid, these membranes maintained higher proton conductivity even under low humidity conditions [24]. As illustrated in Figure 10.7, these composite membranes demonstrated up to a threefold increase in power output at high

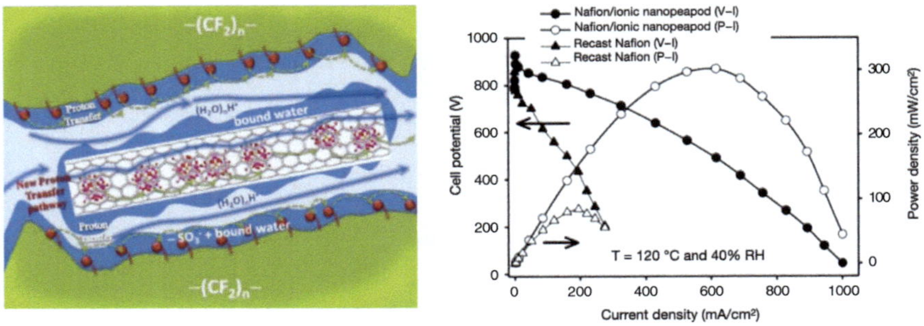

Figure 10.7 Phosphotungstic acid-filled carbon nanotube Nafion composite membrane.

temperatures and low humidity compared to pure Nafion membranes. Similarly, blending cerium oxide nanotubes with Nafion led to noticeable performance gains at low humidity [25].

- **Functionalized GO in PEMs**: GO, with its two-dimensional lamellar structure and abundant hydroxyl and carboxyl functional groups, serves as a versatile component for developing high-performance PEMs. Various strategies have been used to integrate functionalized GO into membranes to improve conductivity, stability, and chemical properties. Jiang's group [26] demonstrated a vacuum filtration approach to deposit GO layers, introducing proton-conducting materials into a GO-based sandwich structure. By chemically modifying GO's hydroxyl groups with silane coupling agents, sulfonic acid groups or cross-links were incorporated, forming effective proton-conducting channels.
Alternative methods include cross-linking GO with sulfonated aromatic compounds using ethylene glycol diglycidyl ether [27], or embedding deoxyribonucleic acid (DNA) between GO layers and securing them via hydrogen bonding or π–π interactions [28]. These techniques capitalize on GO's layered arrangement to achieve high proton conductivity along the horizontal plane, though vertical conductivity typically requires further enhancement.
To address the directional limitations, Cao et al. [29] developed GO-based PEMs with a 3D mesh arrangement using a sol-gel freeze-drying method (Figure 10.8). This structural innovation significantly improved proton conductivity in the vertical direction, enhancing the overall performance of the membrane.

- **Polymer-GO Composite Membranes**: Functionalized GO can also be combined with traditional PEM materials such as Nafion. For instance, spin-coating sulfonated GO layers onto Nafion and subsequent cross-linking results in membranes with superior alcohol-blocking properties without sacrificing proton conductivity [30]. Similarly, grafting sulfonated polymers onto GO surfaces and blending them with Nafion has yielded membranes with enhanced conductivity at low humidity [31]. By forming acid–base pairs between sulfonated GO and polyethylene glycol, researchers achieved improved stability, water retention, and proton conductivity under challenging conditions [32].

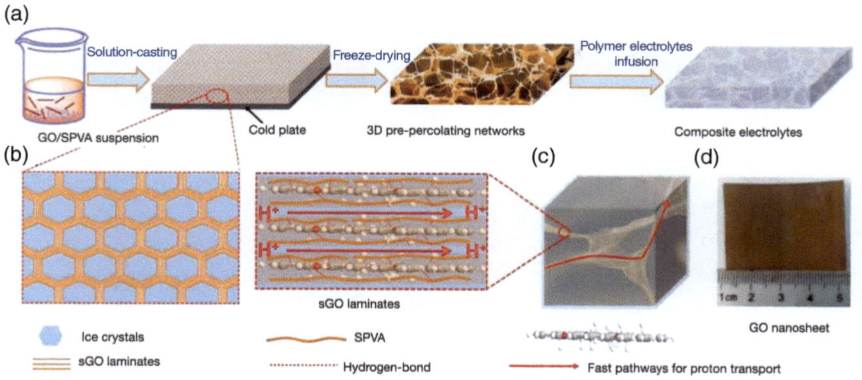

Figure 10.8 Functionalized graphene oxide 3D mesh arrangement of proton exchange membranes. Source: Cao et al. [29]/John Wiley & Sons.

- **Electrospun Nanofiber Reinforcement**: Electrospinning techniques have further advanced PEM development. Electrospun nanofibers, when blended with polymers such as Nafion, create continuous proton transport channels along the fiber–polymer interface. The strong electric fields during electrospinning cause negatively charged sulfonic acid groups to align on the fiber surface, improving the arrangement of these groups in the surrounding matrix. Li et al. [33] prepared poly(vinylidene fluoride) (PVDF) nanofibers and grafted Nafion onto their surfaces. This composite membrane exhibited excellent compatibility between fibers and polymer matrix, significantly boosting proton conductivity and overall fuel cell performance. Chen et al. [34] employed SnO_2 inorganic nanofibers, treating them with sulfuric acid to activate surface hydroxyl groups. The resulting composite membrane, formed by blending these fibers with sulfonated polymers, displayed improved proton conductivity, reduced methanol permeability, and achieved a maximum power output of $150\,mW/cm^2$ in methanol fuel cell tests (Figure 10.9).

10.5 Modulating the Structure of Ordered Microporous Materials

Ordered microporous materials, including mesoporous silica, MOFs, and COFs, are promising candidates for PEMs due to their well-defined pore channels and tunable structures [35–41]. MOFs, in particular, are highly favored because of their long-range ordered pores and hydrogen-bonding networks, which facilitate efficient proton conduction. Their structural flexibility, chemical tunability, and relatively straightforward synthesis make MOFs a viable option for large-scale production of proton conductors [42–46]. The advantages of MOFs as proton conductors include:

1) Efficient adsorption of proton carriers and provision of active sites for proton transport.

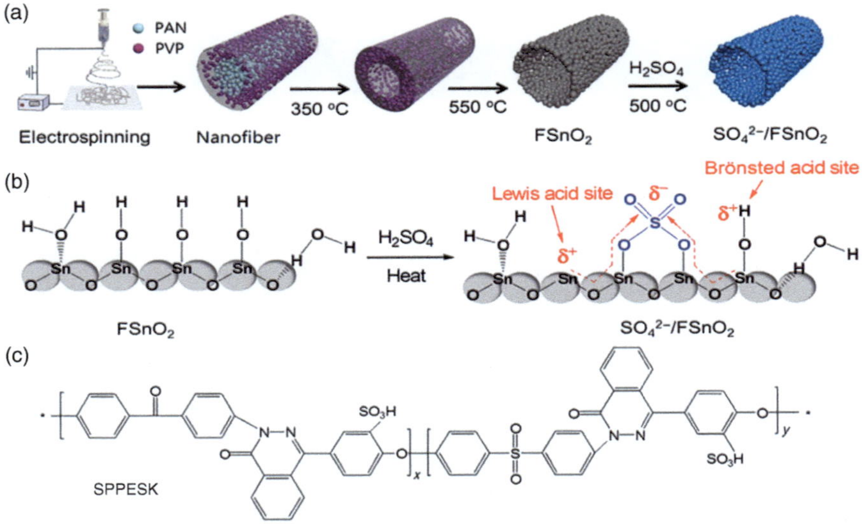

Figure 10.9 Inorganic nanofiber SO_4^{2-}/SnO_2 composite proton exchange membrane with sulfonated polymer. SPPESK, sulfonated poly(phenyl ether sulfone ketone).

2) Well-ordered pore structures that create effective proton transport pathways.
3) Simple synthesis processes that enable large-scale production tailored to specific performance requirements.

For example, Phang et al. developed UiO-66$(SO_3H)_2$ by oxidizing UiO-66$(SH)_2$ with thiols, covalently attaching sulfonic acid groups to the framework. This material exhibited a high proton conductivity of 84 mS/cm at 80 °C and 90% RH, with stable long-term performance [47]. Subhabrata et al. [48] modified UiO-66-NH_2 using a postsynthesis amide coupling method, resulting in two homologous compounds, PSM 1 and PSM 2. As shown in Figure 10.10, the primary difference between these MOFs is the length of the alkyl chains attached to the sulfonic acid groups. PSM 1, with a shorter alkyl chain, achieved a proton conductivity of

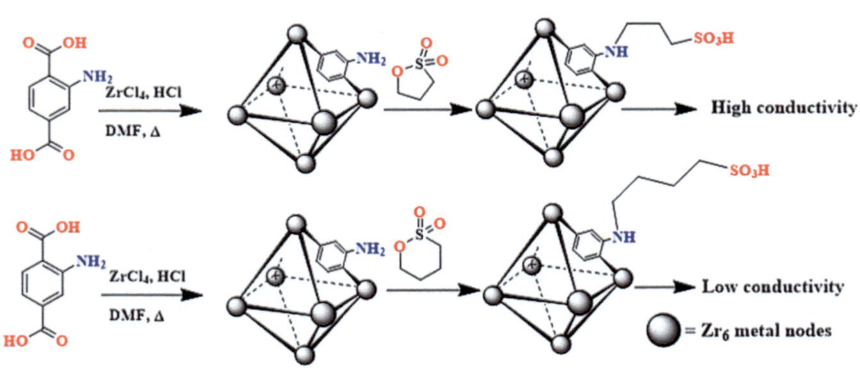

Figure 10.10 Synthesis of UiO-66-NH2 and Its Post-Synthetic Modification to PSM 1 (Top) and PSM 2 (Bottom).

164 mS/cm at 80 °C—comparable to Nafion—while PSM 2, with a longer alkyl chain, reached only 4.6 mS/cm. Theoretical pKa values (3.47 for PSM 1 and 4.91 for PSM 2) and activation energies (0.107 eV for PSM 1 and 0.292 eV for PSM 2) supported the conclusion that proton conduction in MOFs follows the Grotthuss mechanism, in which water facilitates proton hopping through hydrogen-bonded networks.

COFs are another class of crystalline porous materials with rigid, well-defined nano-channels that support proton transport under various humidity conditions [49–56]. Unlike amorphous or flexible networks, COFs' crystalline structure offers stable proton conduction pathways and good water retention. However, their application in PEM fuel cells is challenged by poor processability and structural integrity [57]. Jiang Zhongyi's group at Tianjin University reported a highly crystalline IPC-COF nanosheet (NUS-9) synthesized via interfacial polymerization (Figure 10.11a). These nanosheets were assembled into defect-free, robust

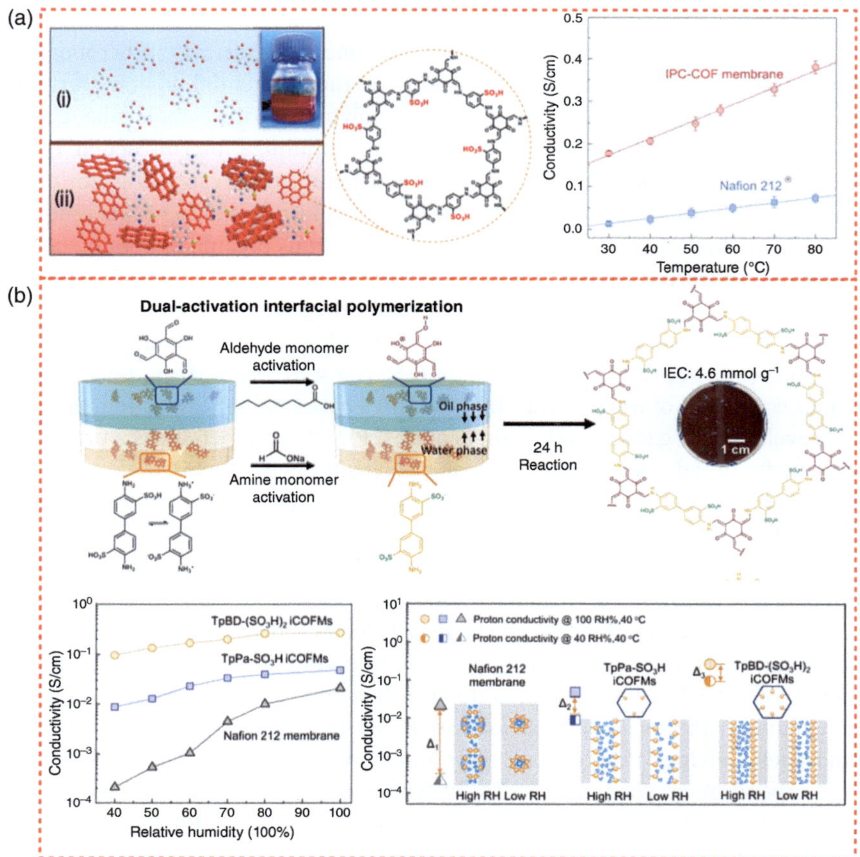

Figure 10.11 Preparation of self-supported sulfonated covalent organic frameworks materials by interfacial polymerization. Source: a) Cao et al. [58]/with permission of John Wiley & Sons, b) Wang et al. [59]/Springer Nature/CC BY 4.0.

IPC-COF membranes using vacuum-assisted self-assembly, resulting in consistent proton conductivity and fuel cell performance across a broad humidity range (30%–98% RH) [58].

Further studies have shown that the crystalline and rigid ionic nanochannels within IPC-COF membranes enable them to retain water effectively and achieve proton conductivity primarily governed by the Grotthuss mechanism. Building on this approach, Jiang and co-workers [59] developed ionic covalent organic framework membranes (iCOFMs) materials using a dual-activated interfacial polymerization technique. By employing a Brønsted acid to activate the aldehyde monomer in the organic phase and a Brønsted base to activate the ionic amine monomer in the aqueous phase, the reaction at the aqueous–organic interface was significantly accelerated. This process yielded highly crystalline iCOFMs with an ultra-high ion exchange capacity of 4.6 mmol/g, as illustrated in Figure 10.11b. The resulting self-supported membranes exhibited an exceptional proton conductivity of 0.66 S/cm, making them well suited for ion transport and separation applications.

In parallel, self-contained microporous polymers of intrinsic microporosity (PIMs) have gained considerable attention for their potential in ion transport and PEM applications. Unlike conventional polymers, whose molecular chains tend to bend or twist to reduce free volume and minimize intermolecular pores, PIMs feature a rigid backbone and a twisted structure that resists efficient chain stacking. This intrinsic property leads to large free volumes, making PIMs uniquely capable of facilitating ion transport. Unlike other porous organic materials such as hyper-cross-linked polymers, conjugated microporous polymers, and COFs, PIMs typically lack a cross-linked network. As a result, they are often soluble in organic solvents, allowing them to be processed into thin films, coatings, and even electrospun fibers [60]. A prominent example is PIM-1, a self-polymerizing microporous polymer composed of trapezoidal PIM structures based on helical bis-indanes. Initially developed for gas separation applications, PIM-1 has emerged as a promising candidate for broader membrane technologies, including PEMs [61].

The unique microporous structure and substantial free volume of PIM materials make them highly conducive to ion transport, positioning them as viable candidates for fuel cell applications. Literature on PIM-based PEMs highlights several promising approaches. For instance, sulfonated PIM polymers have been used as cation exchange membranes, where they form well-defined sub-nanochannels that enhance electrical conductivity [62]. Additionally, the intrinsic microporous structure of PIMs has been exploited to adsorb significant amounts of phosphoric acid, enabling improved performance in PEMs [63]. Tröger base-derived PIMs membranes, relying on acid–base pair interactions, have shown potential for high-temperature applications by effectively reducing phosphoric acid loss [64].

Du et al. [65] synthesized soluble Tröger base-based PIMs for PEMs by copolymerizing 4,4′-diamino-3,3′-dimethylbiphenyl, 4-(2-hydroxyethoxy)-1,3-phenylenediamine dihydrochloride, and dimethoxymethane. The resulting membranes exhibited excellent thermal and mechanical stability, with a swelling rate of less than 8% at 60 °C and proton conductivity of 58.3 mS/cm at 80 °C following hydrochloric acid treatment. Tang et al. [66] advanced this approach by reporting PIM-based membranes with ultramicropores averaging 3.3 Å in radius. This fine pore structure induced a siphoning effect on phosphoric acid, leading to phosphoric acid retention three orders of magnitude higher than that of conventional polybenzimidazole (PBI)-based phosphoric acid-doped membranes. Consequently, these membranes achieved stable operation across a wide temperature range (−20 to 200 °C).

In contrast to conventional PEM materials such as Nafion and sulfonated aromatic hydrocarbon-based polymer electrolytes—where ion-conducting regions are formed through poorly defined microphase separation—PIMs offer a more controlled alternative. Traditional PEMs often develop ion channels of nanometer-scale dimensions upon hydration, which, while supporting ion transport, can compromise ion selectivity and increase hydrogen or methanol permeability. In response, new design approaches emphasize limiting polymer free volume voids, increasing polymer backbone rigidity, reducing thermal chain movement, and carefully managing ion–polymer interactions [67]. PIMs leverage their inherent pore structure, promoting selective ion transport and molecular sieving, making them a compelling choice for next-generation PEMs.

PIMs, while promising, present certain challenges due to their semi-rigid, non-cross-linked network structure and the high concentration of charged functional groups. These features make them susceptible to severe solubilization and reduced selectivity [68]. Efforts to rigidify the polymer framework to reduce free volume voids and enhance selectivity can inadvertently impede rapid ion transport. Overcoming this trade-off requires strategies that both lower the transmembrane ion transport energy barriers and capitalize on pore wall–ion interactions to achieve efficient, low-resistance ion movement.

One approach is the use of rigid covalent triazine-frameworked microporous membranes, as demonstrated by Zuo et al. [69]. As illustrated in Figure 10.12, these membranes are structurally engineered to separate absorbed water within the microcavities from the hydrophobic polymer backbone. This design facilitates selective ion transport through robust sub-nanometer channels. The close proximity of ions to the pore walls enables multiple interactions that promote efficient diffusion by a ligand-like mechanism. Furthermore, the molecular design of the dinitrile monomer allows precise tuning of channel dimensions and chemistries, yielding a membrane with both ultra-high ion transport rates and excellent selectivity.

10.6 Construction of Novel Proton Transport Channels

In conventional PEMs, proton transport channels are typically randomly distributed, resulting in uniform proton conductivity in both the horizontal and vertical directions. However, enhancing proton conduction specifically along the vertical membrane direction can significantly improve fuel cell performance. Researchers have therefore explored approaches to construct vertically aligned proton transport channels.

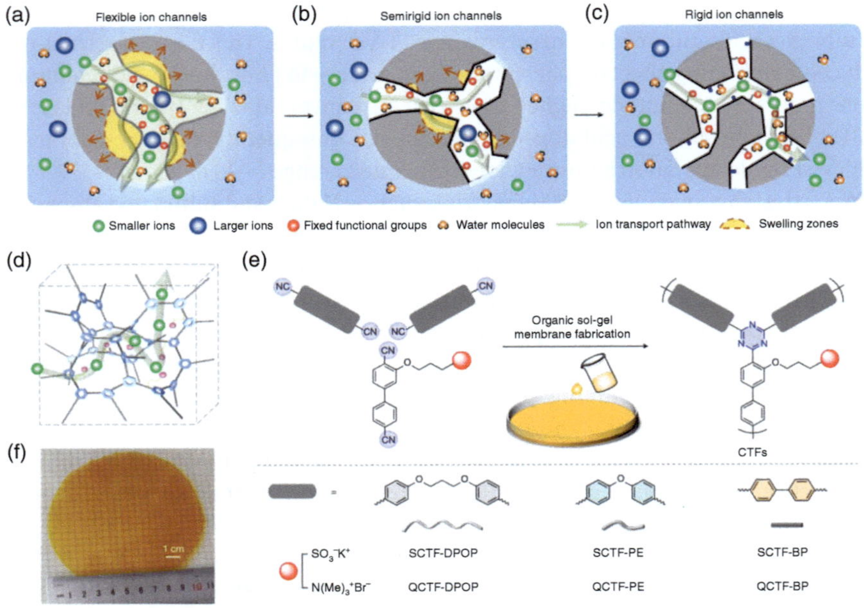

Figure 10.12 Schematic representation of membranes with varying ion channels and the preparation of rigid triazine framework microporous membranes. Source: Zuo et al. [69]/with permission of Springer Nature.

One method involves blending functionalized magnetic nanoparticles into the polymer matrix and applying a magnetic field during the membrane fabrication process. The magnetic particles align along the field lines, forming oriented proton transport pathways [70–75]. For example, Liu et al. [76] synthesized a ferrocyanic coordination polymer that interacts with phosphotungstic acid to produce magnetic properties. When subjected to an ultra-strong magnetic field, this material forms channels aligned along the field lines, as shown in Figure 10.13. This alignment allows protons to traverse the membrane perpendicularly with higher conductivity.

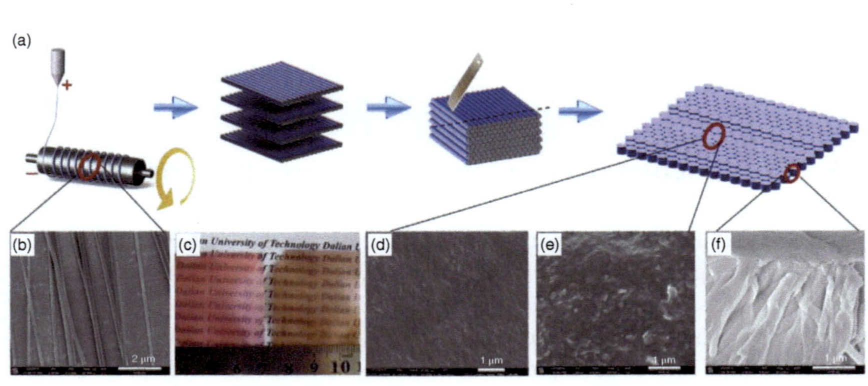

Figure 10.13 Multilayer directionally aligned fiber stacking and cutting for membrane fabrication. Source: Gong et al. [77]/with permission of Elsevier.

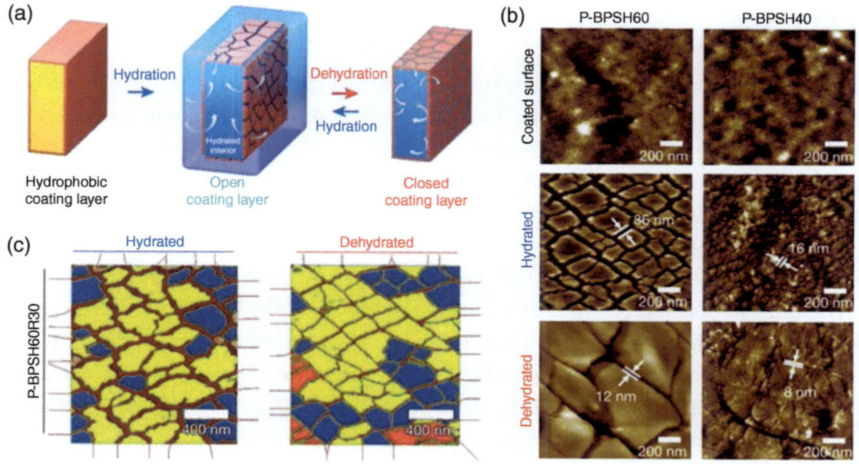

Figure 10.14 Proton exchange membrane with a cracked, self-adjusting surface. Source: Park et al. [78]/with permission of Springer Nature.

Furthermore, the ferrocyanic groups in the polymer engage in free-radical redox reactions, which mitigate free-radical damage, thereby significantly improving the membrane's durability.

To address the challenge of maintaining proton conductivity under low-humidity conditions, various water-retaining materials—such as phosphotungstic acid, water-retaining microcapsules, and sulfonated zeolites—have been incorporated into membranes. However, Park et al. [78] drew inspiration from nature, particularly the behavior of cactus stomata, which close under dry conditions to retain water. They designed a "water-controlling" mechanism on the surface of PEMs. This concept, illustrated in Figure 10.14, uses a hydrophobic nanoscale coating that cracks open when exposed to high humidity, allowing water and protons to pass through. Under low-humidity conditions, the membrane shrinks, causing the coating to reseal and retain water within the membrane.

10.7 Conclusion

In summary, various strategies ranging from molecular structure design and cross-linking methods to the incorporation of functionalized nanomaterials, optimization of ordered microporous structures, and the application of magnetic and electric fields have significantly improved the performance of PEMs. These approaches have enhanced proton conductivity, mechanical stability, and overall durability, but challenges remain, including the complexity of preparation, stringent production conditions, and high costs. For instance, while block copolymers and sulfonated COF membranes offer notable conductivity and structural advantages, their synthesis and scalability pose practical difficulties. Addressing these limitations through innovative modification and cost-effective preparation methods is key to advancing PEM applications in fuel cells.

References

1 Peighambardoust, S.J., Rowshanzamir, S., and Amjadi, M. (2010). Review of the proton exchange membranes for fuel cell applications. *International Journal of Hydrogen Energy* 35 (17): 9349–9384.

2 Sapkota, P., Boyer, C., Dutta, R. et al. (2020). Planar polymer electrolyte membrane fuel cells: powering portable devices from hydrogen. *Sustainable Energy & Fuels* 4 (2): 439–468.

3 Feng, S. and Voth, G.A. (2011). Proton solvation and transport in hydrated Nafion. *The Journal of Physical Chemistry B* 115 (19): 5903–5912.

4 Elliott, J.A. and Paddison, S.J. (2007). Modelling of morphology and proton transport in PFSA membranes. *Physical Chemistry Chemical Physics* 9 (21): 2602–2618.

5 Zawodzinski, T.A., Springer, T.E., Uribe, F. et al. (1993). Characterization of polymer electrolytes for fuel cell applications. *Solid State Ionics* 60 (1): 199–211.

6 Kim, D.S., Robertson, G.P., and Guiver, M.D. (2008). Comb-shaped poly(arylene ether sulfone)s as proton exchange membranes. *Macromolecules* 41 (6): 2126–2134.

7 Lafitte, B. and Jannasch, P. (2007). Proton-conducting aromatic polymers carrying hypersulfonated side chains for fuel cell applications. *Advanced Functional Materials* 17 (15): 2823–2834.

8 Asano, N., Aoki, M., Suzuki, S. et al. (2006). Aliphatic/aromatic polyimide ionomers as a proton conductive membrane for fuel cell applications. *Journal of the American Chemical Society* 128 (5): 1762–1769.

9 Norsten, T.B., Guiver, M.D., Murphy, J. et al. (2006). Highly fluorinated comb-shaped copolymers as proton exchange membranes (PEMs): improving PEM properties through rational design. *Advanced Functional Materials* 16 (14): 1814–1822.

10 Liu, B., Robertson, G.P., Kim, D.S. et al. (2007). Aromatic poly(ether ketone)s with pendant sulfonic acid phenyl groups prepared by a mild sulfonation method for proton exchange membranes. *Macromolecules* 40 (6): 1934–1944.

11 Long, Z., Miyake, J., and Miyatake, K. (2020). Proton exchange membranes containing densely sulfonated quinquephenylene groups for high performance and durable fuel cells. *Journal of Materials Chemistry A* 8 (24): 12134–12140.

12 Kins, C.F., Sengupta, E., Kaltbeitzel, A. et al. (2014). Morphological anisotropy and proton conduction in multiblock copolyimide electrolyte membranes. *Macromolecules* 47 (8): 2645–2658.

13 Takamuku, S. and Jannasch, P. (2012). Multiblock copolymers with highly sulfonated blocks containing di- and tetrasulfonated arylene sulfone segments for proton exchange membrane fuel cell applications. *Advanced Energy Materials* 2 (1): 129–140.

14 Ahn, S.M., Kim, T.H., Yuk, J. et al. (2022). Perfluorocyclobutyl-containing multiblock copolymers to induce enhanced hydrophilic/hydrophobic phase separation and high proton conductivity at low humidity. *Journal of Membrane Science* 641: 119892.

15 Qian, P., Wang, H., Zhang, L. et al. (2022). An enhanced stability and efficiency of SPEEK-based composite membrane influenced by amphoteric side-chain polymer for vanadium redox flow battery. *Journal of Membrane Science* 643: 120011.

16 Kwon, S., Lee, B., and Kim, T.H. (2017). High performance blend membranes based on densely sulfonated poly(fluorenyl ether sulfone) block copolymer and imidazolium-functionalized poly(ether sulfone). *International Journal of Hydrogen Energy* 42 (31): 20176–20186.

17 Chen, W., Dong, T., Xiang, Y. et al. (2022). Ionic crosslinking-induced nanochannels: nanophase separation for ion transport promotion. *Advanced Materials* 34 (3): 2108410.

18 Matsushita, S. and Kim, J.D. (2018). Organic solvent-free preparation of electrolyte membranes with high proton conductivity using aromatic hydrocarbon polymers and small cross-linker molecules. *Solid State Ionics* 316: 102–109.

19 Yaguchi, K., Chen, K., Endo, N. et al. (2010). Crosslinked membranes of sulfonated polyimides for polymer electrolyte fuel cell applications. *Journal of Power Sources* 195 (15): 4676–4684.

20 Chen, S., Zhang, X., Chen, K. et al. (2011). Cross-linked miscible blend membranes of sulfonated poly(arylene ether sulfone) and sulfonated polyimide for polymer electrolyte fuel cell applications. *Journal of Power Sources* 196 (23): 9946–9954.

21 Wu, W., Li, Y., Liu, J. et al. (2018). Molecular-level hybridization of Nafion with quantum dots for highly enhanced proton conduction. *Advanced Materials* 30 (16): 1707516.

22 He, G., Nie, L., Han, X. et al. (2014). Constructing facile proton-conduction pathway within sulfonated poly(ether ether ketone) membrane by incorporating poly(phosphonic acid)/silica nanotubes. *Journal of Power Sources* 259: 203–212.

23 He, G., Zhao, J., Hu, S. et al. (2014). Functionalized carbon nanotube via distillation precipitation polymerization and its application in Nafion-based composite membranes. *ACS Applied Materials & Interfaces* 6 (17): 15291–15301.

24 Hasani-Sadrabadi, M.M., Dashtimoghadam, E., Majedi, F.S. et al. (2016). Ionic nanopeapods: next-generation proton conducting membranes based on phosphotungstic acid filled carbon nanotube. *Nano Energy* 23: 114–121.

25 Ketpang, K., Oh, K., Lim, S.C. et al. (2016). Nafion-porous cerium oxide nanotubes composite membrane for polymer electrolyte fuel cells operated under dry conditions. *Journal of Power Sources* 329: 441–449.

26 He, G., Chang, C., Xu, M. et al. (2015). Tunable nanochannels along graphene oxide/polymer core-shell nanosheets to enhance proton conductivity. *Advanced Functional Materials* 25 (48): 7502–7511.

27 Mao, X., Xu, M., Wu, H. et al. (2019). Supramolecular calix[n]arenes-intercalated graphene oxide membranes for efficient proton conduction. *ACS Applied Materials & Interfaces* 11 (45): 42250–42260.

28 Yang, P., Wu, H., Khan, N.A. et al. (2020). Intrinsic proton conductive deoxyribonucleic acid (DNA) intercalated graphene oxide membrane for high -efficiency proton conduction. *Journal of Membrane Science* 606: 118136.

29 Cao, L., Wu, H., Yang, P. et al. (2018). Graphene oxide-based solid electrolytes with 3D prepercolating pathways for efficient proton transport. *Advanced Functional Materials* 28 (50): 1804944.

30 He, G., He, X., Wang, X. et al. (2016). A highly proton-conducting, methanol-blocking Nafion composite membrane enabled by surface-coating crosslinked sulfonated graphene oxide. *Chemical Communications* 52 (10): 2173–2176.

31 He, X., He, G., Zhao, A. et al. (2017). Facilitating proton transport in Nafion-based membranes at low humidity by incorporating multifunctional graphene oxide nanosheets. *ACS Applied Materials & Interfaces* 9 (33): 27676–27687.

32 Li, J., Wu, H., Cao, L. et al. (2019). Enhanced proton conductivity of sulfonated polysulfone membranes under low humidity via the incorporation of multifunctional graphene oxide. *ACS Applied Nano Materials* 2 (8): 4734–4743.

33 Li, H.Y. and Liu, Y.L. (2014). Nafion-functionalized electrospun poly(vinylidene fluoride) (PVDF) nanofibers for high performance proton exchange membranes in fuel cells. *Journal of Materials Chemistry A* 2 (11): 3783–3793.

34 Chen, W., Chen, M., Zhen, D. et al. (2020). SO_4^{2-}/SnO_2 solid superacid granular stacked one-dimensional hollow nanofiber for a highly conductive proton-exchange membrane. *ACS Applied Materials & Interfaces* 12 (36): 40740–40748.

35 Hoffmann, F., Cornelius, M., Morell, J. et al. (2006). Silica-based mesoporous organic–inorganic hybrid materials. *Angewandte Chemie International Edition* 45 (20): 3216–3251.

36 Lu, S., Wang, D., Jiang, S.P. et al. (2010). HPW/MCM-41 phosphotungstic acid/mesoporous silica composites as novel proton-exchange membranes for elevated-temperature fuel cells. *Advanced Materials* 22 (9): 971–976.

37 Zhao, Y., Yang, H., Wu, H. et al. (2014). Enhanced proton conductivity of hybrid membranes by incorporating phosphorylated hollow mesoporous silica submicrospheres. *Journal of Membrane Science* 469: 418–427.

38 Zhao, Y., Yang, H., Wu, H. et al. (2014). Enhanced proton conductivity of the hybrid membranes by regulating the proton conducting groups anchored on the mesoporous silica. *Journal of Power Sources* 270: 292–303.

39 Seo, K., Nam, K.H., and Han, H. (2020). Proton transport in aluminum-substituted mesoporous silica channel-embedded high-temperature anhydrous proton-exchange membrane fuel cells. *Scientific Reports* 10 (1): 10352.

40 Wu, B., Pan, J., Ge, L. et al. (2014). Oriented MOF-polymer composite nanofiber membranes for high proton conductivity at high temperature and anhydrous condition. *Scientific Reports* 4 (1): 4334.

41 Liu, S.S., Han, Z., Yang, J.S. et al. (2020). Sulfonic groups lined along channels of metal–organic frameworks (MOFs) for super-proton conductor. *Inorganic Chemistry* 59 (1): 396–402.

42 Liu, Y.R., Chen, Y.Y., Zhuang, Q. et al. (2022). Recent advances in MOFs-based proton exchange membranes. *Coordination Chemistry Reviews* 471: 214740.

43 Rao, Z., Lan, M., Wang, Z. et al. (2022). Effectively facilitating the proton conduction of proton exchange membrane by polydopamine modified hollow metal–organic framework. *Journal of Membrane Science* 644: 120098.

44 D'amato, R., Bondi, R., Moghdad, I. et al. (2021). "Shake 'n bake" route to functionalized Zr-UiO-66 metal–organic frameworks. *Inorganic Chemistry* 60 (18): 14294–14301.

45 Pal, S.C. and Das, M.C. (2021). Superprotonic conductivity of MOFs and other crystalline platforms beyond 10^{-1} s cm^{-1}. *Advanced Functional Materials* 31 (31): 2101584.

46 Shimizu, G.K.H., Taylor, J.M., and Kim, S. (2013). Proton conduction with metal-organic frameworks. *Science* 341 (6144): 354–355.

47 Phang, W.J., Jo, H., Lee, W.R. et al. (2015). Superprotonic conductivity of a UiO-66 framework functionalized with sulfonic acid groups by facile postsynthetic oxidation. *Angewandte Chemie International Edition* 54 (17): 5142–5146.

48 Mukhopadhyay, S., Debgupta, J., Singh, C. et al. (2019). Designing UiO-66-based superprotonic conductor with the highest metal–organic framework based proton conductivity. *ACS Applied Materials & Interfaces* 11 (14): 13423–13432.

49 Shi, B., Pang, X., Li, S. et al. (2022). Short hydrogen-bond network confined on COF surfaces enables ultrahigh proton conductivity. *Nature Communications* 13 (1): 1–9.

50 Wang, Y., Ren, L., Wang, J. et al. (2022). In-situ growth of anionic covalent organic frameworks efficaciously enhanced the monovalent selectivity of anion exchange membranes. *Journal of Membrane Science* 659: 120818.

51 Wu, Z., He, Y., Liu, L. et al. (2022). Fast hydroxide conduction via hydrogen-bonding network confined in benzimidazolium-functionalized covalent organic frameworks. *ACS Applied Materials & Interfaces* 14 (38): 43861–43867.

52 Guo, Y., Zou, X., Li, W. et al. (2022). High-density sulfonic acid-grafted covalent organic frameworks with efficient anhydrous proton conduction. *Journal of Materials Chemistry A* 10 (12): 6499–6507.

53 Singh, R. and Kim, D. (2022). Ultrafast ion-transport at hierarchically porous covalent-organic membrane interface for efficient power production. *Nano Energy* 92: 106690.

54 Ying, Y., Peh, S.B., Yang, H. et al. (2022). Ultrathin covalent organic framework membranes via a multi-interfacial engineering strategy for gas separation. *Advanced Materials* 34 (25): 2104946.

55 Hou, S., Ji, W., Chen, J. et al. (2021). Free-standing covalent organic framework membrane for high-efficiency salinity gradient energy conversion. *Angewandte Chemie International Edition* 60 (18): 9925–9930.

56 Tang, J.K., Zhu, C.Y., Jiang, T.W. et al. (2020). Anion exchange-induced single-molecule dispersion of cobalt porphyrins in a cationic porous organic polymer for enhanced electrochemical CO_2 reduction via secondary-coordination sphere interactions. *Journal of Materials Chemistry A* 8 (36): 18677–18686.

57 Xu, H., Tao, S., and Jiang, D. (2016). Proton conduction in crystalline and porous covalent organic frameworks. *Nature Materials* 15 (7): 722–726.

58 Cao, L., Wu, H., Cao, Y. et al. (2020). Weakly humidity-dependent proton-conducting COF membranes. *Advanced Materials* 32 (52): 2005565.

59 Wang, X., Shi, B., Yang, H. et al. (2022). Assembling covalent organic framework membranes with superior ion exchange capacity. *Nature Communications* 13 (1): 1–9.

60 Budd, P.M., Ghanem, B.S., Makhseed, S. et al. (2004). Polymers of intrinsic microporosity (PIMs): robust, solution-processable, organic nanoporous materials. *Chemical Communications* 2: 230–231.

61 Mckeown, N.B. and Budd, P.M. (2006). Polymers of intrinsic microporosity (PIMs): organic materials for membrane separations, heterogeneous catalysis and hydrogen storage. *Chemical Society Reviews* 35 (8): 675–683.

62 Zuo, P., Li, Y., Wang, A. et al. (2020). Sulfonated microporous polymer membranes with fast and selective ion transport for electrochemical energy conversion and storage. *Angewandte Chemie International Edition* 59 (24): 9564–9573.

63 Wang, P., Li, X., Liu, Z. et al. (2022). Construction of highly conductive PBI-based alloy membranes by incorporating PIMs with optimized molecular weights for high-temperature proton exchange membrane fuel cells. *Journal of Membrane Science* 659: 120790.

64 Wang, T., Jin, Y., Mu, T. et al. (2022). Tröger's base polymer blended with poly(ether ketone cardo) for high temperature proton exchange membrane fuel cell applications. *Journal of Membrane Science* 654: 120539.

65 Du, X., Yuan, Y., Dong, T. et al. (2021). Polymer electrolyte membranes from microporous Troger's base polymers for fuel cells. *ACS Applied Energy Materials* 4 (11): 13327–13334.

66 Tang, H., Geng, K., Wu, L. et al. (2022). Fuel cells with an operational range of –20°C to 200°C enabled by phosphoric acid-doped intrinsically ultramicroporous membranes. *Nature Energy* 7 (2): 153–162.

67 Park, H.B., Kamcev, J., Robeson, L.M. et al. (2017). Maximizing the right stuff: the trade-off between membrane permeability and selectivity. *Science* 356 (6343): eaab0530.

68 Ye, C., Wang, A., Breakwell, C. et al. (2022). Development of efficient aqueous organic redox flow batteries using ion-sieving sulfonated polymer membranes. *Nature Communications* 13 (1): 3184.

69 Zuo, P., Ye, C., Jiao, Z. et al. (2023). Near-frictionless ion transport within triazine framework membranes. *Nature* 617 (7960): 299–305.

70 Brijmohan, S.B. and Shaw, M.T. (2007). Magnetic ion-exchange nanoparticles and their application in proton exchange membranes. *Journal of Membrane Science* 303 (1): 64–71.

71 Tong, J.Y., Guo, Q., and Wang, X.X. (2009). Properties and structure of SPEEK proton exchange membrane doped with nanometer CeO_2 and treated with high magnetic field. *Express Polymer Letters* 3 (12): 821–831.

72 Hasanabadi, N., Ghaffarian, S.R., and Hasani-Sadrabadi, M.M. (2011). Magnetic field aligned nanocomposite proton exchange membranes based on sulfonated poly (ether sulfone) and Fe_2O_3 nanoparticles for direct methanol fuel cell application. *International Journal of Hydrogen Energy* 36 (23): 15323–15332.

73 Hasanabadi, N., Ghaffarian, S.R., and Hasani-Sadrabadi, M.M. (2013). Nafion-based magnetically aligned nanocomposite proton exchange membranes for direct methanol fuel cells. *Solid State Ionics* 232: 58–67.

74 Hasani-Sadrabadi, M.M., Majedi, F.S., Coullerez, G. et al. (2014). Magnetically aligned nanodomains: application in high-performance ion conductive membranes. *ACS Applied Materials & Interfaces* 6 (10): 7099–7107.

75 Huang, Y., Xiao, C., Huang, Q. et al. (2018). Magnetic field induced orderly arrangement of Fe_3O_4/GO composite particles for preparation of Fe_3O_4/GO/PVDF membrane. *Journal of Membrane Science* 548: 184–193.

76 Liu, X., Li, Y., Xue, J. et al. (2019). Magnetic field alignment of stable proton-conducting channels in an electrolyte membrane. *Nature Communications* 10 (1): 842.

77 Gong, X. et al. (2017). Aligned electrospun nanofibers as proton conductive channels through thickness of sulfonated poly (phthalazinone ether sulfone ketone) proton exchange membranes. *Journal of Power Sources* 358: 134–141. https://doi.org/10.1016/j.jpowsour.2017.05.022.

78 Park, C.H., Lee, S.Y., Hwang, D.S. et al. (2016). Nanocrack-regulated self-humidifying membranes. *Nature* 532 (7600): 480–483.

11

Highly Sulfonated Poly(Ether Ether Ketone) Nanofibers Constructed Plasmonic Transport Channels

11.1 Introduction

Commercially available perfluorosulfonic acid–based proton exchange membranes (PEMs) (e.g. DuPont's Nafion® series) exhibit strong mechanical properties, chemical stability, and high proton conductivity under high-humidity conditions. Despite these advantages, their high cost, significant fuel permeability, and diminished proton conductivity at elevated temperatures and low humidity restrict their broader application [1]. This has motivated the search for alternative materials, such as sulfonated aromatic hydrocarbon polymers like sulfonated poly(ether ether ketone) (SPEEK). SPEEK membranes offer a combination of good chemical, mechanical, and thermal stability, alongside cost-effective, straightforward preparation. By introducing sulfonate groups into the poly(ether ether ketone) backbone, SPEEK membranes achieve the proton conductivity necessary for fuel cell applications.

In SPEEK membranes, increasing the degree of sulfonation enhances water absorption, which in turn improves proton conductivity. However, excessive sulfonation can lead to overswelling, reduced dimensional stability, and even membrane dissolution, compromising mechanical integrity [2]. Various cross-linking methods—ionic [3], covalent [4], ionic-covalent [5], and end-group cross-linking [6]—have been used to enhance the physical properties of highly sulfonated polymer membranes. For instance, Di Vona demonstrated that thermal cross-linking in the presence of dimethylsulfoxide (DMSO) forms sulfone bridges between sulfonic groups and benzene rings, resulting in reduced swelling and improved mechanical strength [7–9]. However, cross-linking can also reduce the mobility of polymer chains and consume sulfonic acid groups, ultimately lowering proton conductivity.

To overcome these challenges, composite modifications have been introduced to enhance the overall performance of SPEEK membranes. Strategies such as polymer blending [10–12] and the incorporation of functionalized nanofillers—including nanosheets [13–15], nanotubes, nanoclays [16, 17], and nanofibers [18–23]—have

Electrospinning for Proton Exchange Membrane Fuel Cells, First Edition.
Yong Liu and Mohideen Meerasahib Mohamedazeem.

yielded novel composite membranes. Among these approaches, nanofiber composite membranes have gained significant attention due to their ability to form long-range, continuous proton transport channels. For example, Kawakami developed acid-doped nanofiber/sulfonated polyimide composite membranes with rapid proton conduction along the nanofiber surfaces [24], while Pintauro's group fabricated highly sulfonated polymer nanofibers filled with polyurethane, followed by UV-initiated cross-linking for enhanced membrane stability [25–27]. Nevertheless, the lack of proton conductivity in polyurethane fillers partially limited the performance of these composite membranes.

Building on this background, the current chapter focuses on using highly sulfonated SPEEK nanofibers within cross-linked composite membranes to construct efficient proton transport channels. SPEEK nanofiber mats were produced by electrospinning, then infiltrated with a sulfonated SPEEK solution to form dense, defect-free membranes. Upon heat treatment, the sulfonated SPEEK matrix underwent cross-linking, improving dimensional stability and mechanical strength, while the sulfonated SPEEK nanofibers retained their original form. These non-cross-linked regions preserved sulfonate groups and created continuous long-range proton transport channels, ensuring high proton conductivity in the composite membranes.

11.2 Materials Preparation

11.2.1 Preparation of Sulfonated Poly(Ether Ether Ketone) with Different Degrees of Sulfonation

In a typical procedure, 10 g of vacuum-dried polyether ether ketone (PEEK) powder was dissolved in 200 mL of concentrated sulfuric acid at room temperature under magnetic stirring. The temperature was then raised to 50 °C to initiate the sulfonation reaction. After a predetermined reaction time, the process was quenched by cooling with cold water, thereby producing SPEEK with varying degrees of sulfonation. The cooled viscous solution was poured slowly into a large volume of iced water, causing the product to precipitate. The resulting SPEEK was washed with ice water until neutral and then dried under vacuum at 60 °C for 24 h. The degree of sulfonation was determined by ^{1}H nuclear magnetic resonance (NMR) (Bruker Avance HD 400 MHz). The samples prepared at different reaction times exhibited sulfonation degrees of 72%, 80%, and 90%, labeled as SP72, SP80, and SP90, respectively.

11.2.2 Substitution of Counteracting Ions

The sulfonated SPEEK samples (SP72, SP80) were immersed in 2 M NaCl solution at 50 °C for 48 hours to achieve full substitution of H^+ by Na^+ ions. Similarly, SP90 was immersed in 2 M potassium chloride (KCl) solution under identical conditions to replace H^+ with K^+ ions. After counterion substitution, the samples were rinsed

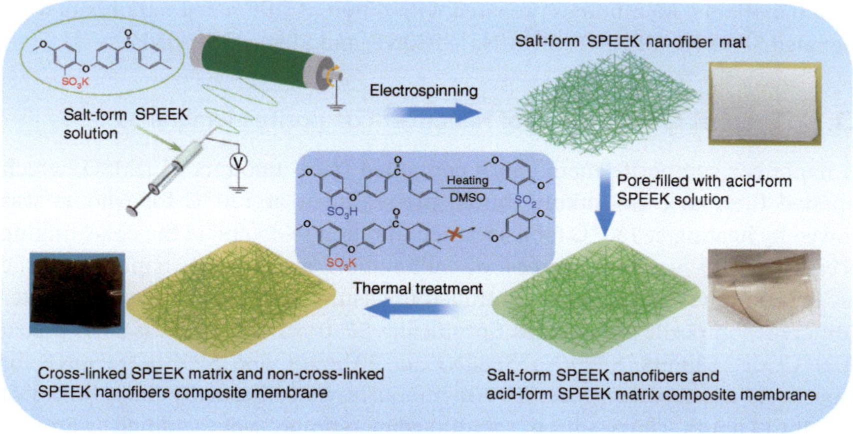

Figure 11.1 Illustration of preparation process of nanofiber composite cross-linked SPEEK membrane. Source: Han et al. [28]/with permission of Elsevier.

with deionized water to remove excess salts and dried under vacuum at 60 °C for 24 hours. The resulting sulfonate forms were denoted as Na^+SP72, Na^+SP80, and K^+SP90.

11.2.3 Preparation of Sulfonated Poly(Ether Ether Ketone) Nanofibers

To produce sulfonated SPEEK nanofibers, a 20 wt% spinning solution was prepared by dissolving sulfonated SPEEK in dimethylformamide (DMF). This solution was drawn into a syringe mounted on a propulsion pump. Electrospinning was carried out at a controlled room temperature of 25 °C and relative humidity (RH) of 50%. The solution was dispensed at a rate of 60 μL/h, and a voltage of +15 kV was applied to the needle connected to a high-voltage power supply. Nanofibers were collected on a grounded rotating cylindrical collector, rotating at 500 rpm, positioned 15 cm from the needle tip. The resulting nanofibers were dried in an oven at 60 °C for 24 hours. Composite membranes were subsequently prepared following the steps illustrated in Figure 11.1. The nanofibers were labeled as Na^+SP72NF, Na^+SP80NF, and K^+SP90NF, corresponding to the counterion-substituted SPEEK used.

11.2.4 Preparation of Nanofiber Composite Membranes

A casting solution was prepared by dissolving sulfonated SPEEK (SP90) in a water/DMSO mixture (mass ratio 15:1) to a concentration of 11 wt%. A predetermined mass of sulfonated SPEEK nanofibers was evenly placed on a horizontally positioned glass plate. The casting solution was then poured over the nanofibers, ensuring complete infiltration into the mats. Solvent evaporation was carried out at room temperature for 24 hours, followed by a drying step at 60 °C for 12 hours to minimize solvent residue. The resulting composite membrane contained approximately

20 wt% nanofibers. Membranes prepared with different sulfonated nanofibers were designated SP90/Na^+SP72NF, SP90/Na^+SP80NF, and SP90/K^+SP90NF.

11.2.5 Thermal Cross-Linking of Nanofiber Composite Membranes

The nanofiber composite membranes contained trace amounts of DMSO, which facilitated thermal cross-linking. Initial pretreatment at 120 °C for 6 hours was followed by heating at 180 °C for a specified duration to complete the cross-linking process. Due to varying degrees of sulfonation among the nanofibers, the cross-linking time was carefully controlled to ensure comparable dissolution rates for different composite membranes. Specifically, SP90/Na^+SP72NF was cross-linked at 180 °C for 15 hours, SP90/Na^+SP80NF for 20 hours, and SP90/K^+SP90NF for 25 hours. As a reference, pure SP90 membranes were thermally cross-linked at 180 °C for 12 hours. The resulting cross-linked membranes were acidified by immersion in 1 M HCl solution for 48 hours, then rinsed thoroughly with deionized water to remove residual acid. These cross-linked membranes were labeled as C-SP90, C-SP90/SP72NF, C-SP90/SP80NF, and C-SP90/SP90NF.

11.3 Results and Discussion

11.3.1 Preparation of Sulfonated Poly(Ether Ether Ketone) and Characterization of Nanofibers

SPEEK was synthesized via a postsulfonation method. The sulfonation of PEEK is an electrophilic substitution reaction that occurs preferentially at benzene ring positions with higher electron density. Concentrated sulfuric acid served both as the solvent and the sulfonation reagent. The degree of sulfonation (DS) was adjusted by controlling the reaction temperature and duration. To prevent uneven sulfonation during the polymer dissolution phase—when sulfonation can occur simultaneously—PEEK was initially dissolved at a lower temperature to slow the reaction. Once fully dissolved, the temperature was increased to the desired reaction level for a set period, resulting in a more uniform DS. The DS of SPEEK was confirmed using ^{1}H NMR analysis, as illustrated in Figure 11.2.

The sulfonation degree of SPEEK was determined using ^{1}H NMR spectroscopy (Figure 11.2). A distinct chemical shift at 7.52 ppm corresponds to the proton on the benzene ring adjacent to the sulfonic acid group (HE). By integrating the proton signals from the benzene ring and applying Eq. (11.1), the DS was calculated. The corresponding ion exchange capacities (IECs) were derived using Eq. (11.2). The results are summarized in Table 11.1.

$$\frac{\mathrm{DS}}{12-2\mathrm{DS}}=\frac{A_{\mathrm{H_E}}}{\sum A_{\mathrm{H_{A,B,B',C,C',D,D'}}}} \tag{11.1}$$

$$\mathrm{IEC}=\frac{1000\mathrm{DS}}{M_{\mathrm{PEEK}}+80.06\mathrm{DS}} \tag{11.2}$$

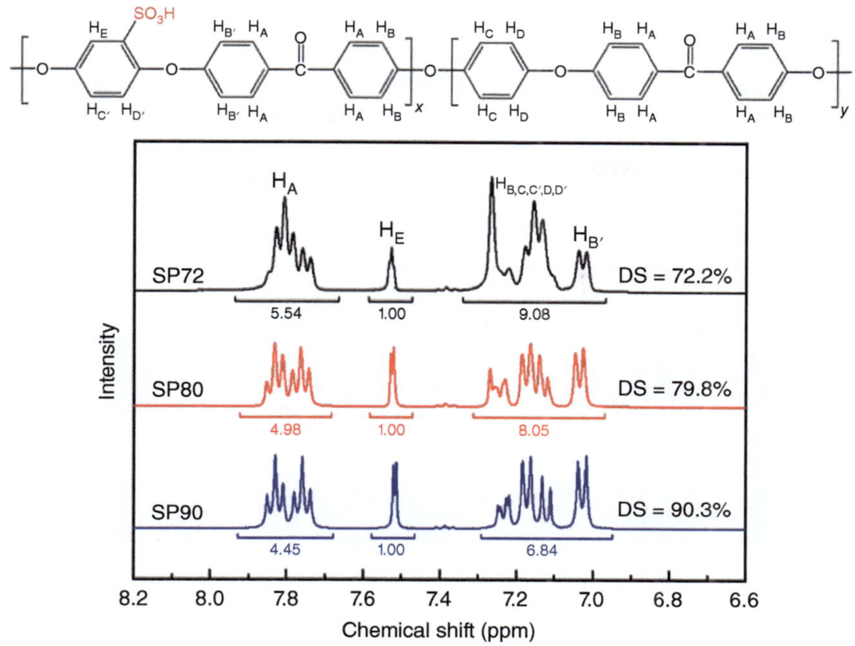

Figure 11.2 Structure of SPEEK molecular structure ($x + y = n$, $x/(x + y) = \text{DS}$) and the corresponding ^{1}H NMR spectra with different sulfonation degree.

Table 11.1 Degree of sulfonation and IEC values of the synthesized SPEEK.

SPEEK	Reaction time at 50 °C (h)	IEC (mmol/g)	DS (%)
SP72	2.5	2.08	72.2
SP80	3.5	2.27	79.8
SP90	10	2.50	90.3

Sulfonated SPEEK nanofibers were then prepared via electrostatic spinning. As indicated in Table 11.2, the stability of the different salt forms of SPEEK in water varied. The K^+ and Na^+ sulfonated forms of SPEEK exhibited significantly lower water absorption and swelling compared to the H^+ sulfonated form. Among these, K^+-salt SPEEK showed superior water stability over Na^+-salt SPEEK. Consequently, K^+SP90NF nanofibers were chosen to construct highly sulfonated nanofiber composite membranes, as they retained their fiber morphology during solution casting. The morphology of the electrospun SPEEK nanofibers was examined via scanning electron microscope (SEM) (Figure 11.3). The fibers displayed a smooth, bead-free surface, with diameters ranging from 40 to 120 nm, averaging approximately 60 nm.

Table 11.2 The water uptake and swelling ratio of different salt-form SPEEK membranes.

H^+ or Na^+ or K^+ form SPEEK membranes	Soaked in water (20 °C 4 h)		Soaked in water (80 °C, 4 h)
	Water uptake (%)	Swelling ratio (%)	
H^+SP90	284.5	68.6	Dissolved
Na^+SP90	91.1	29.2	Dissolved
K^+SP90	58.2	21.8	Dissolved

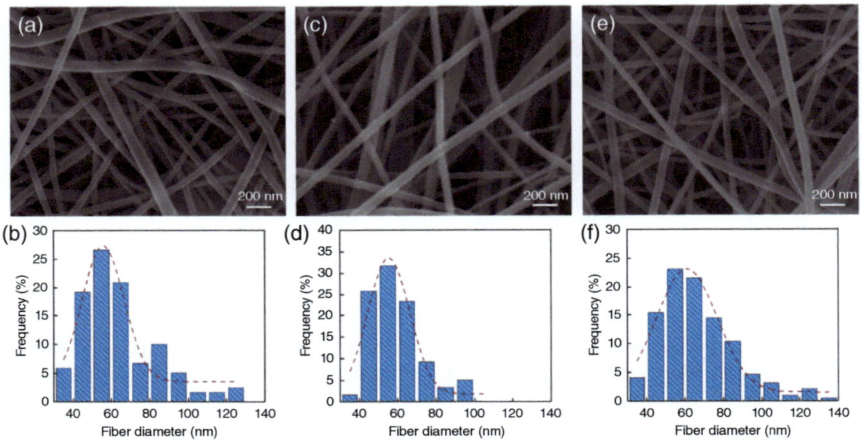

Figure 11.3 SEM micrographs of electrospun SPEEK nanofibers (a) Na^+SP72NF, (c) Na^+SP80NF, (e) K^+SP90NF, and the corresponding fiber diameter distribution (b), (d) and (f), respectively. Source: Han et al. [28]/with permission of Elsevier.

11.3.2 Cross-Sectional Morphology of Fiber Composite Membranes

A semi-transparent composite membrane, approximately 50 μm thick and containing about 20 wt% fibers, was obtained by slowly casting H^+SP90 SPEEK solution onto SPEEK nanofiber mats in their salt form. SEM images of the cross-sectional morphology prior to thermal cross-linking are shown in Figure 11.4a. The SP90/Na^+SP72NF (Figure 11.4b) and SP90/Na^+SP80NF (Figure 11.4c) composite membranes displayed clearly defined and evenly distributed fibers. In contrast, the SP90/K^+SP90NF composite membrane (Figure 11.4d) exhibited less distinct fiber morphology. This difference arose because the highly sulfonated K^+SP90 SPEEK fibers tended to dissolve partially during the casting process, causing slight structural damage. Despite this, the fibers remained distinguishable within the composite matrix. Previous attempts using Na^+SP90 fibers were less successful as they were prone to excessive water absorption and swelling, complicating the preservation of fiber morphology during film formation.

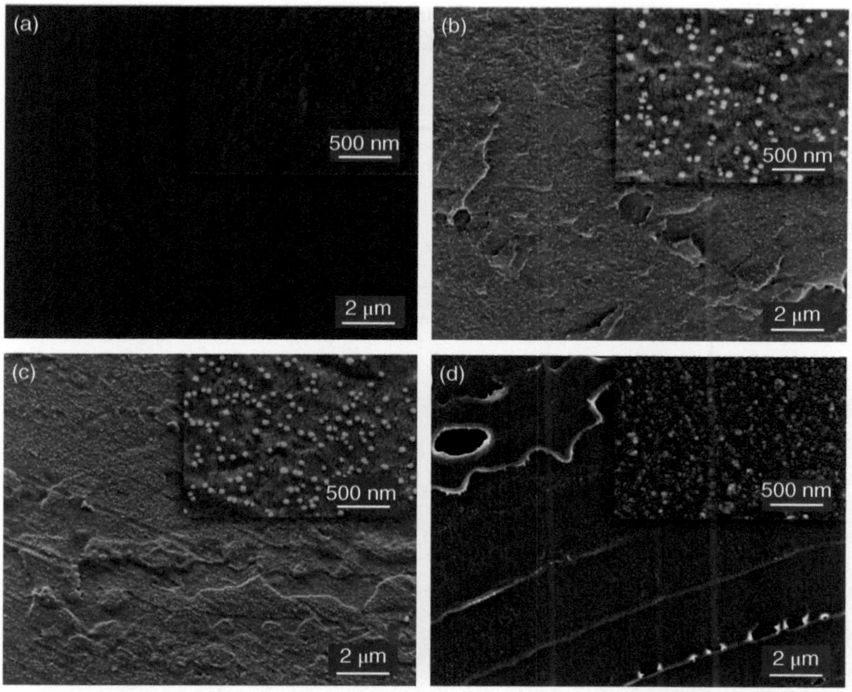

Figure 11.4 SEM micrographs of the cross-section of pristine and composite SPEEK membranes before cross-linked thermal treatment: (a) pristine SPEEK SP90, (b) SP90/Na^+SP72NF, (c) SP90/Na^+SP80NF, (d) SP90/K^+SP90NF. Source: Han et al. [28]/with permission of Elsevier.

Thermal cross-linking was used to improve the mechanical stability of the highly sulfonated membranes. During the preparation of the nanofiber composite membranes, a trace amount of DMSO left in the casting solution initiated a thermal cross-linking reaction of $-SO_3H$ groups, forming $-SO_2$-bridges. After thermal treatment, the IEC values of both pure SP90 and the composite membranes decreased notably, as shown in Table 11.3, with pure SP90 dropping from 2.50 to 2.05 mmol/g after 12 hours of cross-linking. In the composite membranes, because the SPEEK fibers did not undergo significant cross-linking, prolonging the cross-linking time and increasing the matrix cross-linking density helped maintain overall membrane stability. Some exchange of Na^+ or K^+ ions from the SPEEK fibers with H^+ ions in the matrix occurred during casting, resulting in minor cross-linking within the fibers. This low cross-linking density aided in maintaining the fiber structure.

Figure 11.5 illustrates the cross-sectional morphology after thermal cross-linking. The fiber structures became less distinct, likely due to the increased molecular mobility at elevated temperatures and the interaction between sulfonic acid groups on the matrix and the fiber surfaces. This cross-linking blurred the interface between fibers and matrix, resulting in a more unified composite structure.

Table 11.3 The IEC of thermal cross-linked composite SPEEK membranes.

Cross-linked SPEEK	Thermal treatment time at 180 °C (h)	Initial IEC[a)] (mmol/g)	IEC[b] (mmol/g)
C-SP90	12	2.50	2.05 ± 0.04
C-SP90/SP72NF	15	2.42	1.97 ± 0.07
C-SP90/SP80NF	20	2.45	1.89 ± 0.06
C-SP90/SP90NF	25	2.50	1.78 ± 0.09

a) Nominal IEC value of the contribution of each polymer.
b) Determined by titration.

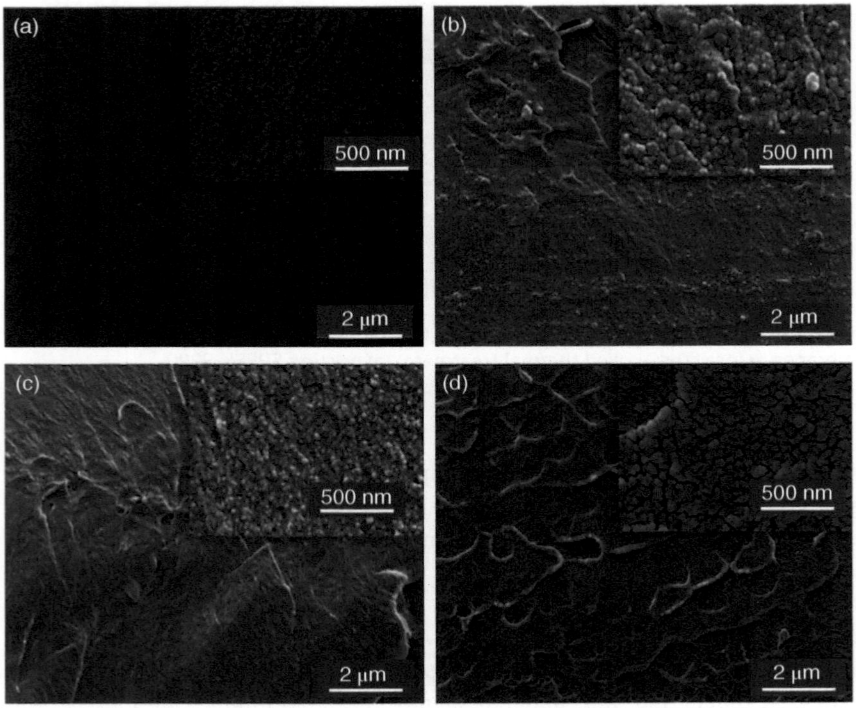

Figure 11.5 SEM micrographs of the cross-section of pristine and composite SPEEK membranes after cross-linked thermal treatment: (a) pristine SPEEK C-SP90, (b) C-SP90/SP72NF, (c) C-SP90/ SP80NF, (d) C-SP90/SP90NF. Source: Han et al. [28]/with permission of Elsevier.

11.3.3 Chemical Structure Characterization of Composite Membranes

The Fourier transform infrared (FT-IR) spectra of the cross-linked nanofiber composite membranes (Figure 11.6a) revealed characteristic peaks of SPEEK: the carbonyl group (C=O) at 1645 cm^{-1}, benzene ring stretching at 1587 cm^{-1}, and C–C stretching at 1465 cm^{-1}. Peaks at 1076 and 1010 cm^{-1} correspond to the symmetric

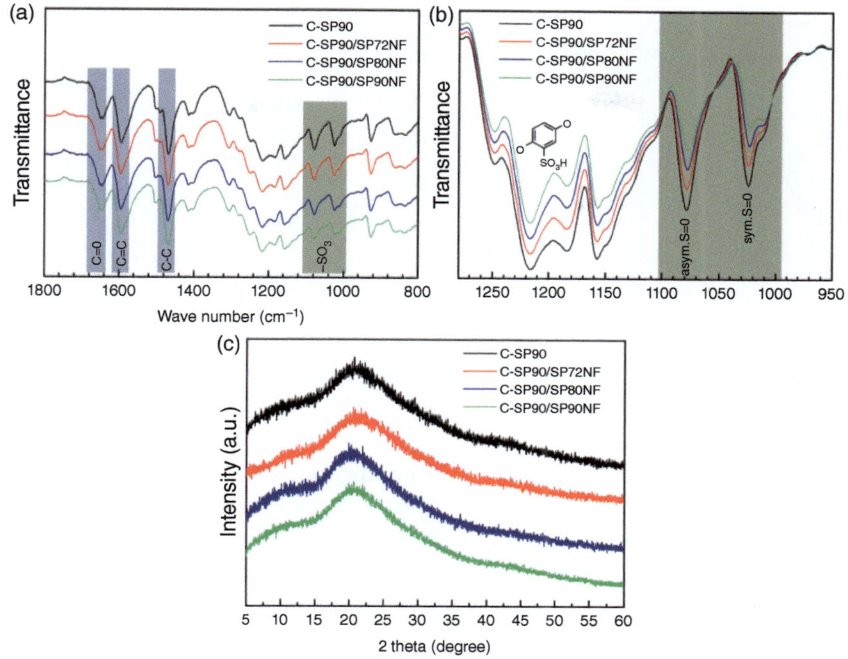

Figure 11.6 (a) FT-IR spectra, (b) enlarged part of FT-IR spectra, and (c) XRD patterns of thermal cross-linked pristine SPEEK and composite SPEEK membranes.

and asymmetric stretching vibrations of the sulfonic acid group's S=O bond, while the peak at 1217 cm^{-1} indicated 1,2,4-phenylene ring substitution [29, 30]. Composite membranes with a higher cross-linking degree exhibited weakened $-SO_3H$ peaks due to reduced sulfonic acid group content. The X-ray diffraction (XRD) patterns (Figure 11.6c) showed that the high number of sulfonic acid groups disrupted the polymer's chain stacking, resulting in a broad peak around $2\theta = 21°$, characteristic of SPEEK's amorphous structure.

11.3.4 Mechanical Properties and Thermal Stability

The mechanical properties of PEMs are crucial for ensuring fuel cell longevity. Thermally cross-linked SPEEK membranes displayed a balance of toughness and rigidity due to their cross-linked networks. Stress–strain curves (Figure 11.7a) revealed tensile stress values ranging from 57 to 68 MPa and Young's modulus from 0.93 to 1.31 GPa. Membranes with higher cross-linking densities exhibited greater stiffness; for instance, the C-SP90/SP90NF film had a Young's modulus of 1.31 GPa, outperforming the C-SP90 film's 0.93 GPa. While C-SP90/SP90NF's elongation at break (20%) was lower than C-SP90's (38%), its tensile stress was slightly higher at 68 MPa compared to 60 MPa.

Thermal stability, evaluated by thermogravimetric analysis (TGA) (Figure 11.7b), showed three stages of weight loss: Volatilization of water and small molecules

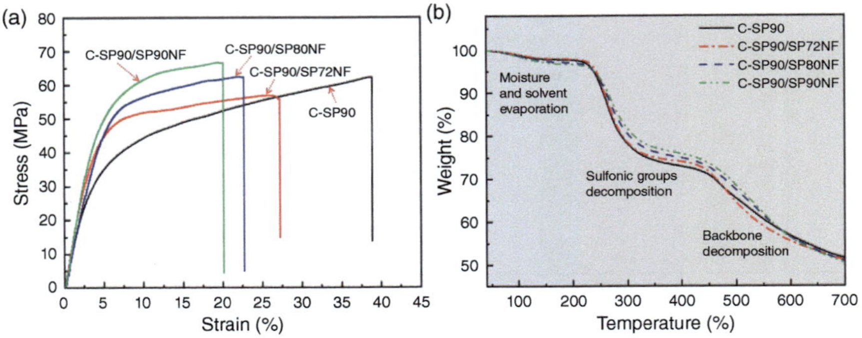

Figure 11.7 (a) Stress–strain and (b) TGA curves of thermal cross-linked pristine SPEEK and composite SPEEK membranes.

(30–200 °C), degradation of sulfonic acid groups (220–400 °C), and degradation of the polymer backbone (450–700 °C). The TGA data revealed that the weight loss associated with sulfonate group degradation decreased in the order: C-SP90 > C-SP90/SP72NF > C-SP90/SP80NF > C-SP90/SP90NF. This sequence correlates with increasing cross-linking density, as cross-linking reduces the number of $-SO_3H$ groups available for degradation.

11.3.5 Water Absorption and Solubility

The water absorption capacity of PEMs, heavily influenced by their IEC and degree of sulfonation, plays a critical role in proton conduction. Table 11.3 illustrates that un-cross-linked, highly sulfonated SP90 membranes suffered from excessive water uptake (284.5%) and swelling (68.6%), resulting in poor dimensional stability. In contrast, Figures 11.8 show that the cross-linked membranes achieved significant reductions in water absorption and swelling. These cross-linked membranes maintained moderate swelling levels—around 20% at 40 °C and 30% at 80 °C—and exhibited improved dimensional stability. The reduction in swelling is

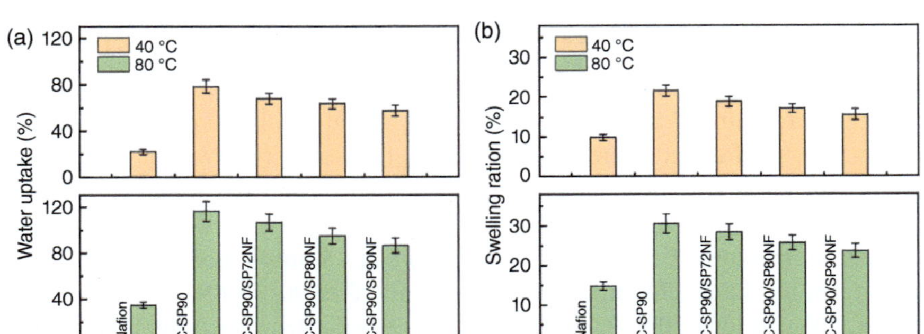

Figure 11.8 (a) Water uptake and (b) swelling ratio of Nafion, thermal cross-linked pristine SPEEK and composite SPEEK membranes at various temperatures.

attributed to the formation of covalent bonds during cross-linking, which produce a more robust, compact polymer network. This structure reduces the free volume available for water absorption and restricts the mobility of polymer chains, making the membrane less permeable to water molecules. While cross-linked membranes still demonstrate higher water absorption and swelling compared to Nafion, their performance remains manageable due to the stabilizing effects of cross-linking.

11.3.6 Membrane Proton Conductivity

Proton conductivity measurements of cross-linked SPEEK membranes and recast Nafion under fully hydrated conditions (Figure 11.9a) reveal a temperature-dependent increase in conductivity. This increase can be explained by enhanced water activity and polymer chain mobility at higher temperatures. The Arrhenius plots in Figure 11.9b indicate activation energy values (Ea) between 6.11 and 8.39 kJ/mol, consistent with the Grotthuss mechanism of proton transport. This mechanism involves protons hopping between water molecules via hydrogen bonds, resulting in a low energy barrier and limited temperature dependence [31, 32].

Although proton conductivity typically correlates with IEC, the cross-linked nanofiber composite membranes demonstrated superior conductivity despite having lower IEC values compared to C-SP90 membranes. For instance, the C-SP90/SP90NF membrane achieved a proton conductivity of approximately 200 mS/cm at 80 °C, surpassing both Nafion (140 mS/cm) and C-SP90 (135 mS/cm). Furthermore, the proton conductivity of the composite membranes increased with the sulfonation degree of the incorporated nanofibers—C-SP90/SP80NF and C-SP90/SP72NF, which displayed proton conductivities of 172 and 137 mS/cm, respectively, at 80 °C. The enhanced proton conductivity is attributed to the structural advantage of the nanofibers, which create long-range, effective water channels through abundant sulfonic acid groups. These channels connect isolated hydrated regions within the matrix, thereby facilitating efficient proton transport (Figure 11.10).

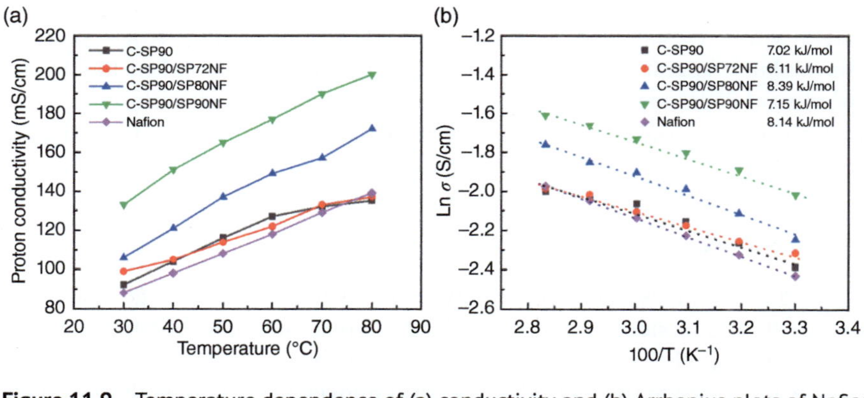

Figure 11.9 Temperature dependence of (a) conductivity and (b) Arrhenius plots of Nafion, thermal cross-linked pristine SPEEK and composite SPEEK membranes.

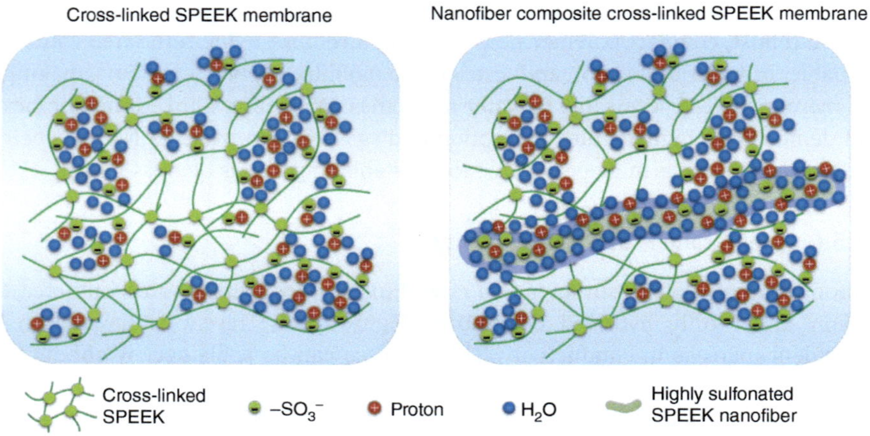

Figure 11.10 Schematic illustration of the proton transport of thermal cross-linked nanofiber composite membranes.

At 80 °C, the proton conductivity of the PEM increased with higher RH (Figures 11.11 and 11.12a). While conductivity was low at 51% and 74% RH, it rose significantly at 94% and 100% RH, confirming that water content strongly influences proton transport. The C-SP90/SP90NF membrane outperformed C-SP90 at lower RH due to its high sulfonated fiber continuity, which supports the formation of long-range water channels. However, all thermally cross-linked SPEEK membranes had lower conductivities than did Nafion under low RH conditions, mainly because Nafion's unique phase separation structure enhances proton conduction at reduced humidity. At 94% RH, C-SP90/SP90NF achieved a proton conductivity of 145 mS/cm, slightly exceeding Nafion's 126 mS/cm.

Small-angle X-ray scattering (SAXS) analysis (Figure 11.12b) revealed that C-SP90/SP90NF had a distinct scattering peak at 1.40 nm^{-1}, corresponding to ion clusters averaging 4.5 nm in size. In contrast, C-SP90 exhibited a weaker, less defined scattering peak, indicating smaller and more discrete ion clusters. These findings underscore the role of nanofibers in enhancing the aggregation of hydrophilic ion clusters within cross-linked membranes. While cross-linking restricts polymer chain mobility and sulfonic acid group aggregation, un-cross-linked highly sulfonated nanofibers maintain flexible chains that form larger ion clusters. These nanofibers connect isolated ionic clusters in the matrix, improving proton transport and overall composite membrane performance.

11.3.7 Antioxidant Stability

Antioxidant stability is a critical property for PEMs in proton exchange membrane fuel cell (PEMFC) systems, where peroxyl radicals (e.g. HOO· and HO·) generated at the catalyst electrode can degrade the membrane and reduce its operational lifetime. To assess resistance to free radical attack, the antioxidant stability of the thermally cross-linked membranes was evaluated using Fenton's reagent. Un-cross-linked

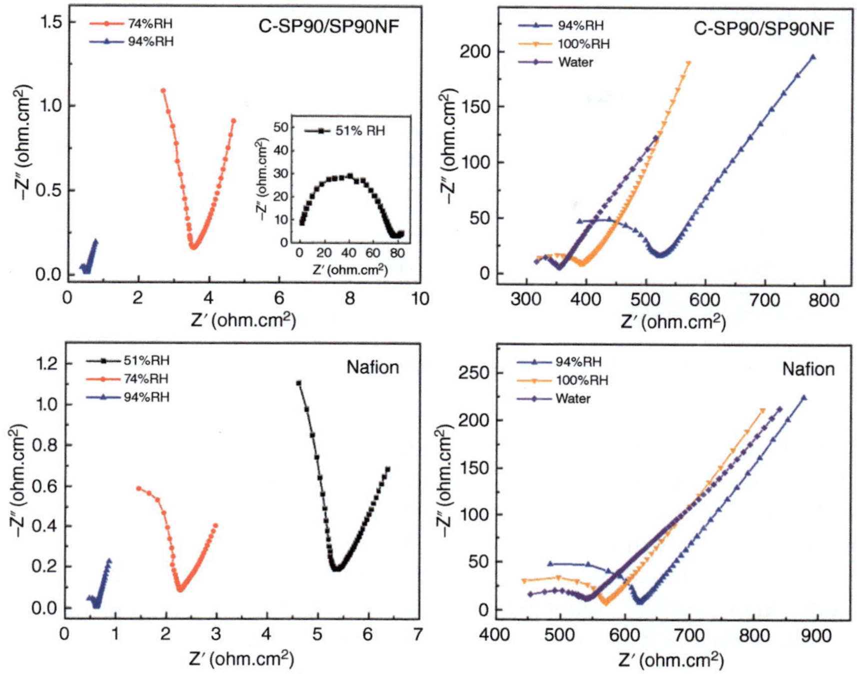

Figure 11.11 Nyquist plots from impedance data of thermal cross-linked composite SPEEK and Nafion membranes at 80 °C and varying humidity.

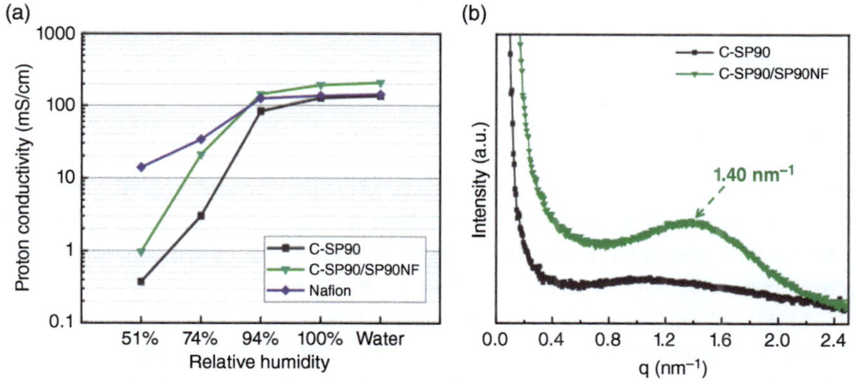

Figure 11.12 (a) Proton conductivity of membranes at 80 °C with different relative humidity and (b) SAXS of thermal cross-linked pristine SPEEK and composite SPEEK membranes.

Table 11.4 Oxidative stability of the prepared cross-linked SPEEK membranes.

Cross-linked SPEEK	W_L (%)	τ_1 (min)	τ_2 (min)
C-SP90	1.72	105	165
C-SP90/SP72NF	1.84	115	180
C-SP90/SP80NF	1.69	125	205
C-SP90/SP90NF	1.57	140	235

τ_1: Elapsed time for the membrane to start to break in Fenton's reagent at 80 °C.

τ_2: Elapsed time for the membrane to dissolve completely in Fenton's reagent at 80 °C.

SPEEK membranes, due to their high sulfonation degree and solubility in hot water, could not be tested under these conditions. Thermally cross-linked SPEEK membranes were treated with a 3 wt% H_2O_2 solution containing 2 ppm $FeSO_4$ at 80 °C for 1 hour. As indicated in Table 11.4, these cross-linked membranes exhibited minimal weight loss (less than 2%), demonstrating a significant enhancement in antioxidant stability compared to the un-cross-linked variants. Furthermore, the rupture time (τ_1) and disappearance time (τ_2) in Fenton's reagent extended with increased cross-linking density. For instance, the τ_1 and τ_2 of C-SP90 were 105 and 165 minutes, respectively. By contrast, the values for C-SP90/SP90NF, heat-treated for a longer duration at 180 °C, increased to 140 and 235 minutes, respectively. This improvement highlights that increasing cross-link density can effectively enhance membrane stability. Moreover, membranes with higher cross-linking density exhibited reduced water absorption, which likely contributed to lower permeability of peroxyl radicals, thereby improving their antioxidant properties [33].

11.3.8 Fuel Cell Performance

The performance of thermally cross-linked SPEEK membranes (C-SP90 and C-SP90/SP90NF) and recast Nafion membranes was evaluated in a single-cell H_2/Air PEMFC test at 80 °C and 100% RH as displayed in Figure 11.13. Membrane electrode assembly (MEA) were assembled using membranes with thicknesses of 51 (C-SP90), 47 (C-SP90/SP90NF), and 58 μm (Nafion). The polarization curves reveal open circuit voltage (OCVs) above 0.95 V for all MEAs, indicating high catalytic activity and low fuel permeability. C-SP90/SP90NF membranes exhibited superior proton transport due to the interconnection of hydrated ion channels by continuous, highly sulfonated nanofibers. As a result, the ohmic resistance of C-SP90/SP90NF was lower than that of C-SP90. At higher current densities, the voltage of C-SP90/SP90NF decreased more slowly than that of C-SP90 or Nafion, yielding a peak power density of 485 mW/cm^2 compared to 334 and 385 mW/cm^2 for C-SP90 and Nafion, respectively. These results demonstrate that C-SP90/SP90NF membranes offer higher proton conductivity and superior fuel cell performance under hydrated conditions.

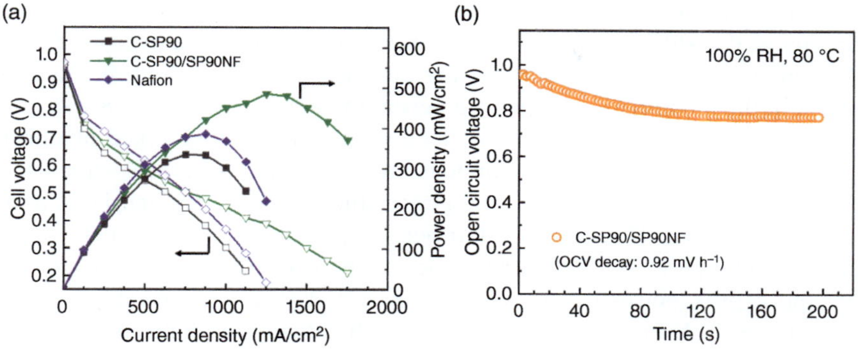

Figure 11.13 (a) H_2/air PEMFC single-cell performance of MEAs using C-SP90, C-SP90/SP90NF, and recast Nafion membranes at 80 °C and under 100% RH condition. (b) OCV changes in the durability test of the C-SP90/SP90NF membrane.

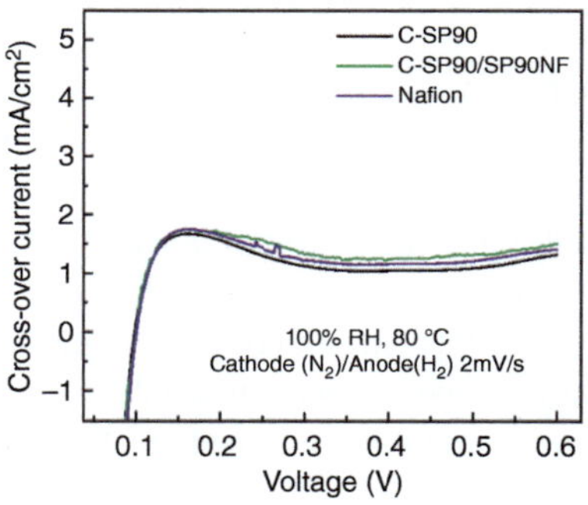

Figure 11.14 Linear sweep voltammograms for hydrogen crossover on the MEAs of C-SP90, and C-SP90/SP90NF and recast Nafion membranes.

The hydrogen permeation rates were measured using linear sweep voltammetry (LSV) at 80 °C and 100% RH. As shown in Figure 11.14, C-SP90 and Nafion membranes exhibited low hydrogen permeation currents of 1.04 and 1.15 mA/cm^2 at 0.4 V, respectively. C-SP90/SP90NF membranes displayed a slightly higher hydrogen permeation current of 1.25 mA/cm^2, attributable to their high proton conductivity. Chemical durability was further assessed through a 200-hour OCV durability test. As shown in Figure 11.13b, the OCV decay rate of C-SP90/SP90NF was only 0.92 mV/h, with the OCV stabilizing at 0.78 V. The absence of any sharp drops or abrupt changes indicates that the cross-linked composite membranes maintain good chemical stability.

11.4 Conclusion

In this chapter, we successfully prepared a series of composite membranes combining highly sulfonated SPEEK nanofibers with a sulfonated SPEEK matrix. Through thermal cross-linking in the presence of DMSO, we obtained composite membranes with a cross-linked network structure, where cross-linking occurred primarily in the SPEEK matrix while the nanofibers remained non-cross-linked. This approach significantly improved the dimensional stability and mechanical strength of the membranes, achieving a maximum tensile strength of 68 MPa and maintaining water swelling rates below 30% at 80 °C. The un-cross-linked nanofiber regions established long-range, continuous proton transport channels, enhancing proton conductivity to 200 mS/cm—outperforming Nafion (140 mS/cm) and pure SPEEK membranes (135 mS/cm). Furthermore, these composite membranes demonstrated superior fuel cell performance, with the C-SP90/SP90NF membrane achieving a peak power density of 485 mW/cm^2, 1.45 times greater than that of the pure SPEEK membrane. These findings highlight the promising potential of thermally cross-linked nanofiber composite membranes for advanced fuel cell applications.

References

1 Prykhodko, Y., Fatyeyeva, K., Hespel, L. et al. (2021). Progress in hybrid composite Nafion®-based membranes for proton exchange fuel cell application. *Chemical Engineering Journal* 409: 127329.

2 Kim, Y.S. (2021). Polymer electrolytes with high ionic concentration for fuel cells and electrolyzers. *ACS Applied Polymer Materials* 3 (3): 1250–1270.

3 Won, M., Kwon, S., and Kim, T.H. (2015). High performance blend membranes based on sulfonated poly(arylene ether sulfone) and poly(p-benzimidazole) for PEMFC applications. *Journal of Industrial and Engineering Chemistry* 29: 104–111.

4 Mikhailenko, S.D., Wang, K., Kaliaguine, S. et al. (2004). Proton conducting membranes based on cross-linked sulfonated poly(ether ether ketone) (SPEEK). *Journal of Membrane Science* 233 (1): 93–99.

5 Lei, L., Zhu, X., Xu, J. et al. (2017). Highly stable ionic-covalent cross-linked sulfonated poly(ether ether ketone) for direct methanol fuel cells. *Journal of Power Sources* 350: 41–48.

6 Han, J., Kim, K., Kim, J. et al. (2019). Cross-linked highly sulfonated poly(arylene ether sulfone) membranes prepared by in-situ casting and thiol-ene click reaction for fuel cell application. *Journal of Membrane Science* 579: 70–78.

7 Di Vona, M.L., Sgreccia, E., Licoccia, S. et al. (2009). Analysis of temperature-promoted and solvent-assisted cross-linking in sulfonated poly(ether ether ketone) (SPEEK) proton-conducting membranes. *The Journal of Physical Chemistry B* 113 (21): 7505–7512.

8 Di Vona, M.L., Alberti, G., Sgreccia, E. et al. (2012). High performance sulfonated aromatic ionomers by solvothermal macromolecular synthesis. *International Journal of Hydrogen Energy* 37 (10): 8672–8680.

9 Knauth, P., Pasquini, L., Maranesi, B. et al. (2013). Proton mobility in sulfonated polyetheretherketone (SPEEK): influence of thermal crosslinking and annealing. *Fuel Cells* 13 (1): 79–85.

10 Vinothkannan, M., Rhan Kim, A., Suk Nahm, K. et al. (2016). Ternary hybrid (SPEEK/SPVdF-HFP/GO) based membrane electrolyte for the applications of fuel cells: profile of improved mechanical strength, thermal stability and proton conductivity. *RSC Advances* 6 (110): 108851–108863.

11 Kim, A.R., Vinothkannan, M., and Yoo, D.J. (2017). Sulfonated-fluorinated copolymer blending membranes containing SPEEK for use as the electrolyte in polymer electrolyte fuel cells (PEFC). *International Journal of Hydrogen Energy* 42 (7): 4349–4365.

12 Kim, A.R., Poudel, M.B., Chu, J.Y. et al. (2023). Advanced performance and ultra-high, long-term durability of acid-base blended membranes over 900 hours containing sulfonated PEEK and quaternized poly(arylene ether sulfone) in H2/O2 fuel cells. *Composites Part B: Engineering* 254: 110558.

13 Liu, Y., Lu, Y., Haragirimana, A. et al. (2020). Immobilized phosphotungstic acid for the construction of proton exchange nanocomposite membranes with excellent stability and fuel cell performance. *International Journal of Hydrogen Energy* 45 (35): 17782–17794.

14 Zhai, S., Song, H., Jia, X. et al. (2021). Fabrication of water-insoluble phosphotungstic acid-carbon nitride nanohybrids for promoting proton transport of nanocomposite proton exchange membranes. *Journal of Power Sources* 506: 230195.

15 Yin, Z., Geng, H., Yang, P. et al. (2021). Improved proton conduction of sulfonated poly (ether ether ketone) membrane by sulfonated covalent organic framework nanosheets. *International Journal of Hydrogen Energy* 46 (52): 26550–26559.

16 Gao, J., Dong, X., Tian, Q. et al. (2023). Carbon nanotubes reinforced proton exchange membranes in fuel cells: an overview. *International Journal of Hydrogen Energy* 48 (8): 3216–3231.

17 Woo, S.H., Lee, S.Y., Yoon, Y.G. et al. (2022). Functionalized nanoclays for improved properties of composite proton exchange membranes. *Journal of Power Sources* 549: 232083.

18 Boaretti, C., Pasquini, L., Sood, R. et al. (2018). Mechanically stable nanofibrous sPEEK/Aquivion® composite membranes for fuel cell applications. *Journal of Membrane Science* 545: 66–74.

19 Liu, G., Tsen, W.C., Jang, S.C. et al. (2019). Mechanically robust and highly methanol-resistant sulfonated poly(ether ether ketone)/poly(vinylidene fluoride) nanofiber composite membranes for direct methanol fuel cells. *Journal of Membrane Science* 591: 117321.

20 Wang, L., Deng, N., Liang, Y. et al. (2020). Metal-organic framework anchored sulfonated poly(ether sulfone) nanofibers as highly conductive channels for hybrid proton exchange membranes. *Journal of Power Sources* 450: 227592.

21 Yuan, Q., Fu, Z., Wang, Y. et al. (2020). Coaxial electrospun sulfonated poly (ether ether ketone) proton exchange membrane for conductivity-strength balance. *Journal of Membrane Science* 595: 117516.

22 Zhu, B., Sui, Y., Wei, P. et al. (2021). NH_2-UiO-66 coated fibers to balance the excellent proton conduction efficiency and significant dimensional stability of proton exchange membrane. *Journal of Membrane Science* 628: 119214.

23 Li, L., Liu, X., Guo, Y. et al. (2022). Reasonable construction of proton conducting channel via biomimetic caterpillar-like alumina fiber to improve the properties of its composite proton exchange membrane. *International Journal of Hydrogen Energy* 47 (69): 29915–29924.

24 Ibaraki, T., Tanaka, M., and Kawakami, H. (2019). Fast surface proton conduction on acid-doped polymer nanofibers in polymer electrolyte composite membranes. *Electrochimica Acta* 296: 1042–1048.

25 Choi, J., Lee, K.M., Wycisk, R. et al. (2008). Nanofiber network ion-exchange membranes. *Macromolecules* 41 (13): 4569–4572.

26 Choi, J., Lee, K.M., Wycisk, R. et al. (2010). Sulfonated polysulfone/POSS nanofiber composite membranes for PEM fuel cells. *Journal of the Electrochemical Society* 157 (6): B914.

27 Ballengee, J.B., Haugen, G.M., Hamrock, S.J. et al. (2013). Properties and fuel cell performance of a nanofiber composite membrane with 660 equivalent weight perfluorosulfonic acid. *Journal of the Electrochemical Society* 160 (4): F429–F435.

28 Han, D., Zhang, J., Wang, C. et al. (2024). Construction of proton transport channels in cross-linked composite membrane with highly sulfonated SPEEK nanofibers. *International Journal of Hydrogen Energy* 52: 11–24. https://doi.org/10.1016/j.ijhydene.2023.07.285.

29 Banerjee, S. and Kar, K.K. (2017). Impact of degree of sulfonation on microstructure, thermal, thermomechanical and physicochemical properties of sulfonated poly ether ether ketone. *Polymer* 109: 176–186.

30 Beyraghi, F., Mirfarsi, S.H., Rowshanzamir, S. et al. (2020). Optimal thermal treatment conditions for durability improvement of highly sulfonated poly(ether ether ketone) membrane for polymer electrolyte fuel cell applications. *International Journal of Hydrogen Energy* 45 (24): 13441–13458.

31 Sun, F., Qin, L.L., Zhou, J. et al. (2020). Friedel-Crafts self-crosslinking of sulfonated poly(etheretherketone) composite proton exchange membrane doped with phosphotungstic acid and carbon-based nanomaterials for fuel cell applications. *Journal of Membrane Science* 611: 118381.

32 Yin, C., Li, J., Zhou, Y. et al. (2018). Enhancement in proton conductivity and thermal stability in nafion membranes induced by incorporation of sulfonated carbon nanotubes. *ACS Applied Materials & Interfaces* 10 (16): 14026–14035.

33 Borup, R., Meyers, J., Pivovar, B. et al. (2007). Scientific aspects of polymer electrolyte fuel cell durability and degradation. *Chemical Reviews* 107 (10): 3904–3951.

12

Polydopamine-Modified Halloysite Nanotube/Sulfonated Poly(Ether Ether Ketone) Cross-Linked Composite Membrane

12.1 Introduction

The properties of sulfonated aromatic hydrocarbon polymers are strongly influenced by their degree of sulfonation. Polymers with a high sulfonation degree contain numerous sulfonic acid groups, providing abundant proton transport sites and yielding high proton conductivity. However, this also leads to excessive water uptake (WU) and swelling, which compromise the membrane's mechanical stability. Conversely, polymers with a lower degree of sulfonation exhibit reduced water absorption and swelling, resulting in better mechanical stability. However, their lower sulfonic acid group content often hinders the formation of a continuous hydrophilic network, thus limiting proton conductivity [1]. As a result, the performance of proton exchange membranes (PEMs) often involves a trade-off between proton conductivity and mechanical stability due to variations in sulfonation degree.

To address this issue, researchers have pursued various strategies. Some have focused on synthesizing novel polymer structures through molecular design, while others have employed cross-linking methods—such as covalent and ionic bonding—to stabilize the polymer matrix [2]. These approaches, though effective, are often complex. A simpler and more widely studied alternative is the incorporation of inorganic nanoparticles and other additives into sulfonated polymers to produce hybrid membranes, thereby enhancing the overall performance of PEMs.

The preparation of hybrid membranes typically involves blending inorganic nanofillers with sulfonated polymer solutions. This approach improves the thermal and chemical stability of the membrane, enhances mechanical strength, and reduces fuel permeability. Various inorganic fillers have been reported in the literature, including poly(sesquicarbazone), porous silica [3–6], eclogite nanotubes, carbon nanotubes, graphene oxide, carbon nitride nanosheets [7, 8], alumina, tin oxide nanofibers, and metal-organic framework porous materials [9, 10].

However, many of these inorganic fillers lack inherent proton conductivity. Additionally, the interfaces between the fillers and the polymer matrix are often weak and poorly compatible. This can lead to difficulties in achieving uniform

Electrospinning for Proton Exchange Membrane Fuel Cells, First Edition.
Yong Liu and Mohideen Meerasahib Mohamedazeem.

dispersion, with the fillers tending to agglomerate rather than form a consistent membrane structure. Such aggregation can obstruct proton transport channels, ultimately reducing both proton conductivity and mechanical strength in the resulting hybrid membranes [11, 12]. To overcome these challenges, the surface modification of inorganic fillers is typically required. This enhances their compatibility with the polymer matrix and enables the incorporation of proton-conducting groups, improving the performance of the hybrid membranes.

Incorporating nanoparticles into the structure of composite or hybrid membranes can effectively balance the challenges associated with high sulfonation levels, such as excessive water absorption and swelling. For example, nanoparticles used as fillers can occupy the membrane's structural matrix, hindering the free diffusion of water molecules. This controlled restriction reduces the membrane's overall WU and, at the same time, enhances its mechanical strength and stability. Additionally, the presence of nanoparticles helps optimize polymer chain arrangement, thereby lowering the swelling rate (SR) [13–15].

Gang et al. [16] demonstrated this approach by incorporating g-C_3N_4 nanosheets into a sulfonated poly(ether ether ketone) (SPEEK) matrix. At higher g-C_3N_4 content, stronger interactions between the nanosheets and the polymer restricted the mobility of the SPEEK backbone. This produced a denser polymer matrix that effectively limited swelling. While many fillers lack inherent proton conductivity and may show poor compatibility with the polymer matrix, functionalized fillers can address these shortcomings. By modifying the surface of these fillers with functional groups such as $-SO_3H$, $-PO_3H_2$, or $-NH_2$, researchers have introduced proton-conducting pathways and enhanced compatibility with the polymer backbone. Liu et al. [17] achieved a 45% increase in proton conductivity over pristine membranes by incorporating highly sulfonated carbon nanoparticles, while Rao et al. [18] developed co-doped carbon nanotubes with phosphoric acid-functionalized nanotubes and graphene oxide, achieving a proton conductivity of 282 mS/cm at 90 °C and 95% relative humidity (RH).

Halloysite nanotubes (HNTs), characterized by their hollow tubular structures and large specific surface areas, hold promise for PEMs. However, as HNTs themselves do not conduct protons, they require modification. Surface modifications, such as polymer coatings, introduce hydrophilic amine or sulfonic acid groups to HNTs, improving their dispersion in polymer films and enhancing proton conductivity. For instance, Liu et al. [7] prepared sulfonated halloysite nanotubes (SHNTs) by coating them with sulfonated polystyrene. When combined with SPEEK, the resulting composite membrane exhibited improved compatibility, interfacial adhesion, and a dense, homogeneous structure. Moreover, SHNTs encouraged phase separation within the membrane, further enhancing proton transport.

Polydopamine (PDA) is widely used for material surface modifications due to its adhesive properties and the versatility of its functional groups. PDA coatings on HNTs can introduce amine groups, which facilitate the loading of heteropolyacids like phosphotungstic acid or the formation of acid–base pairs with the matrix's sulfonic acid groups. This approach can significantly enhance proton transport in composite membranes.

In this study, dopamine was used to prepare HNTs via a self-polymerization process. These dopamine-modified halloysite nanotubes (DHNTs) were blended with highly sulfonated SPEEK to form SP/DHNTs nanocomposite membranes. Subsequent thermal cross-linking, catalyzed by dimethylsulfoxide (DMSO), minimized water absorption and membrane swelling. At the interface, abundant amine groups on DHNT surfaces formed acid–base pairs with the SPEEK sulfonic acid groups. The resulting acid–base pairs facilitated proton transport via the Grotthuss mechanism, creating low-energy proton transport pathways. Additionally, these interfacial interactions restricted water penetration, further reducing swelling. As a result, SP/DHNTs membranes required shorter cross-linking times yet exhibited water absorption and SRs comparable to those of pure SPEEK membranes with longer cross-linking durations. This allowed the retention of more sulfonic acid groups within the composite membrane's matrix. The high aspect ratio of DHNTs allowed the formation of long-range, continuous proton transport channels at the interface with SPEEK. These channels interconnected isolated hydrated ion regions within the matrix, enhancing proton transport throughout the membrane. Consequently, the SP/DHNTs composite membranes exhibited both improved mechanical stability and superior proton conductivity.

12.2 Materials Preparation

12.2.1 Polydopamine Coating of Halloysite Nanotubes

In 100 mL of deionized water, 0.3 g of HNT powder was ultrasonically dispersed. Using a cell disruptor, the dispersion was treated under ice bath conditions at 80% power for 20 minutes, with 5-second pulses alternating with 2-second rests, to ensure thorough dispersion of the HNTs. Subsequently, 0.3 g of dopamine hydrochloride was added to the dispersion, and the pH was adjusted to 8.5 by adding Tris. The mixture was stirred vigorously and maintained at 30 °C for 24 hours. After the reaction, the resulting solid was separated via vacuum filtration. To remove any residual Tris and unreacted dopamine monomers, the product was alternately washed several times with deionized water and ethanol. Finally, the DHNTs were dried in a vacuum oven at 60 °C.

12.2.2 Sulfonation of Poly(Ether Ether Ketone)

Poly(ether ether ketone) (PEEK) powder was vacuum dried at 80 °C for 24 hours before use. Dried PEEK measuring 10 g was added to 100 mL of concentrated sulfuric acid, stirred at 20 °C, and left to react for 12 hours. The reaction temperature was then raised to 50 °C and continued for another 10 hours. To terminate the reaction, the solution was immediately cooled with cold water. The reaction mixture was slowly poured into a large volume of ice water, causing precipitation of the fibrous SPEEK product. The precipitate was washed multiple times with deionized water until neutral, then dried in a vacuum oven at 60 °C for 24 hours. The degree of sulfonation of SPEEK was determined to be 88% by ^{1}H nuclear magnetic resonance (NMR).

12.2.3 Preparation of Thermally Cross-Linked Composite Membranes

A certain amount of DHNTs was ultrasonically dispersed in DMSO before being mixed with the SPEEK casting solution (using DMSO as the solvent). The resulting mixture was both magnetically stirred and ultrasonically treated to ensure homogeneity. After allowing the solution to rest for 2 hours to remove any bubbles, it was cast onto a clean glass plate and dried at 60 °C for 24 hours. This process yielded composite membranes with varying DHNTs contents, designated SP/DHNTs-x, where x (4, 8, or 12) represents the mass percentage of DHNTs in the membrane.

These composite membranes were then thermally cross-linked in two steps: pretreatment at 120 °C for 6 hours, followed by heating to 180 °C for a specified duration to reduce membrane swelling. After cross-linking, the membranes were acidified in 1 M hydrogen chloride (HCl) solution for 24 hours and washed until neutral. For comparison, a pure SPEEK cross-linked membrane was prepared under identical conditions.

12.3 Results and Discussion

12.3.1 Characterization of Dopamine-Modified Halloysite Nanotubes

The polydopamine coating process for HNTs is depicted in Figure 12.1a. Initially, HNT powder is dispersed in water, and the coating process is carried out at pH 8.5 and 30 °C. During this step, dopamine undergoes oxidative self-polymerization, adhering to the nanotube surfaces. This results in the formation of a polydopamine layer that isolates individual HNTs, preventing aggregation. Consequently, the dispersion of HNTs in the sulfonated poly(ether ether ketone) (SPEEK) solution is significantly improved. Additionally, the polydopamine coating introduces abundant amine groups, which form acid–base pairs with sulfonic acid groups, thus establishing long-range proton transport channels at the interface between DHNTs and the SPEEK matrix. As shown in Figure 12.1b, the polydopamine coating induces a visible color change, turning the white HNT powder black—an indication of dopamine's oxidative polymerization into a black polydopamine layer.

The morphology of both HNTs and DHNTs was examined using scanning electron microscopy (SEM) and transmission electron microscopy (TEM) (Figure 12.2). Following dopamine reaction, the tubular structure of the HNTs is retained, but the initially smooth surfaces become rough and textured, indicating the presence of a polydopamine layer. This surface modification enhances the hydrophilicity of DHNTs, leading to better dispersion within the SPEEK solution and minimizing aggregation. SEM images confirm that dopamine predominantly polymerizes on the HNT surface, with minimal formation of dopamine microspheres. High-resolution TEM images further reveal that the polydopamine layer on the DHNT surface is approximately 9 nm thick, providing a robust, uniform coating.

The elemental composition of the DHNTs was analyzed using SEM-based energy dispersive X-ray spectroscopy (SEM-EDS) and elemental mapping (Figure 12.3). The mapping results show a uniform distribution of carbon (C), nitrogen (N),

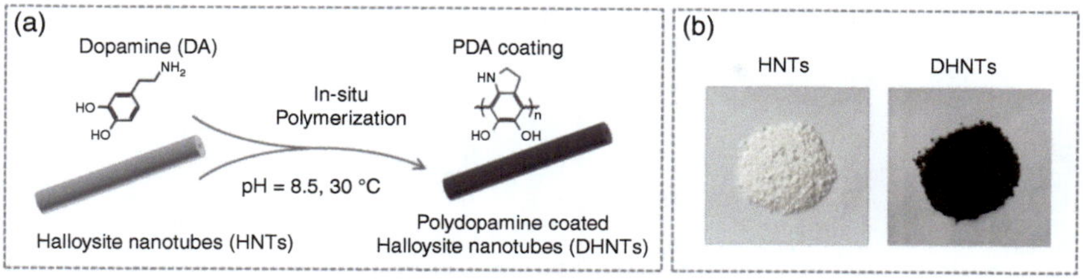

Figure 12.1 (a) Schematic representation of the synthesis of DHNTs nanohybrids and (b) visual appearance of HNT and DHNTs powders.

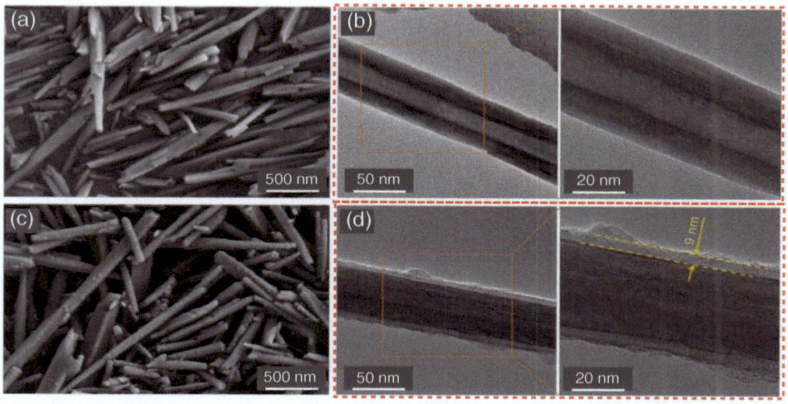

Figure 12.2 SEM and TEM images of (a, b) HNTs, (c, d) DHNTs.

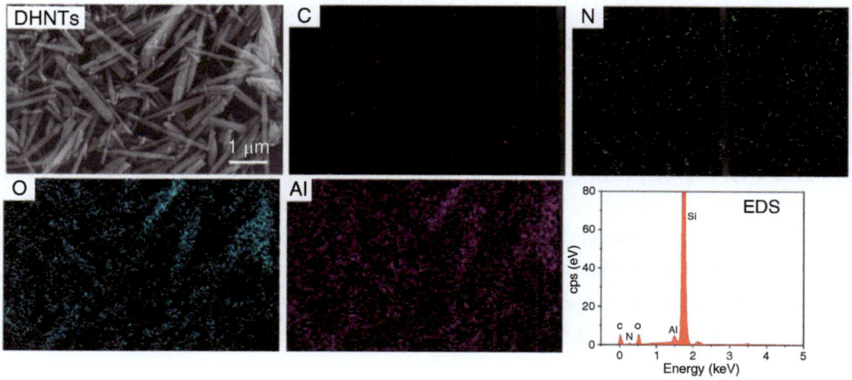

Figure 12.3 Elemental mapping of C, N, O and Al in DHNTs nanohybrids with a SEM-EDS.

and oxygen (O)—key elements within the polydopamine coating. This uniform distribution confirms that the polydopamine layer is evenly applied across the HNT surface. The EDS maps further corroborate the presence of these elements, providing additional evidence for the successful formation of the polydopamine coating.

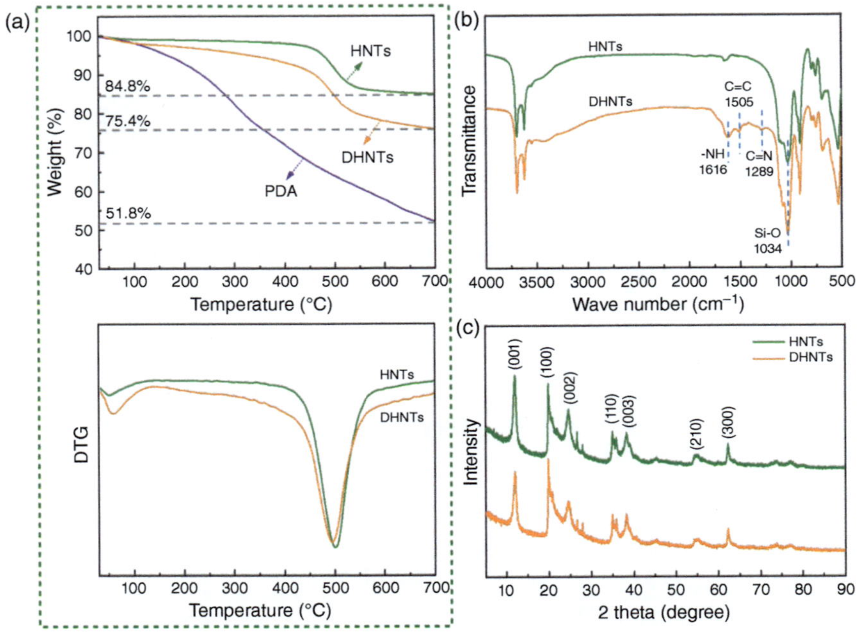

Figure 12.4 (a) TGA curves, (b) FT-IR spectra and (c) XRD patterns of HNTs and DHNTs.

To quantify the mass of the polydopamine coating on DHNTs, thermogravimetric analysis (TGA) was conducted on HNTs, DHNTs, and polydopamine particles, as illustrated in Figure 12.4a. Both HNTs and DHNTs show similar thermal degradation behavior; however, DHNTs exhibit slightly higher weight loss between 300 and 700 °C, which is attributed to the thermal degradation of the polydopamine coating. From the TGA data, the polydopamine coating layer on DHNTs is calculated to account for approximately 28.5% of the total mass.

Figure 12.4b displays the Fourier transform infrared (FT-IR) spectra of both HNTs and DHNTs. A prominent absorption peak at 1034 cm^{-1} corresponds to the stretching vibration of the Si–O bonds in halloysite. In the spectrum of DHNTs, additional peaks are observed at 1616, 1505, and 1289 cm^{-1}, which are attributed to the –NH bending vibration, aromatic ring benzene ring (C=C) stretching vibration, and carbon–nitrogen (C–N) bond stretching vibration of polydopamine, respectively. These features provide clear evidence of the successful polydopamine coating on the HNT surface.

The X-ray diffraction (XRD) patterns of HNTs and DHNTs are shown in Figure 12.4c. According to Joint Committee on Powder Diffraction Standards (JCPDS) card No. 29-1487, the strong diffraction peaks at $2\theta = 12.2°$, 20.1°, 24.7°, 35.1°, 38.5°, 54.7°, and 62.5°, as well as the weaker peaks at 73.8° and 77.1°, corresponding to the crystalline structure of pure HNTs. The XRD patterns of DHNTs remain essentially unchanged compared to pure HNTs, indicating that the crystalline structure of HNTs is preserved after the polydopamine coating. This suggests that the polydopamine layer is amorphous and does not interfere with the inherent crystallinity of the HNTs.

12.3.2 Discussion on the Sulfonation Process of Poly(Ether Ether Ketone)

Sulfonation of PEEK typically relies on concentrated sulfuric acid as the sulfonating agent. Unlike stronger agents such as fuming sulfuric acid or chlorosulfonation, concentrated sulfuric acid selectively undergoes electrophilic substitution on the electron-rich benzene rings of PEEK, typically introducing only one sulfonic acid group per active ring. Because sulfonation reactions are reversible, an excess of concentrated sulfuric acid is needed. Additionally, it is crucial to fully dry the PEEK and conduct the reaction under a nitrogen atmosphere to suppress reverse sulfonation.

PEEK is an aromatic polymer with multiple benzene rings, whose rigid structure and aromaticity cause the polymer chains to stack closely, forming strong intramolecular π–π interactions. This results in a semicrystalline structure that provides excellent mechanical properties and chemical stability. However, these same characteristics also make PEEK a high-melting, insoluble polymer, creating challenges in processing. Only a limited number of solvents can dissolve PEEK, and the sulfonation process is closely tied to dissolution. Initially, sulfonation occurs on the surface of PEEK particles, disrupting the orderly stacking of polymer chains. This disruption allows the concentrated sulfuric acid to gradually penetrate the particles, causing them to swell and dissolve. The process is inherently slow, typically requiring several hours, which leads to variations in sulfonation degree between particles that dissolve early and those that dissolve later.

Some studies suggest that using methanesulfonic acid as a preliminary solvent, followed by the addition of concentrated sulfuric acid, results in SPEEK with a more uniform sulfonation degree [20]. However, in our experiment, we found that the PEEK obtained from Vitrex did not dissolve in methanesulfonic acid, even at elevated temperatures of up to 80 °C. This may be attributed to the higher crystallinity of the PEEK material, which limits the ability of methanesulfonic acid to penetrate the polymer and facilitate dissolution. Fortunately, sulfonation progresses slowly at room temperature [208], so differences in sulfonation degree due to dissolution time were not pronounced during the shorter dissolution process.

To address this, we ground the PEEK into fine powder and used mechanical stirring under strong shear forces to enhance dissolution in concentrated sulfuric acid. By accelerating the dissolution process, we minimized differences in sulfonation degree. Once the PEEK was fully dissolved, the temperature was raised to 50 °C, and the reaction was allowed to proceed for 10 hours, yielding a high-sulfonation SPEEK. Using ^{1}H NMR (Figure 12.5), we calculated the sulfonation degree of SPEEK to be 88%, corresponding to an ion exchange capacity (IEC) of 2.45 mmol/g.

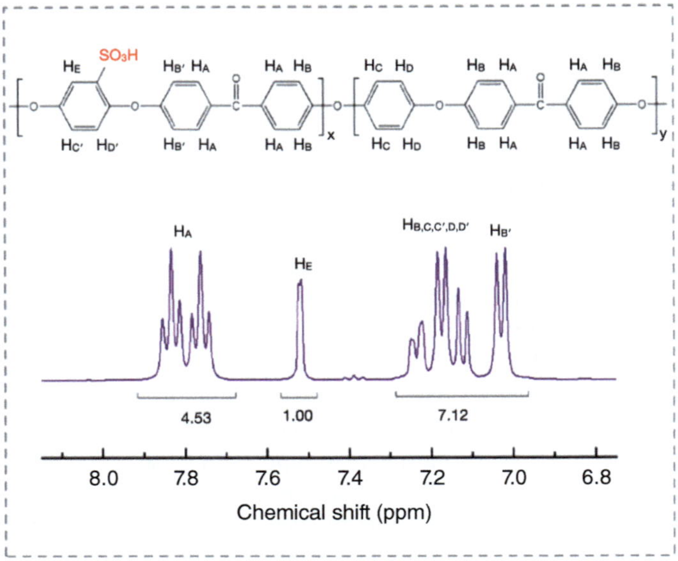

Figure 12.5 Structure and ^{1}H NMR spectra of SPEEK.

12.3.3 Morphological Characterization of Composite Membranes

Figure 12.6 displays the cross-sectional SEM images of pure SPEEK membranes and SP/DHNTs composite membranes. The pure SPEEK membrane exhibits a relatively smooth and flat fractured surface, as seen in Figure 12.6a,d. In contrast, the SP/DHNTs composite membranes show rougher fractured surfaces due to the incorporation of DHNT nanorods with high aspect ratios, as illustrated in Figure 12.6b,c,g. These images also indicate that the DHNTs are well dispersed throughout the SPEEK matrix, with no noticeable aggregation.

In comparison, the composite membranes made with uncoated HNTs and SPEEK exhibit significant aggregation, as shown in Figure 12.6k,l. This aggregation is attributed to the lack of surface functionalization on the HNTs, which reduces their compatibility with the polymer matrix. Conversely, the surface functional groups (–NH and –OH) present on DHNTs can form acid–base pairs or hydrogen bonds with the sulfonic acid groups in SPEEK. This improved interfacial compatibility prevents aggregation and results in a more uniform distribution.

Figure 12.7 further supports this observation, showing SEM mapping and elemental distribution for the SP/DHNTs-8 composite membrane. The mapping results confirm that DHNTs are uniformly distributed within the SPEEK matrix, ensuring consistent composite morphology and potentially enhancing the membrane's overall performance.

12.3.4 Chemical and Physical Properties of the Composite Membranes

Figure 12.8a presents the FT-IR spectra of pure SPEEK membranes and SP/DHNTs composite membranes. All membranes show characteristic peaks typical of SPEEK,

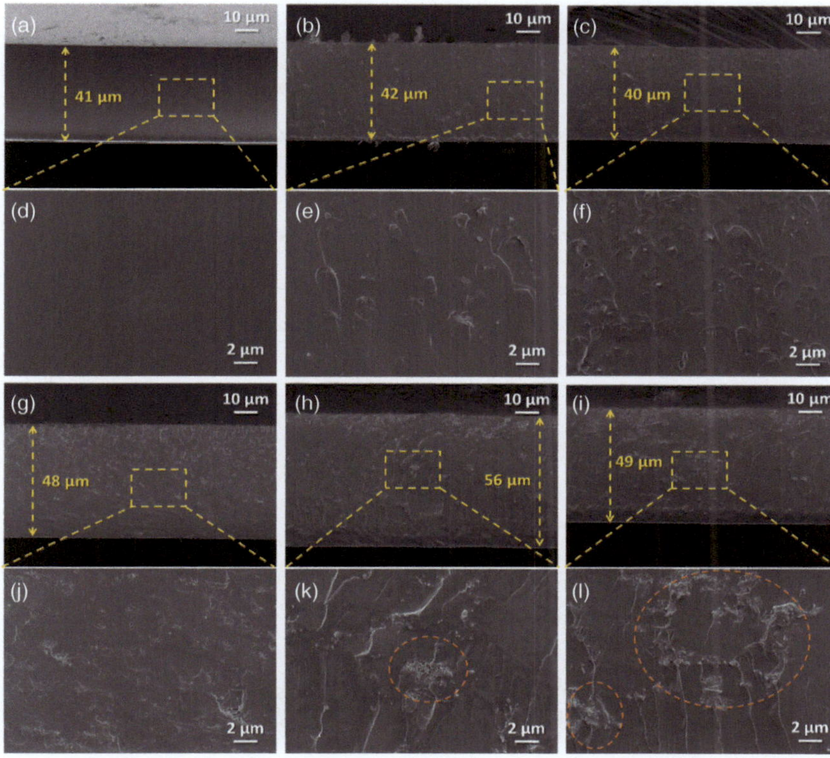

Figure 12.6 SEM images of the cross-section of (a, d) pristine SPEEK, (b, e) SP/DHNTs-4, (c, f) SP/DHNTs-8, (g, j) SP/DHNTs-12, (h, k) SP/HNTs-8 and (i, l) SP/HNTs-12 composite membranes.

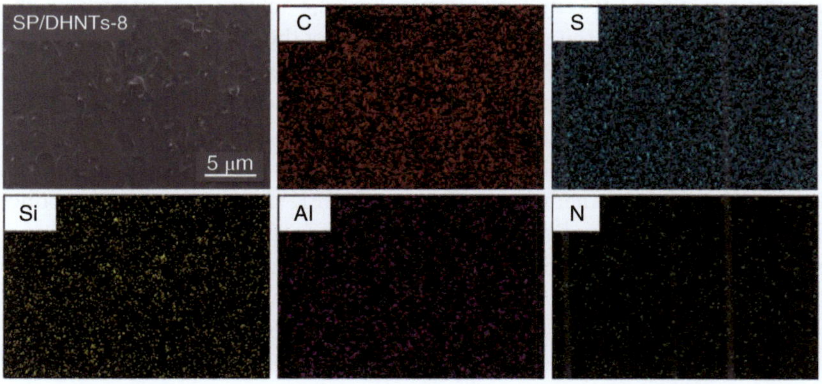

Figure 12.7 Elemental mapping of C, S, Si, Al, and N in SP/DHNTs-8 composite membrane by SEM-EDX (energy-dispersive X-ray spectroscopy).

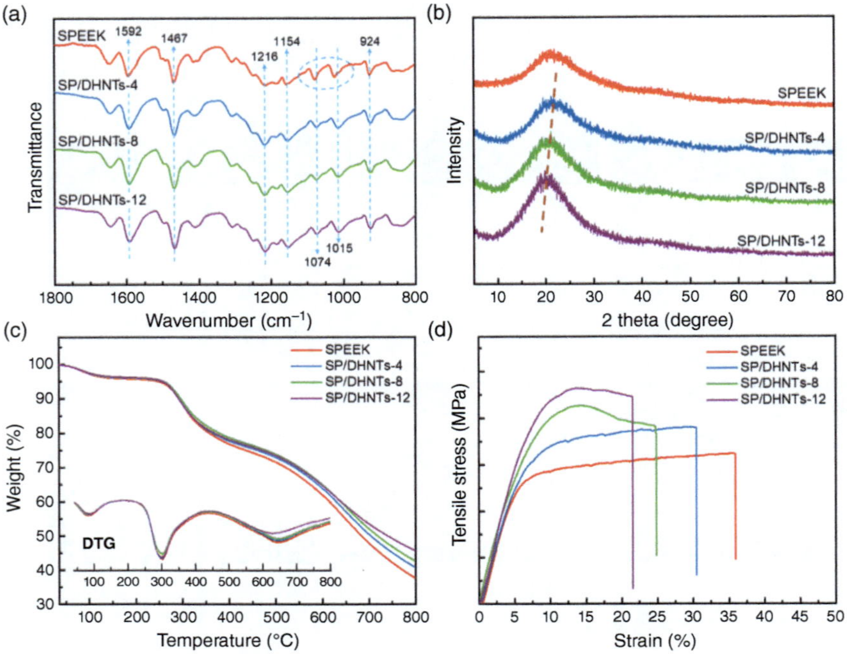

Figure 12.8 (a) FT-IR spectra, (b) XRD patterns, (c) TGA curves and (d) stress–strain curves of pristine SPEEK, SP/DHNTs-4, SP/DHNTs-8, and SP/DHNTs-12 composite membranes.

including the carbonyl (C=O) stretch at 1645 cm^{-1}, the C=C stretch at 1592 cm^{-1}, and the carbon–carbon single bond (C–C) stretching vibration at 1467 cm^{-1}. Peaks associated with the sulfonic acid groups in SPEEK appear at 1076 and 1010 cm^{-1}, corresponding to symmetric and asymmetric S=O stretching vibrations. Additionally, a peak at 1216 cm^{-1} represents the 1,2,4-substitution pattern on the benzene ring. As shown in the spectra, the intensity of the $-SO_3H$ peaks is lower in the highly cross-linked pure SPEEK membrane compared to the low-cross-linked SP/DHNTs composite membranes, reflecting the reduced content of free sulfonic acid groups in the pure SPEEK membrane.

Figure 12.8b displays the XRD patterns of the pure SPEEK membranes and the SP/DHNTs composite membranes. In pure SPEEK membranes, the introduction of numerous sulfonic acid groups disrupts the regular stacking of polymer chains, resulting in an amorphous structure characterized by a broad peak near $2\theta = 21°$. In comparison, the composite membranes containing DHNTs exhibit a more pronounced diffraction peak at $2\theta = 20.1°$, corresponding to the crystalline structure of HNTs. This causes a slight leftward shift in the diffraction peak of the SP/DHNTs composite membranes relative to the pure SPEEK membranes. Moreover, as the DHNT content increases, the intensity of this diffraction peak becomes stronger, indicating that the structural contribution of the DHNTs becomes more prominent.

Table 12.1 Mechanical properties of as-prepared membranes.

Membranes	Tensile strength (MPa)	Young's modulus (MPa)	Elongation at break (%)
Pristine SPEEK	64.9 ± 4.2	939 ± 53	35.9 ± 5.1
SP/DHNTs-4	76.2 ± 3.1	897 ± 68	30.5 ± 3.8
SP/DHNTs-8	74.7 ± 2.2	1083 ± 76	23.6 ± 3.5
SP/DHNTs-12	72.1 ± 3.4	1012 ± 87	21.5 ± 2.5

To achieve long-term operation, PEMs must demonstrate robust thermal and mechanical stability. Both pure SPEEK membranes and SP/DHNTs composite membranes exhibit three distinct thermal degradation stages, as shown in Figure 12.8c. The first stage, occurring below 150 °C, is primarily attributed to the evaporation of water associated with the sulfonic acid groups in SPEEK. The second stage, between 250 and 400 °C, corresponds to the decomposition of $-SO_3H$ groups, while the third stage, above 450 °C, reflects the degradation of the PEEK backbone. The inclusion of DHNTs in the composite membrane does not significantly alter its thermal stability, as HNTs themselves are thermally stable and their content in the composite membrane is relatively low. Furthermore, the thermal decomposition of polydopamine is not distinctly visible in the TGA curve. As the DHNT content increases, the final residual mass of the composite membrane also rises.

Mechanical performance data for both membrane types are shown in Figure 12.8d and summarized in Table 12.1. The addition of DHNT nanotubes enhances the tensile strength of the composite membrane compared to the pure SPEEK membrane, thanks to the reinforcement provided by the nanoparticles [209]. For instance, the original SPEEK membrane exhibits a tensile stress of 64.9 MPa, a Young's modulus of 939 MPa, and an elongation at break of 35.9%. In the SP/DHNTs composite membranes, both tensile stress and Young's modulus increase, reflecting improved mechanical robustness. However, the elongation at break decreases as the DHNT content rises. When the DHNT concentration becomes too high, excessive nanoparticle aggregation can introduce structural defects. These defects weaken the membrane's mechanical integrity, increasing its susceptibility to fracture under stress.

12.3.5 Ion Exchange Capacities, Water Absorption Rate, and Swelling Rate

The IEC reflects the number of moles of sulfonic acid groups per gram of membrane material. It is a key factor influencing the membrane's water absorption, SR, and proton conductivity. As shown in Table 12.2, the initial sulfonation degree of pure SPEEK is 88%, resulting in an IEC of 2.45 mmol/g. This high sulfonation level, while beneficial for proton conductivity, leads to significant swelling and poor mechanical stability. A simple thermal cross-linking treatment can improve stability but consumes sulfonic acid groups. As a result, the IEC of the membrane decreases after cross-linking, as confirmed by acid–base titration.

Table 12.2 The IEC of pristine composite SPEEK membranes.

Membranes	Thermal treatment time at 180 °C (h)	Initial IEC[a)] (mmol/g)	IEC[b)] (mmol/g)
SPEEK	12	2.45	1.95 ± 0.03
SP/DHNTs-4	10	2.35	1.91 ± 0.04
SP/DHNTs-8	8	2.26	1.92 ± 0.03
SP/DHNTs-12	6	2.16	1.89 ± 0.05

a) Calculated IEC values.
b) Determined by titration.

In SP/DHNTs composite membranes, two primary factors affect the IEC value: (i) the amount of DHNTs incorporated and (ii) the cross-linking duration. Since DHNTs lack sulfonic acid groups, increasing their concentration in the composite reduces the overall density of $-SO_3H$ groups, thereby lowering the IEC. For instance, when the DHNT content ranges from 0 to 12 wt%, the IEC value of the composite membrane theoretically decreases from 2.45 to 2.26 mmol/g. Furthermore, the amine groups in the polydopamine-coated HNTs interact with the $-SO_3H$ groups in SPEEK, leading to additional consumption of sulfonic acid groups and further reducing the IEC.

Thermal cross-linking, catalyzed by DMSO, also consumes $-SO_3H$ groups as it forms sulfone cross-links within the SPEEK matrix. The longer the cross-linking time, the higher the degree of cross-linking, which results in fewer available $-SO_3H$ groups [178]. This leads to a lower IEC, reduced water absorption, and a lower SR, all of which can impact proton conductivity. Therefore, it is critical to carefully control the cross-linking time in SP/DHNTs membranes, particularly as DHNT content increases. For example, when DHNTs are present at 0–12 wt%, the cross-linking time should be reduced from 12 hours to 6 hours to maintain a balance between stability and proton conductivity. Acid–base titration is then used to determine the final IEC values of the cross-linked membranes.

To ensure effective proton transport within PEMs, a carefully balanced WU is essential. Sulfonic acid groups within the membrane attract water, creating hydrogen-bond networks that facilitate proton conduction through both Grotthuss and Vehicle mechanisms. However, excessive WU can lead to a high SR, undermining the membrane's dimensional stability and mechanical strength. Overabsorption of water may also dilute the H^+ concentration, reducing proton conductivity.

In sulfonated aromatic polymer membranes, the content of $-SO_3H$ groups—reflected by the IEC—is the primary determinant of water absorption. Incorporating DHNTs into SPEEK and thermally cross-linking the SPEEK matrix can both reduce water absorption and improve membrane stability. As shown in Figure 12.9a, a thermally cross-linked pure SPEEK membrane exhibits a WU of 58.2% and an SR of 17.6% at 25 °C. In contrast, an un-cross-linked SPEEK membrane with 88% sulfonation severely swells and loses its structural integrity upon immersion in water at 25 °C, indicating the necessity of cross-linking. The thermal cross-linking process

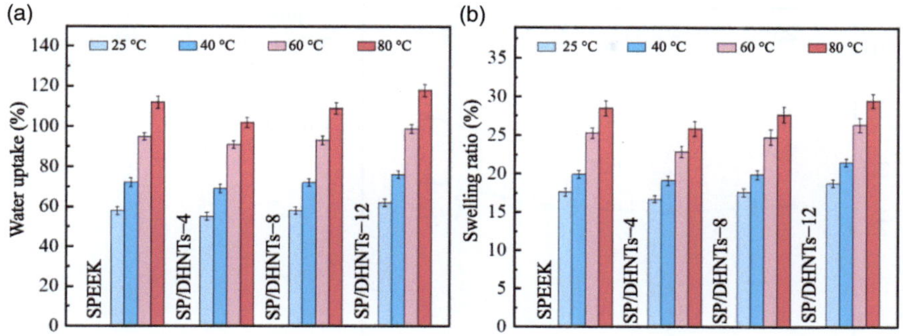

Figure 12.9 (a) Water uptake and (b) swelling ratio of composite membranes at various temperatures.

creates sulfone covalent bonds between polymer chains, establishing a robust three-dimensional network. As cross-linking density increases, the membrane becomes more compact and structurally stable, reducing the free volume available for water absorption.

The formation of cross-linked bonds also restricts the mobility of polymer chains, thereby decreasing water permeability and further limiting both WU and SR. However, as Figure 12.9a,b demonstrates, higher temperatures lead to increased WU and SR. For example, the WU of a cross-linked pure SPEEK membrane rises from 71.8% at 40 °C to 112.3% at 80 °C, while the SR increases from 19.9% to 28.4%. This trend arises because elevated temperatures relax the polymer chains, weaken molecular entanglements, and enlarge free volume, thereby allowing more water molecules to penetrate the membrane [187].

The addition of DHNTs to SPEEK for composite membrane preparation reduces both the WU and SR of the membrane. However, DHNTs alone do not provide sufficient dimensional stability. Without additional stabilization, the composite membrane continues to swell excessively and may dissolve in water. Therefore, thermal cross-linking remains necessary to achieve stability, although the cross-linking time can be shortened. By reducing the duration of thermal cross-linking, it is possible to limit the SR while retaining more sulfonic acid groups within the composite membrane, thereby maintaining better proton conductivity. As shown in Figure 12.9b, all membranes exhibit a SR below 30% after immersion in water at 80 °C, regardless of cross-linking time. For example, the SP/DHNTs-12 composite membrane, cross-linked for only 6 hours, achieves WU and SR values (118.1% and 29.5%, respectively) similar to those of a pure SPEEK membrane cross-linked for 12 hours (112.3% and 28.4%, respectively). The observed decrease in WU and SR for the composite membranes containing DHNTs is attributable to several factors:

1) **Reduced Hydrophilicity**: DHNTs are less hydrophilic than highly sulfonated SPEEK. Incorporating DHNTs reduces the availability of spaces within the polymer matrix for water molecules, thus limiting WU.
2) **Restricted Polymer Chain Mobility**: The inclusion of DHNTs nanofillers constrains the mobility of the SPEEK polymer chains. This restricted mobility

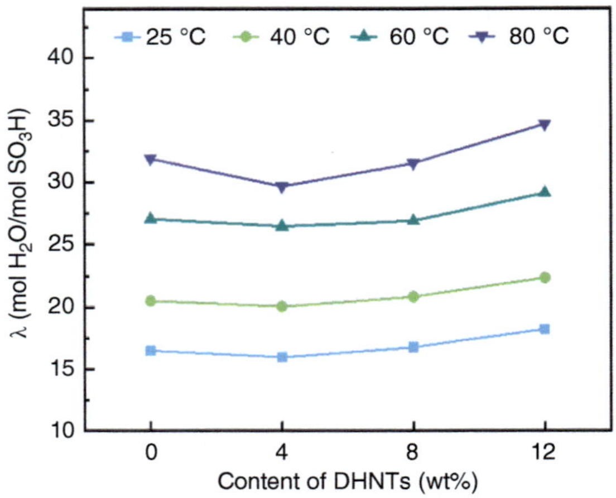

Figure 12.10 The number of water molecules adsorbed per unit of acidic group at different temperatures of the membranes.

reduces the permeability of the membrane to water molecules, resulting in lower water absorption and swelling.

3) **Ionic Cross-Linking Effect**: Amine groups on the surface of DHNTs can form ionic bonds with the sulfonic acid groups in SPEEK. These ionic interactions create a cross-linking effect that further minimizes swelling.
4) **Reinforced Three-Dimensional Network**: During thermal cross-linking, DHNTs are embedded into the SPEEK matrix's cross-linked network. These rigid nanorods, with their high aspect ratio, act as structural anchors that enhance the robustness of the three-dimensional network. As a result, the membrane becomes more dimensionally stable, with significantly reduced WU and SRs.

Proton conductivity, which depends on effective water absorption, can be evaluated through the λ value, defined as the number of water molecules per sulfonic acid group. The λ value, calculated using WU and IEC values, offers insight into how water absorption affects proton conduction. A low λ value indicates insufficient WU, impeding the formation of a continuous hydrogen-bond network between hydrated ions and reducing proton conductivity. Conversely, a high λ value can dilute the H^+ concentration, also decreasing proton conductivity. As shown in Figure 12.10, the λ values of pure SPEEK and SP/DHNTs composite membranes range from 16.5 to 18.2 at 25 °C and increase to 29.6–34.7 at 80 °C. These λ values confirm that the prepared membranes absorb sufficient water to support efficient proton conduction.

12.3.6 Proton Conductivity

Proton conductivity is the key performance indicator for PEMs, directly determining the efficiency of fuel cells. As shown in Figure 12.11a, the proton conductivities of pure SPEEK cross-linked membranes and SP/DHNTs cross-linked composite membranes were measured at various temperatures. Proton conductivity increases as temperature rises, which is attributed to enhanced polymer chain mobility and increased water molecule activity. This temperature dependence is consistent with the Arrhenius equation.

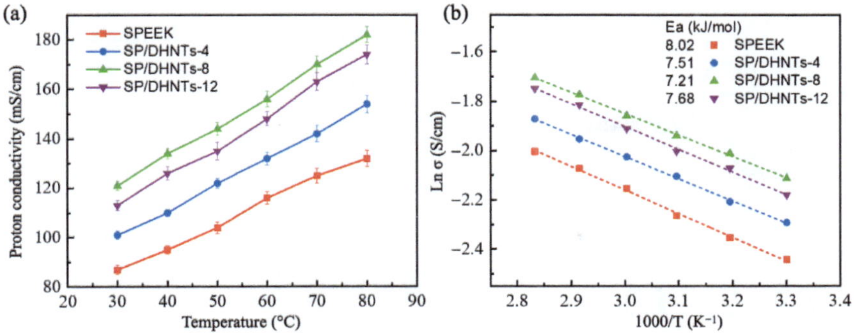

Figure 12.11 (a) Temperature dependence of proton conductivity and (b) Arrhenius plots of pristine SPEEK and SP/DHNTs composite membranes.

The SP/DHNTs cross-linked composite membranes demonstrate significantly higher proton conductivities than do pure SPEEK cross-linked membranes. For instance, at 80 °C and 100% RH, the proton conductivity of the pure SPEEK membrane reaches approximately 132 mS/cm. In comparison, the proton conductivities of SP/DHNTs-4, SP/DHNTs-8, and SP/DHNTs-12 composite membranes increase to 154, 182, and 174 mS/cm, respectively. Figure 12.12 illustrates the proton conduction mechanism in the composite membranes. Several factors contribute to the improved proton conductivity in the SP/DHNTs membranes:

1) **Higher Sulfonic Acid Group Density**: The SP/DHNTs composite membranes exhibit a lower cross-linking density compared to pure SPEEK membranes. This results in a higher concentration of sulfonic acid groups, providing more sites for proton hopping and thus improving proton conduction.
2) **Amine Groups on DHNTs Surfaces**: The DHNTs' polydopamine coating contains abundant amine groups. These groups form acid–base pairs with the sulfonic acid groups in SPEEK. During thermal cross-linking, sulfonic acid groups in these acid–base pairs are not consumed by the cross-linking reaction, preserving a higher number of active proton-conducting groups at the interface. This leads to a surface rich in proton transport media.
3) **Grotthuss Mechanism Facilitation**: The Grotthuss mechanism involves protons hopping across distances of about 0.376 nm. The proximity of the amine groups on DHNTs to the sulfonic acid groups in SPEEK creates an environment conducive to this hopping. These short-range interactions establish new proton pathways, enhancing the proton diffusion network.
4) **Continuous Proton Transfer Channels**: While the SPEEK matrix's hydrated ion regions are typically isolated, the high-aspect-ratio nanorods of DHNTs introduce long-range continuous proton transport channels. These channels interconnect previously separated hydrated ion regions, facilitating continuous proton transfer throughout the composite membrane.

Although DHNTs themselves do not transport protons, their proper incorporation improves the membrane's structure. However, excessive DHNT content

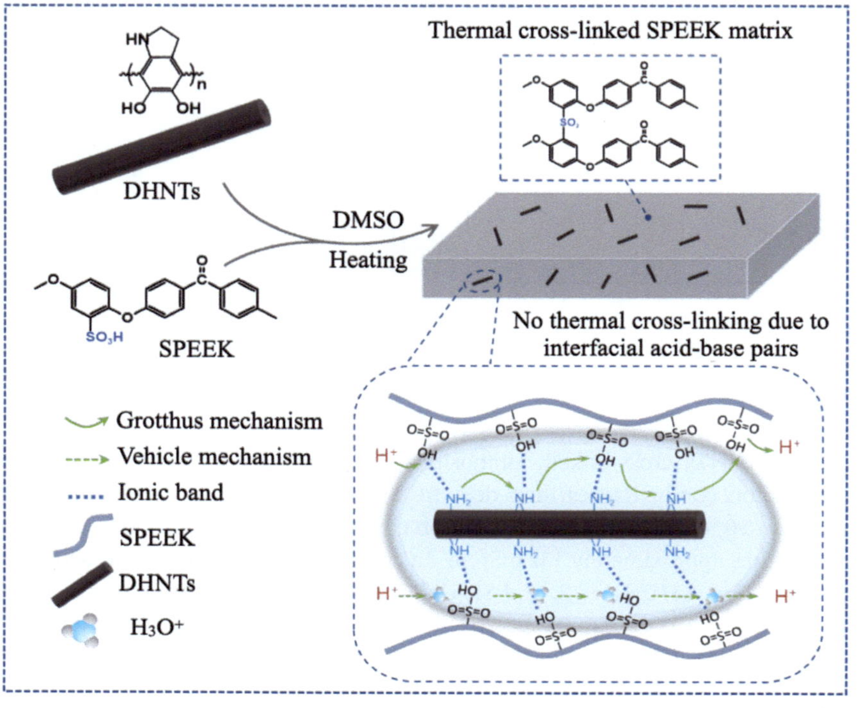

Figure 12.12 Schematic illustration for the proton transportation mechanism of the SP/DHNTs composite membrane.

can reduce proton conductivity by diluting the sulfonic acid group density and causing aggregation. For example, SP/DHNTs-12 exhibits slightly lower proton conductivity than does SP/DHNTs-8 due to DHNT agglomeration, as evident in the SEM images of the membrane's cross-section (see Figure 12.12). This aggregation blocks the formation of uniform, continuous proton pathways. Despite this, SP/DHNTs-12 still outperforms SP/DHNTs-4 and pure SPEEK membranes in terms of proton conductivity.

To further validate the formation of proton transport channels, small-angle X-ray scattering (SAXS) was conducted on the membranes. As shown in Figure 12.13, the pure SPEEK membrane exhibits a weak scattering peak at 1.15 nm^{-1}, corresponding to an average ion cluster size of 5.46 nm. This indicates the presence of discrete ionic domains. By contrast, the SP/DHNTs-8 composite membrane displays a stronger scattering peak at 0.91 nm^{-1}, reflecting an increased ion cluster size of 6.90 nm. These results demonstrate that DHNTs play a key role in connecting the hydrophilic ionic domains within the composite membrane. By bridging the discrete ionic regions in the SPEEK matrix, DHNTs facilitate the formation of larger, long-range proton transport channels.

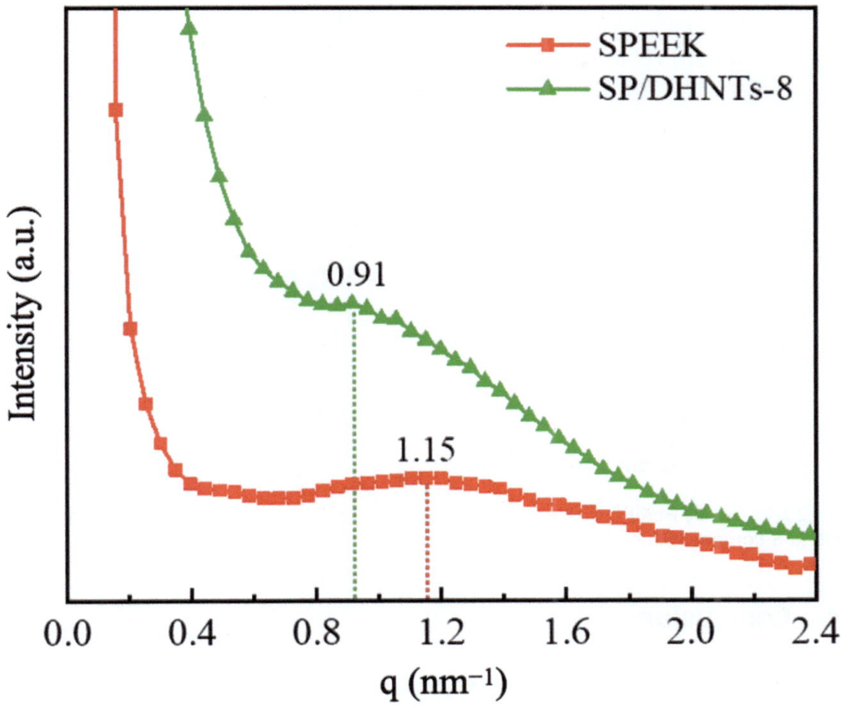

Figure 12.13 SAXS of pristine SPEEK and SP/DHNTs-8 composite membranes.

12.3.7 Fuel Cell Performance

To evaluate fuel cell performance, pure SPEEK membranes and SP/DHNTs-8 composite membranes were used to fabricate membrane electrode assembly (MEA) for single-cell H_2/O_2 testing. The resulting current-voltage polarization curves, recorded at 80 °C and 100% RH, are shown in Figure 12.14. Both MEAs achieved open-circuit voltages greater than 0.98 V, indicating high catalytic activity and low H_2 permeability. For the SP/DHNTs-8 composite membrane, the uniform dispersion of DHNTs in the SPEEK matrix and the acid–base pair interactions between DHNTs and sulfonic acid groups led to the formation of long-range proton transport channels. This enhanced proton transport reduced the membrane's ohmic resistance. The polarization curves revealed that as current density increased, the SP/DHNTs-8 membrane exhibited a slower voltage drop than the pure SPEEK membrane. Consequently, the maximum power density of the SP/DHNTs-8 membrane reached 406 mW/cm^2, compared to 308 mW/cm^2 for the pure SPEEK membrane. The superior proton conductivity of SP/DHNTs-8 was a key factor contributing to its improved fuel cell performance.

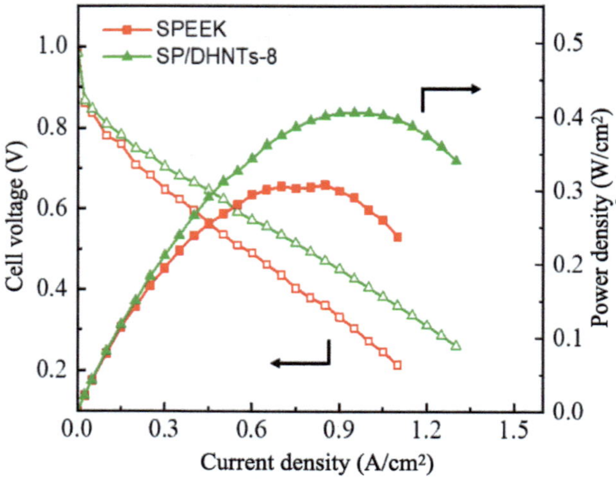

Figure 12.14 H_2/O_2 PEMFC single-cell performance of MEAs using pristine SPEEK and SP/DHNTs-8 composite membranes at 80 °C, 100% RH.

12.4 Conclusion

In this study, dopamine was polymerized onto HNTs to create DHNTs, which were then incorporated into high-sulfonation SPEEK to prepare SP/DHNTs composite membranes. Thermal cross-linking stabilized the membranes by reducing water absorption and swelling below 30%. The DHNTs were well dispersed, forming acid–base pairs with sulfonic acid groups, which introduced efficient proton transfer pathways. These long-range proton channels improved conductivity and enhanced fuel cell performance. Compared to pure SPEEK membranes, SP/DHNTs-8 exhibited higher proton conductivity (182 mS/cm) and greater peak power density (406 mW/cm^2) at 80 °C and 100% RH. Thus, the combination of DHNTs and optimized cross-linking provides a balance between mechanical stability and superior proton conductivity.

References

1 Yang, W., Zhang, J., and Wang, Y. (2013). Charged membranes prepared by SPEEK of very low degree of sulfonation. *Journal of Applied Polymer Science* 128 (5): 2875–2883.

2 Mirfarsi, S.H., Parnian, M.J., Rowshanzamir, S. et al. (2022). Current status of cross-linking and blending approaches for durability improvement of hydrocarbon-based fuel cell membranes. *International Journal of Hydrogen Energy* 47 (27): 13460–13489.

3 Liu, S., Liu, L., Wang, Y. et al. (2023). Inorganic-organic composite membranes containing amino-functionalized mesoporous silica loaded phosphotungstic acid with enhanced fuel cell performance and stability. *International Journal of Hydrogen Energy* 48 (25): 9436–9450.

4 Pourzare, K., Mansourpanah, Y., Farhadi, S. et al. (2020). Improving the efficiency of Nafion-based proton exchange membranes embedded with magnetically aligned silica-coated Co_3O_4 nanoparticles. *Solid State Ionics* 351: 115343.

5 Yang, X.B., Zhao, L., Goh, K. et al. (2020). A phosphotungstic acid coupled silica-Nafion composite membrane with significantly enhanced ion selectivity for vanadium redox flow battery. *Journal of Energy Chemistry* 41: 177–184.

6 Kumari, M., Sodaye, H.S., and BINDAL R C. (2018). Cross-linked sulfonated poly(ether ether ketone)-poly ethylene glycol/silica organic–inorganic nanocomposite membrane for fuel cell application. *Journal of Power Sources* 398: 137–148.

7 Liu, X., He, S., Song, G. et al. (2016). Proton conductivity improvement of sulfonated poly(ether ether ketone) nanocomposite membranes with sulfonated halloysite nanotubes prepared via dopamine-initiated atom transfer radical polymerization. *Journal of Membrane Science* 504: 206–219.

8 Bai, H., Zhang, H., He, Y. et al. (2014). Enhanced proton conduction of chitosan membrane enabled by halloysite nanotubes bearing sulfonate polyelectrolyte brushes. *Journal of Membrane Science* 454: 220–232.

9 Zhang, Y., Wang, H., Qian, P. et al. (2021). Hybrid proton exchange membrane of sulfonated poly(ether ether ketone) containing polydopamine-coated carbon nanotubes loaded phosphotungstic acid for vanadium redox flow battery. *Journal of Membrane Science* 625: 119159.

10 Sun, X., Zhang, T., Liu, X. et al. (2020). Multi-functionalized acid-base double-shell nanotubes are incorporated into the proton exchange membrane to cope with low humidity conditions. *International Journal of Hydrogen Energy* 45 (55): 30673–30688.

11 Kim, T., Sihn, Y., Yoon, I.H. et al. (2021). Monolayer hexagonal boron nitride nanosheets as proton-conductive gas barriers for polymer electrolyte membrane water electrolysis. *ACS Applied Nano Materials* 4 (9): 9104–9112.

12 Lu, Y., Liu, Y., Li, N. et al. (2020). Sulfonated graphitic carbon nitride nanosheets as proton conductor for constructing long-range ionic channels proton exchange membrane. *Journal of Membrane Science* 601: 117908.

13 Wu, H., Cao, Y., Li, Z. et al. (2015). Novel sulfonated poly (ether ether ketone)/phosphonic acid-functionalized titania nanohybrid membrane by an in situ method for direct methanol fuel cells. *Journal of Power Sources* 273: 544–553.

14 Hou, J., Sun, X., Gu, X. et al. (2019). Construction of continuous proton-conduction channels through polyvinylimidazole nanotubes to enhance proton conductivity of polymer electrolyte membrane. *Journal of Applied Polymer Science* 136 (8): 47106.

15 Zhang, P., Li, W., Wang, L. et al. (2020). Polydopamine-modified sulfonated polyhedral oligomeric silsesquioxane: an appealing nanofiller to address the trade-off between conductivity and stabilities for proton exchange membrane. *Journal of Membrane Science* 596: 117734.

16 Gang, M., He, G., Li, Z. et al. (2016). Graphitic carbon nitride nanosheets/ sulfonated poly(ether ether ketone) nanocomposite membrane for direct methanol fuel cell application. *Journal of Membrane Science* 507: 1–11.

17 Liu, L., Lu, Y., Pu, Y. et al. (2021). Highly sulfonated carbon nano-onions as an excellent nanofiller for the fabrication of composite proton exchange membranes with enhanced water retention and durability. *Journal of Membrane Science* 640: 119823.

18 Rao, Z., Lan, M., Zhu, D. et al. (2022). Synergistically promoted proton conduction of proton exchange membrane by phosphoric acid functionalized carbon nanotubes and graphene oxide. *Journal of Membrane Science* 659: 120810.

13

Construction of Ordered Proton Transport Channels with Phosphotungstic Acid Modified Magnetic Nanoparticles

13.1 Introduction

Sulfonated aromatic hydrocarbon polymers require a moderate degree of sulfonation to control water absorption and swelling. However, this results in lower sulfonic acid group content [1], and these groups are typically attached directly to the polymer backbone. As a consequence, it is challenging to achieve distinct microphase separation, leaving the proton transport channels narrow and poorly connected. This limits the formation of continuous hydrophilic regions, thereby reducing proton conductivity. Enhancing proton transport is therefore critical to advancing the widespread application of these membranes.

In fuel cells, protons must travel through the membrane from the anode to the cathode, a process that faces intrinsic resistance. Reducing membrane thickness can lower this resistance, but very thin membranes often suffer from increased fuel permeability and compromised mechanical strength [2]. Additionally, conventional proton exchange membranes (PEMs) generally feature disordered proton transport channels that are often horizontally oriented. As a result, proton conductivity tends to be similar in both the parallel and perpendicular directions relative to the membrane surface, with perpendicular conductivity often lagging behind [3–5]. In contrast, an ideal PEM would exhibit enhanced perpendicular proton conductivity. Constructing vertical proton transport channels can significantly reduce proton transport resistance, thereby improving overall fuel cell performance [6–9].

Phosphotungstic acid ($H_3PW_{12}O_{40}$) (abbreviated as HPW) is a strong heteropoly acid with a Hammett acidity function (H_0) of −13.16, qualifying it as a superacid. It fully dissociates even at low pH, providing numerous proton transport sites when incorporated into polymer membranes. However, its water solubility presents challenges, as it tends to leach out under hydrated conditions, leading to diminished proton conductivity. Stabilization approaches, such as forming ionic bonds between HPW and amine groups, have proven effective in mitigating leaching and retaining its proton-conducting properties [10–12].

Herein, this chapter focuses on stabilizing HPW by loading it onto polydopamine-coated magnetic nanoparticles (DMNPs). The resulting nanoparticles are mixed

Electrospinning for Proton Exchange Membrane Fuel Cells, First Edition.
Yong Liu and Mohideen Meerasahib Mohamedazeem.

with a sulfonated polyetheretherketone sulfonated poly(ether ether ketone) (SPEEK) solution and cast under a magnetic field to construct vertically oriented proton transport channels. The amine groups on polydopamine form ionic bonds with HPW, reducing leaching and ensuring stable integration. Furthermore, applying a uniform magnetic field during membrane casting aligns the HPW-modified nanoparticles along the field direction, creating continuous vertical proton transport pathways. By combining the proton conductivity and water retention properties of HPW with the structural alignment achieved under a magnetic field, this method aims to significantly enhance proton transport efficiency and fuel cell performance.

13.2 Materials Preparation

13.2.1 Preparation of Magnetic Nanoparticles

Fe_3O_4 nanoparticles were synthesized via a co-precipitation method [13]. First, 100 mL of deionized water was added to a three-necked flask, heated to boiling under a nitrogen atmosphere, and then allowed to cool. Afterward, 5.4 g (20 mmol) of $FeCl_3{\cdot}6H_2O$ and 2.8 g (10.07 mmol) of $FeSO_4{\cdot}7H_2O$ were dissolved into the deoxygenated water, and the mixture was stirred mechanically at 80 °C under nitrogen. Next, 16 mL of $NH_3{\cdot}H_2O$ was added rapidly, and the reaction proceeded for 5 minutes. A 2-mL aliquot of 1 M sodium citrate solution was then introduced to prevent aggregation of the nanoparticles. The suspension was stirred for an additional 2 hours, after which the precipitate was collected using a permanent magnet. The magnetic particles were washed several times with deionized water until neutral, followed by two ethanol washes. The final product was vacuum dried at 60 °C, yielding Fe_3O_4 magnetic nanoparticles (MNPs).

13.2.2 Preparation of Phosphotungstic Acid-Loaded Magnetic Nanoparticles

To prepare DMNPs, 0.2 g of Fe_3O_4 particles were ultrasonically dispersed in 10 mL of deionized water and transferred to 90 mL of pH 8.5 Tris buffer. Then, 0.2 g of dopamine hydrochloride was added, and the mixture was stirred at 30 °C in air for 6 hours. The resulting product was separated magnetically, washed multiple times with deionized water and ethanol, and vacuum dried at 60 °C to obtain DMNPs.

Next, 0.05 g of DMNPs was dispersed in 10 mL of ethanol containing 5 wt% HPW. The mixture was stirred for 12 hours to promote interaction between HPW and the surface amine groups of the DMNPs. The HPW-modified nanoparticles were then separated using a strong magnet, washed with ethanol to remove excess acid, and vacuum dried at 60 °C. This process yielded HPW-loaded MNPs (DMNPs@HPW).

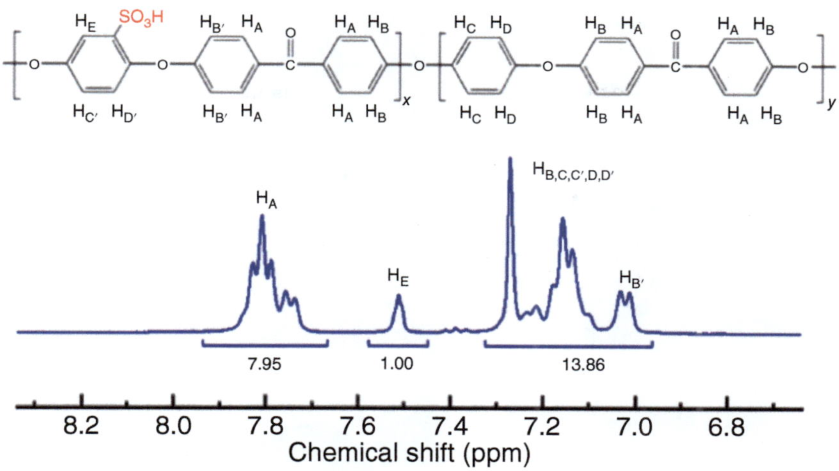

Figure 13.1 Structure and ^{1}H NMR spectra of SPEEK.

13.2.3 Sulfonation of Poly(Ether Ether Ketone) and Preparation of Composite Membranes

Poly(ether ether ketone) (PEEK) powder was vacuum-dried at 80 °C for 24 hours before sulfonation. Subsequently, 10 g of dried PEEK was added to 100 mL of concentrated sulfuric acid, stirred at 20 °C, and allowed to react for 12 hours. The temperature was then raised to 50 °C for 1.5 hours before terminating the reaction by cooling with cold water. The resulting solution was poured into a large volume of ice water to precipitate the sulfonated polymer. The precipitate was washed several times with deionized water until neutral, then vacuum-dried at 60 °C for 24 hours. The degree of sulfonation of the SPEEK was determined to be 50%, as confirmed by ^{1}H nuclear magnetic resonance (NMR) spectroscopy (Figure 13.1).

13.2.4 Preparation of Composite Membranes

To prepare the composite membranes, a 1% (w/w) dispersion of DMNPs@HPW was made in NMP, and 0.3 g of SPEEK was dissolved in 2 g of NMP. The appropriate amount of the DMNPs@HPW dispersion was combined with the SPEEK solution, then mechanically stirred and ultrasonically treated to ensure uniform mixing. After standing for 2 hours to remove air bubbles, the casting solution was ready. This solution was drop-cast onto a clean glass plate and placed within a uniform magnetic field (0.3 T), aligned vertically to the liquid surface, as depicted in Figure 13.2. The film was dried at 60 °C for 24 hours.

The resulting membrane was acidified by immersion in 2 M hydrochloric acid for 24 hours and then rinsed with water until neutral. This produced a composite

Figure 13.2 Optical photographs of (a, b, c) magnetic field generator, (d) SP/DMNPs @HPW-2, (e) M-SP/DMNPs@HPW-2. Source: Han et al. [14]/with permission of Elsevier.

membrane in which DMNPs@HPW particles were aligned within the SPEEK matrix, labeled as M-SP/DMNPs@HPW-x. As a control, another composite membrane was prepared without applying a magnetic field, resulting in randomly distributed DMNPs@HPW particles. This membrane was labeled as SP/DMNPs@HPW-x, where "x" indicates the mass fraction of DMNPs@HPW in the composite membrane.

13.3 Result and Discussion

13.3.1 Preparation and Characterization of DMNPs@HPW

1) **Preparation of HPW-Loaded Magnetic Nanoparticles:** The functionalization process for MNPs and their subsequent HPW loading is illustrated in Figure 13.3. In the first step, MNPs are dispersed in a dopamine solution at pH 8.5. In the presence of oxygen, dopamine undergoes self-polymerization, forming a protective polydopamine coating on the MNP surfaces. This coating prevents environmental degradation and aggregation, while also providing amine groups capable of binding acidic species.

   In the second step, the DMNPs are dispersed in an ethanol solution of HPW. The amine groups on the DMNPs react with HPW through acid–base interactions, resulting in the formation of MNP-polymetallic oxide salt hybrids (DMNPs@HPW). During this process, the amine groups protonate to form

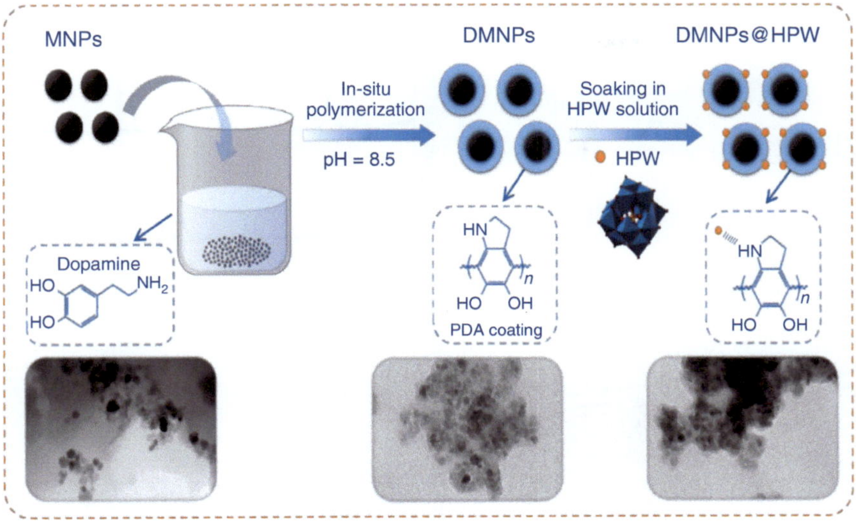

Figure 13.3 Schematic illustration of the preparation of HPW-functionalized MNPs. Source: Han et al. [14]/with permission of Elsevier.

positively charged ammonium ions, which electrostatically interact with the negatively charged HPW species. Because organic cation-polyanion salts have low solubility in water and ethanol [15], excess polyacid is easily removed by ethanol washing. The remaining HPW ions remain stably attached to the polydopamine-modified nanoparticles and do not dissolve in water.

2) **Characterization of DMNPs@HPW**: Fourier transform infrared (FT-IR) analysis of MNPs, DMNPs, and DMNPs@HPW is presented in Figure 13.4a. The broad peak at 580 cm^{-1} corresponds to Fe–O vibrations [16], while the peak at 1630 cm^{-1} arises from surface-adsorbed water. For the DMNPs, the peak at 1255 cm^{-1} indicates phenolic hydroxyl group vibrations, and the peaks at 1490 and 1610 cm^{-1} correspond to benzene ring (C=C) vibrations. After loading HPW, characteristic Keggin-type anion vibrations appear at 1080 cm^{-1} (P–O), 978 cm^{-1} (terminal W=O), 895 cm^{-1} (corner-sharing W–O–W), and 810 cm^{-1} (edge-sharing W–O–W) [17], confirming successful immobilization of HPW on DMNPs.
   X-ray diffraction patterns of MNPs, DMNPs, and DMNPs@HPW are shown in Figure 13.4b. All samples exhibit diffraction peaks consistent with the spinel structure of Fe_3O_4 (JCPDS No. 19-0629), including peaks at $2\theta = 30.1°$, 35.5°, 43.1°, 53.4°, 57.0°, and 62.6°, corresponding to the (220), (311), (411), (422), (511), and (440) crystal planes of Fe_3O_4. No additional crystalline peaks for polydopamine or HPW are observed, indicating that both are present in an amorphous form on the Fe_3O_4 nanoparticles.

Thermogravimetric analysis (TGA) of MNPs, DMNPs, and DMNPs@HPW samples under nitrogen atmosphere provided insight into the loading of polydopamine (PDA) and HPW on the MNPs. As shown in Figure 13.4c, the MNPs exhibited two

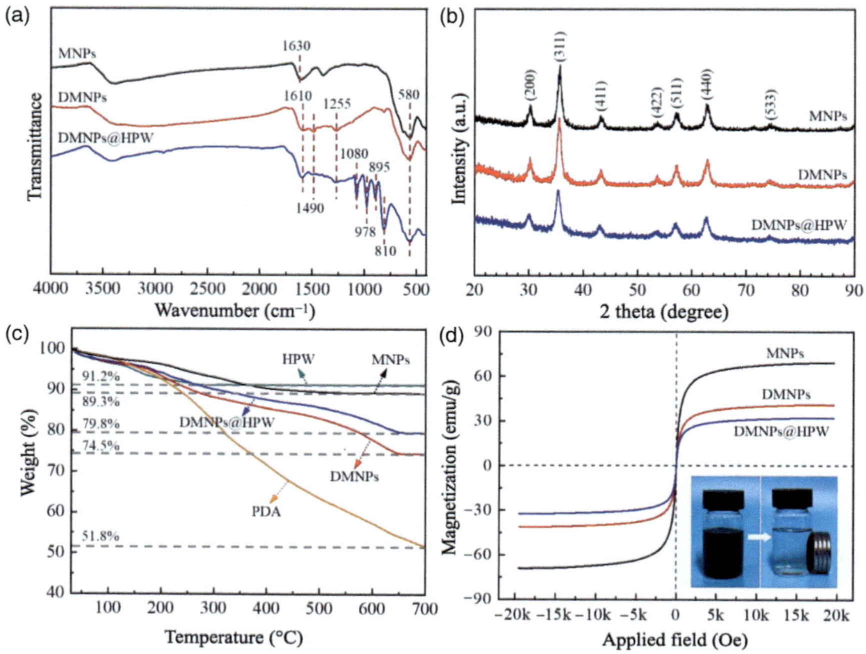

Figure 13.4 (a) FT-IR spectra (b) XRD patterns of MNPs, DMNPs and DMNPs@HPW; (c) TGA spectra of PDA, HPW, MNPs, DMNPs and DMNPs@HPW; (d) Hysteresis loops of MNPs, DMNPs and PWA-DMNPs, inset photo shows DMNPs@HPW dispersed in an aqueous solution and attracted by an external magnet. Source: Han et al. [14]/with permission of Elsevier.

distinct stages of weight loss: a 3% reduction below 140 °C, attributed to moisture removal, and a 7.7% decrease between 220 and 400 °C, likely resulting from impurity decomposition. In contrast, DMNPs and DMNPs@HPW demonstrated additional weight loss due to the decomposition of the PDA polymer. The final residue masses were 74.5% for DMNPs and 79.8% for DMNPs@HPW, translating to PDA and HPW loadings of approximately 39.5% and 31.7%, respectively.

Magnetic properties of the particles were examined using vibrating sample magnetometry (VSM), with hysteresis loops presented in Figure 13.4d. All samples displayed typical superparamagnetic behavior, with no hysteresis, negligible remnant magnetization, and coercivity close to zero. The saturation magnetization of the MNPs was 69.3 emu/g, which decreased to 41.1 emu/g for DMNPs and further to 32.4 emu/g for DMNPs@HPW due to the non-magnetic nature of the PDA coating and HPW. Despite this decrease, the particles maintained excellent magnetic responsiveness, as illustrated by their swift separation under a magnetic field.

Scanning electron microscope (SEM) provided morphological details of the MNPs, DMNPs, and DMNPs@HPW. MNPs exhibited nearly spherical shapes with an average diameter of 21 nm, as shown in Figure 13.5a. Following polydopamine coating and subsequent HPW loading, the diameters increased to approximately 30 nm for DMNPs and 33 nm for DMNPs@HPW, as seen in Figures 13.5b and c.

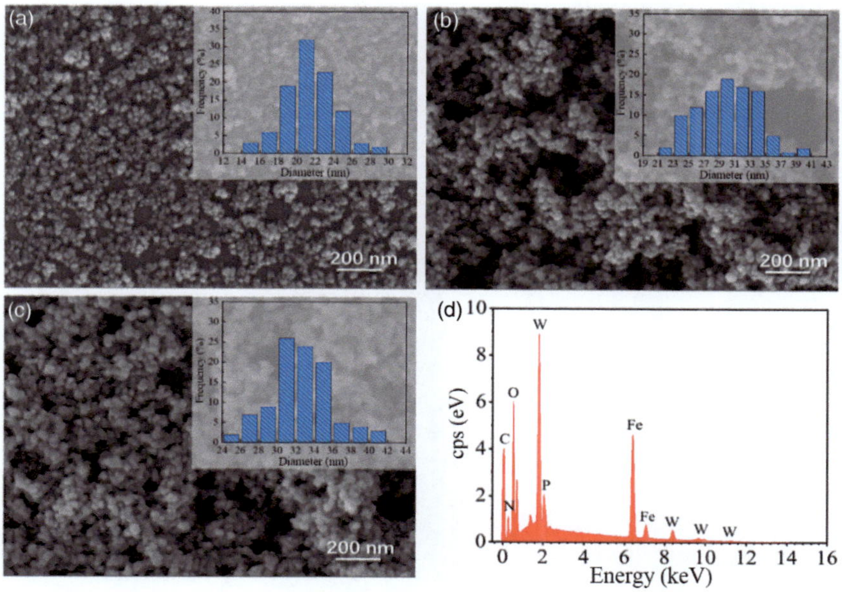

Figure 13.5 SEM micrographs of (a) MNPs, (b) DMNPs and (c) DMNPs@HPW (insets are the diameter distribution of the nanoparticles). (d) EDS spectrum of DMNPs@HPW. Source: Han et al. [14]/with permission of Elsevier.

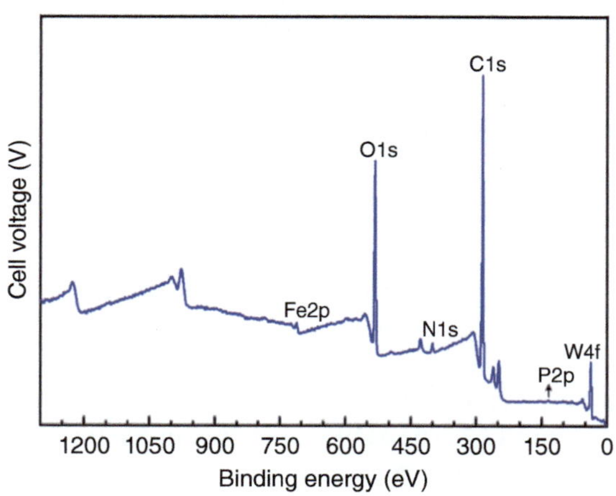

Figure 13.6 XPS spectra of DMNPs@HPW.

Further confirmation of successful HPW loading came from elemental analysis using energy dispersive X-ray spectroscopy (EDS) and X-ray photoelectron spectroscopy (XPS). The EDS spectra revealed distinct peaks for N (0.28 keV), P (2.06 keV), and W (1.80, 8.41, 9.65 keV), corresponding to the nitrogen in PDA and the phosphorus and tungsten in HPW. XPS analysis as shown in Figure 13.6, corroborated these findings, confirming the successful integration of HPW onto the DMNPs.

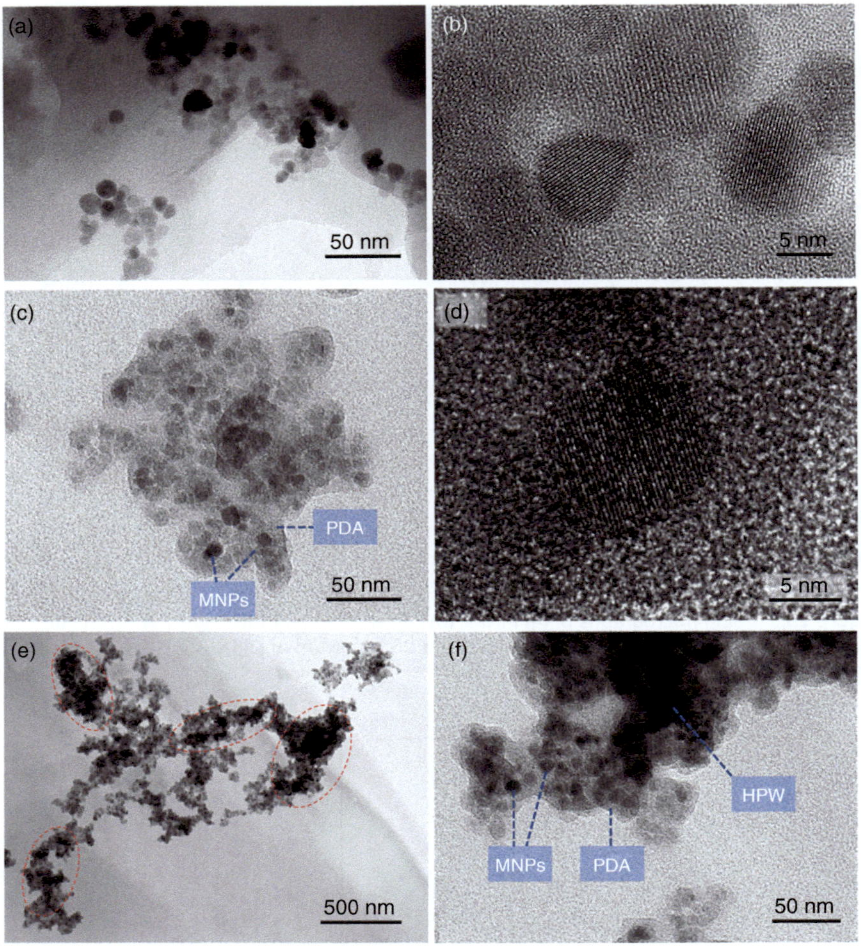

Figure 13.7 TEM images of (a, b) MNPs, (c, d) DMNPs and (e, f) DMNPs@HPW. Source: Han et al. [14]/with permission of Elsevier.

Transmission electron microscopy (TEM) images of the MNPs, DMNPs, and DMNPs@HPW are displayed in Figure 13.7. The MNPs, as shown in Figure 13.7a, appear nearly spherical with good monodispersity, and their lattice structure is clearly evident in Figure 13.7b. The images of DMNPs in Figure 13.7c and d show the dark MNPs uniformly embedded in the lighter PDA polymer matrix. The lattice structure of the MNPs remains visible, with the surrounding PDA polymer clearly distinguishable. Figures 13.7e and f illustrate the incorporation of HPW onto the DMNPs. The presence of darker regions in these images corresponds to tungsten atoms from HPW, which scatter electrons more strongly. These darker areas confirm the successful loading of HPW onto the DMNPs. This evidence from TEM imaging demonstrates that the MNPs are well coated with PDA and that HPW is effectively incorporated into the composite structure.

13.3.2 Morphology of Composite Membrane Sections

The cross-sectional SEM images of pure SPEEK membranes and DMNPs@HPW composite membranes, shown in Figure 13.8, reveal that all membranes exhibit relatively dense structures without significant defects. The pure SPEEK membrane, displayed in Figures 13.8a, features a smooth, dense morphology. In contrast, DMNPs@HPW-2 composite membrane cross-section, seen in Figures 13.8c and d, has a slightly rougher appearance, with the DMNPs@HPW nanofillers dispersed evenly throughout the SPEEK matrix. The surface modification of MNPs by PDA and HPW appears to enhance particle dispersion; however, minor local aggregations are noted (circled in Figure 13.8d), likely stemming from the PDA coating process.

The composite membrane M-SP/DMNPs@HPW-2, prepared under a magnetic field, demonstrates an even more distinctive morphology. As shown in Figures 13.8e and f, the DMNPs@HPW particles align along the direction of the magnetic field,

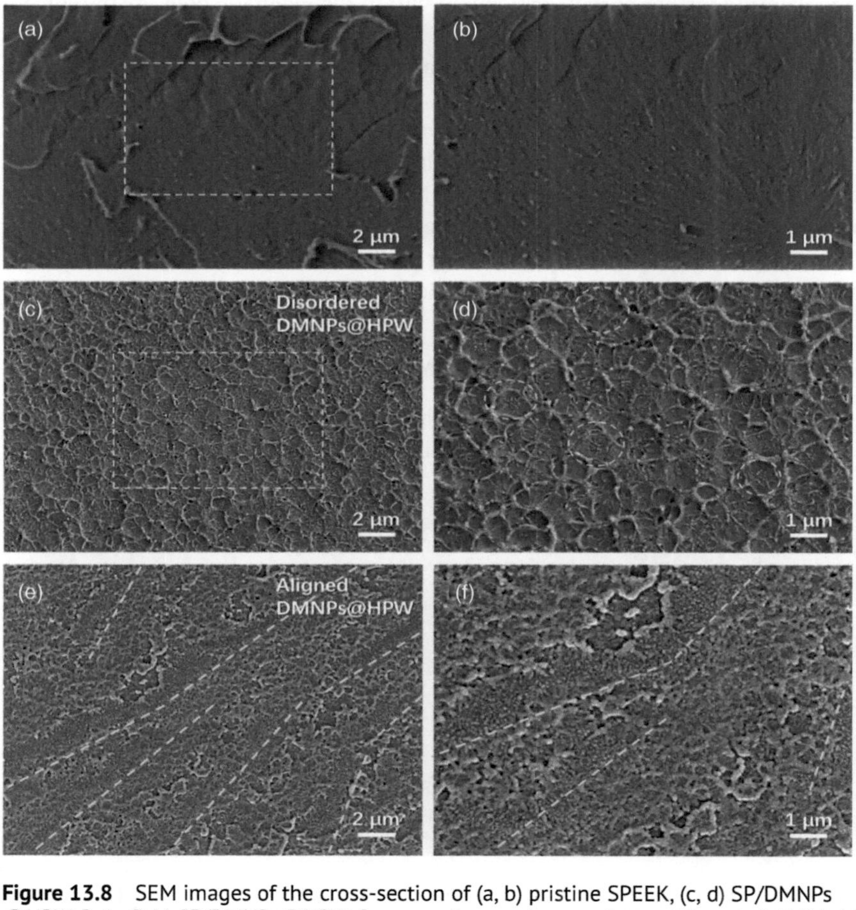

Figure 13.8 SEM images of the cross-section of (a, b) pristine SPEEK, (c, d) SP/DMNPs @HPW-2, (e, f) M-SP/DMNPs@HPW-2 composite membranes. Source: Han et al. [14]/with permission of Elsevier.

creating an ordered internal structure. This alignment clearly indicates that the magnetic field significantly influences the arrangement of DMNPs@HPW within the SPEEK matrix, leading to a more organized filler distribution compared to the randomly dispersed particles in the SP/DMNPs@HPW-2 membrane.

13.3.3 Proton Conduction Pathways and Particle Distribution

More significantly, the DMNPs@HPW particles loaded with the strong protonic acid HPW interconnect to create continuous, long-range proton transfer channels within the SPEEK matrix, which enhances proton conductivity. Within the composite membrane, the DMNPs@HPW particles show an inclined alignment. This inclination likely occurs because, when the magnetic field is first applied, the DMNPs@HPW particles quickly magnetize and align in columnar structures along the magnetic field direction due to dipole–dipole interactions. This arrangement persists until the solvent begins to evaporate, increasing the viscosity of the solution and preventing further particle movement within the magnetic field. As the solvent fully evaporates and the polymer chains undergo random motion, the initially columnar arrangement becomes inclined and distorted, forming channels that are not strictly vertical.

Figure 13.9 highlights the impact of varying DMNPs@HPW content on the composite membranes. It reveals that while the particles effectively form continuous paths at moderate loadings, excessive DMNPs@HPW can lead to aggregation. Such aggregation disrupts the ordered structure and reduces the effectiveness of the proton conduction pathways, emphasizing the need for optimized particle content to maintain both structural order and efficient proton transport.

13.3.4 Characterization of Composite Membranes X-ray Diffraction

The composite membranes were characterized through FT-IR, XRD and TGA analyses to investigate their chemical structure, crystalline properties, and thermal

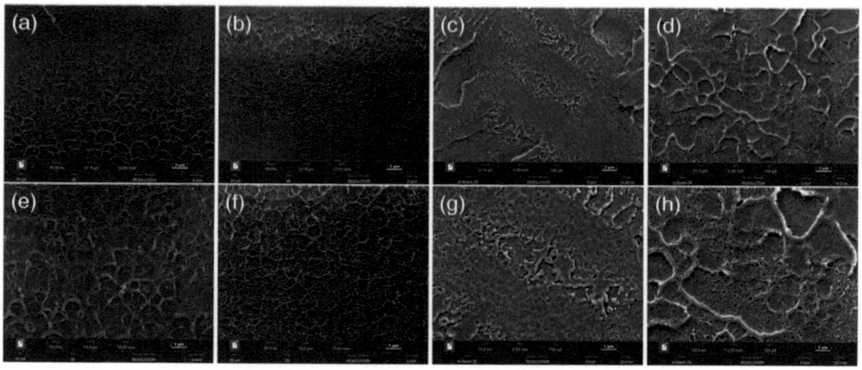

Figure 13.9 SEM images of the cross-section of (a, e) SP/DMNPs@HPW-1, (b, f) SP/DMNPs @HPW-4, (c, g) M-SP/DMNPs@HPW-1 and (d, h) M-SP/DMNPs@HPW-4 composite membranes. Source: Han et al. [14]/with permission of Elsevier.

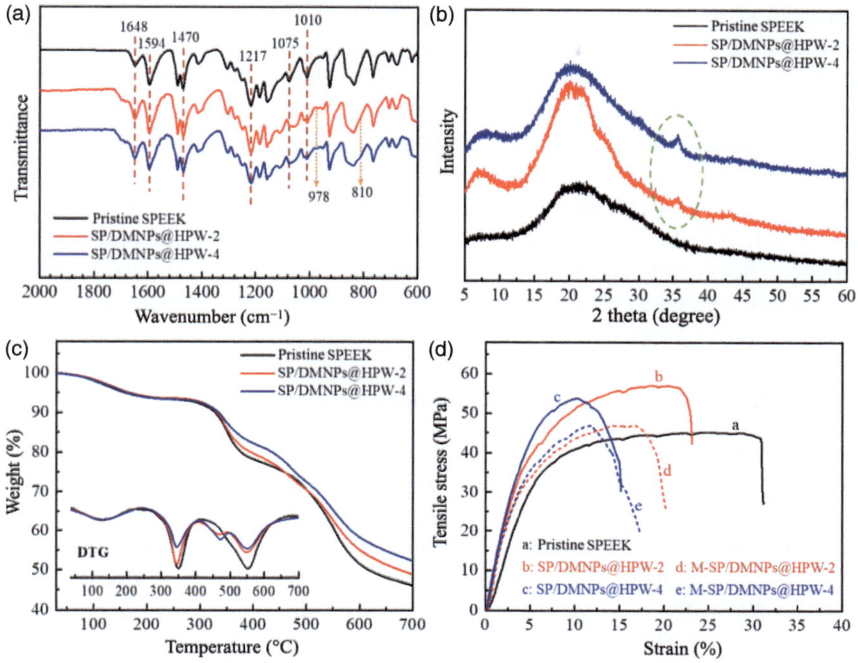

Figure 13.10 (a) FT-IR spectra, (b) XRD patterns, and (c) TGA curves of pristine SPEEK, SP/DMNPs@HPW-2, and SP/DMNPs@HPW-4 composite membranes. (d) Stress–strain curves of pristine SPEEK, SP/DMNPs@HPW-2, SP/DMNPs@HPW-4, M-SP/DMNPs@HPW-2, and M-SP/DMNPs@HPW-4 composite membranes.

stability. The FT-IR spectra of pure SPEEK membranes and SP/DMNPs@HPW composite membranes are presented in Figure 13.10a. For SPEEK, characteristic peaks include a strong absorption peak at $1648\,cm^{-1}$ (C=O stretching), peaks at 1594 and $1470\,cm^{-1}$ (C=C stretching in benzene rings), and peaks at 1217, 1075, and $1010\,cm^{-1}$ (S=O and S–O stretching from $-SO_3H$ groups) [18].

The composite membranes exhibit similar peaks, with additional peaks at 978 and $810\,cm^{-1}$ indicating the presence of HPW. Due to the low DMNPs@HPW content and overlapping absorption bands, these HPW peaks appear weak, and characteristic peaks for dopamine are not clearly visible. An increase in DMNPs@HPW content leads to a gradual decrease in the intensity of the $-SO_3H$ peaks, potentially due to hydrogen bond formation between $-SO_3H$ and functional groups on DMNPs@HPW.

Figure 13.10b shows XRD patterns of the pure SPEEK membrane and SP/DMNPs@HPW composite membranes. The SPEEK membrane displays a broad peak at $2\theta = 21°$, consistent with its amorphous nature due to the irregular arrangement of sulfonated polymer chains. In the composite membranes, a characteristic peak at $2\theta = 35.5°$ corresponds to the (311) plane of Fe_3O_4, indicating the presence of DMNPs@HPW and suggesting that the crystalline structure of the iron oxide particles remains intact.

Thermal stability was evaluated by TGA, with results shown in Figure 13.10c. For the pure SPEEK membrane, weight loss occurs in two stages: below 150 °C due to absorbed water and above 400 °C due to backbone degradation. The SP/DMNPs@HPW composite membranes exhibit similar thermal behavior, but partial weight loss around 450 °C is observed, likely from dopamine degradation. As shown in Figure 13.10d, the residual Masses of SP/DMNPs@HPW-2 and SP/DMNPs@HPW-4 are 49.1% and 52.4%, respectively, slightly higher than the pure SPEEK membrane (46.2%), due to the presence of iron oxide and tungsten elements. These results confirm that the composite membranes maintain good thermal stability below 250 °C, making them suitable for proton exchange membrane fuel cells (PEMFC) applications.

13.3.5 Mechanical Properties of Composite Membranes

The mechanical properties of the pure SPEEK membrane and its composite counterparts were evaluated and are presented in Figure 13.10d and Table 13.1. Incorporating nanoparticles into SPEEK significantly enhanced the tensile strength of the composite membranes due to the reinforcing effect of the fillers. The pure SPEEK membrane exhibited a tensile strength (Ts) of 47.4 MPa, a Young's modulus (Y) of 672 MPa, and an elongation at break (Eb) of 30.3%. In contrast, the SP/DMNPs@HPW-2 and SP/DMNPs@HPW-4 composite membranes showed increased tensile strength (58.7 and 51.9 MPa, respectively) and Young's modulus (1021 and 993 MPa, respectively). However, their elongation at break decreased to 20.5% and 11.7%, likely due to minor nanoparticle agglomeration causing stress-concentrating defects, which reduced the membranes' flexibility and made them more susceptible to rupture at these defect sites.

For the M-SP/DMNPs@HPW-2 and M-SP/DMNPs@HPW-4 composite membranes prepared under a uniform magnetic field, a slight reduction in mechanical strength was observed. The tensile strength of these membranes was measured at 47.8 and 48.1 MPa, with elongation at break values of 15.5% and 10.4%, respectively. The alignment of magnetic particles into long-range proton transport channels may have introduced structural anisotropy, increasing susceptibility to localized rupture under stress. Nonetheless, the mechanical properties of all composite membranes remain within acceptable limits for practical PEMFC applications.

Table 13.1 Mechanical properties of as-prepared membranes.

Membranes	Tensile strength (MPa)	Young's modulus (MPa)	Elongation at break (%)
Pristine SPEEK	47.4 ± 4.2	672 ± 53	30.3 ± 5.1
SP/DMNPs@HPW-2	58.7 ± 3.1	1021 ± 68	20.5 ± 3.8
SP/DMNPs@HPW-4	51.9 ± 2.7	993 ± 126	11.7 ± 2.2
M-SP/DMNPs@HPW-2	47.8 ± 2.2	921 ± 76	15.5 ± 3.5
M-SP/DMNPs@HPW-4	48.1 ± 3.4	884 ± 87	10.4 ± 2.5

DMNP, polydopamine-coated magnetic nanoparticles; SPEEK, sulfonated poly(ether ether ketone).

13.3.6 Ion Exchange Capacities, Water Uptake, and Swelling Ratio

Ion exchange capacities (IECs), water uptake, and swelling ratio are crucial indicators of the composite membranes' performance. The IEC, which quantifies the concentration of hydrophilic functional groups like sulfonic acids or HPW, was measured via acid–base titration. For the pure SPEEK membrane, the IEC is about 1.49 mmol/g, closely aligning with the theoretical value of 1.52 mmol/g, as calculated based on its 50% sulfonation degree from ^{1}H NMR. Incorporating DMNPs@HPW nanoparticles slightly reduced the IEC to a range of 1.44–1.48 mmol/g, primarily due to the increased membrane weight and the acid–base interactions between the sulfonic groups and the polydopamine's basic amine groups. This interaction reduces the availability of free H^+ ions from sulfonic groups, causing the slight decline in IEC.

Water uptake, critical for facilitating proton transport, and swelling rate were assessed as a function of temperature. As shown in Figure 13.11a, the pure SPEEK membrane's water uptake rose from approximately 33.5% at 40 °C to 65.4% at 80 °C, with a corresponding increase in swelling rate from 9.8% to 22.5%. This trend reflects greater polymer chain mobility at elevated temperatures, which weakens chain entanglements and increases free volume, allowing more water to penetrate the membrane.

All DMNPs@HPW composite membranes demonstrated higher water uptake across the temperature range compared to the pure SPEEK membrane. This enhancement is attributed to the strong water-retaining ability of HPW. The SP/DMNPs@HPW-4 composite membrane, containing a higher proportion of DMNPs@HPW, showed the highest water uptake. Despite this increased water uptake, the composite membranes exhibited slightly lower swelling rates than the pure SPEEK membrane. As shown in Figure 13.11b and Table 13.2, the addition of nanoparticles and the acid–base interactions between sulfonic groups and DMNPs@HPW helped reinforce the composite structure and reduce polymer chain mobility. This reinforcement, along with the formation of hydrogen-bonding networks, stabilized the water content while limiting dimensional changes upon water absorption. At 80 °C, the swelling rates of the composite membranes remained around 20%, indicating adequate dimensional stability for PEMFC applications.

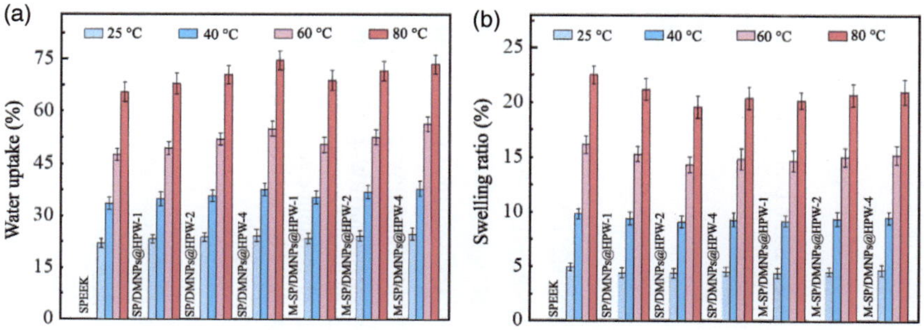

Figure 13.11 (a) Water uptake and (b) swelling ratio of composite membranes at various temperatures.

Table 13.2 Basic properties of the composite membranes.

Code	IEC[a)] (mmol/g)	WU (%)		SR (%)		$\sigma_{//}$[b)] (mS/cm)	
		40 °C	80 °C	40 °C	80 °C	40 °C	80 °C
Pristine SPEEK	1.49 ± 0.04	33.5 ± 1.8	65.4 ± 2.7	9.8 ± 0.4	22.5 ± 0.8	72.4 ± 1.8	121.2 ± 3.1
SP/DMNPs@HPW-1	1.48 ± 0.06	34.8 ± 2.0	67.8 ± 2.9	9.4 ± 0.6	21.2 ± 1.0	82.4 ± 1.3	139.8 ± 2.2
SP/DMNPs@HPW-2	1.47 ± 0.05	35.6 ± 1.7	70.5 ± 2.6	9.1 ± 0.4	19.6 ± 0.9	98.5 ± 1.5	157.4 ± 3.8
SP/DMNPs@HPW-4	1.44 ± 0.08	37.5 ± 1.9	74.6 ± 2.7	9.3 ± 0.6	20.4 ± 1.0	91.8 ± 3.2	150.2 ± 4.1
M-SP/DMNPs@HPW-1	1.48 ± 0.05	35.3 ± 1.9	68.8 ± 3.0	9.2 ± 0.5	20.2 ± 0.7	86.3 ± 1.7	145.9 ± 3.5
M-SP/DMNPs@HPW-2	1.46 ± 0.04	36.8 ± 2.1	71.5 ± 2.8	9.4 ± 0.6	20.7 ± 1.1	102.3 ± 2.0	159.3 ± 2.7
M-SP/DMNPs@HPW-4	1.45 ± 0.07	37.9 ± 2.2	73.5 ± 2.4	9.5 ± 0.7	21.0 ± 1.2	97.3 ± 2.4	153.8 ± 4.2

DMNP, polydopamine-coated magnetic nanoparticles; IEC, ion exchange capacity; SPEEK, sulfonated poly(ether ether ketone); SR, swelling rate; WU, water uptake.

a) Estimated by acid–base titration.

b) In-plane proton conductivity.

13.3.7 Proton Conductivity of the Membranes

Proton conductivity is a critical property of PEMs and directly affects fuel cell performance. As shown in Figure 13.12, proton conductivity was measured in both parallel (σ//) and perpendicular (σ⊥) directions under 100% relative humidity (RH) across various temperatures.

1) **Proton Conductivity in the Parallel Direction:** For all tested membranes, proton conductivity increased with temperature, consistent with known principles. For example, the pure SPEEK membrane's conductivity ranged from 60.8 to 121.2 mS/cm between 30 and 80 °C. This trend arises because elevated temperatures increase the mobility of polymer chain segments and water molecules, thereby improving proton transport. Incorporating DMNPs@HPW nanoparticles into the SPEEK matrix further enhanced conductivity by increasing water absorption and providing additional proton pathways. The SP/DMNPs@HPW composite membranes demonstrated higher proton conductivity, peaking at 157.4 mS/cm at 80 °C for the SP/DMNPs@HPW-2 membrane. A similar enhancement was observed in the M-SP/DMNPs@HPW membranes, where proton conductivity reached 159.3 mS/cm at 80 °C. However, excessive nanoparticle loading led to slight aggregation, which hindered proton transfer and caused a minor decrease in conductivity. Despite this, the composite membranes consistently outperformed pure SPEEK.

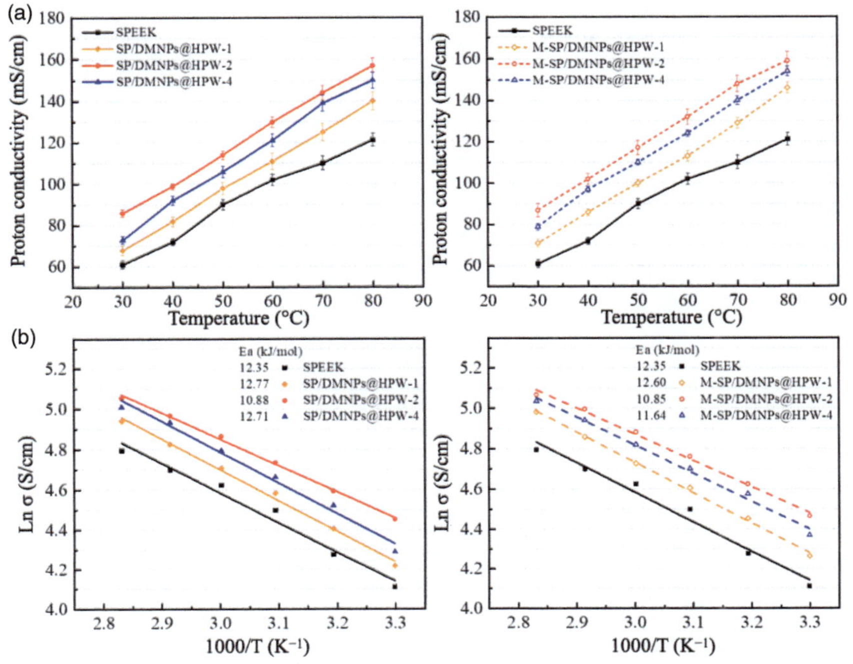

Figure 13.12 (a) Temperature dependence in-plane conductivity and (b) Arrhenius plots of pristine SPEEK and DMNPs@HPW composite membranes.

2) **Proton Conductivity in the Perpendicular Direction:** As shown in Figure 13.12, the σ⊥ values of the composite membranes were higher than those of pure SPEEK, with the M-SP/DMNPs@HPW membranes showing the most pronounced increase. At 80 °C, the pure SPEEK membrane's $\sigma\perp$ value was 82.2 mS/cm, whereas SP/DMNPs@HPW-2 and M-SP/DMNPs@HPW-2 reached 113.4 and 129.2 mS/cm, representing 38% and 57% improvements, respectively. These enhancements result from the nanoparticles aligning along the magnetic field during membrane preparation, forming structured proton transport pathways.
3) **Proton Conductivity Anisotropy and Channel Structure:** Figure 13.13 illustrates the anisotropy ratio (σ///σ⊥) of proton conductivity. For all membranes, this ratio exceeded 1.0, indicating greater conductivity in the parallel direction. In the pure SPEEK membrane, this is attributed to the horizontal orientation of natural proton transport channels within the matrix. However, the M-SP/DMNPs@HPW membranes prepared under a magnetic field showed a lower anisotropy ratio (1.23 compared to 1.47 for pure SPEEK), reflecting their improved perpendicular conductivity. The alignment of DMNPs@HPW under the magnetic field induced continuous water channels that connected previously isolated paths, reducing the directional discrepancy in proton transport. This improved connectivity is responsible for the reduced anisotropy and the enhanced overall proton conduction in the composite membranes.

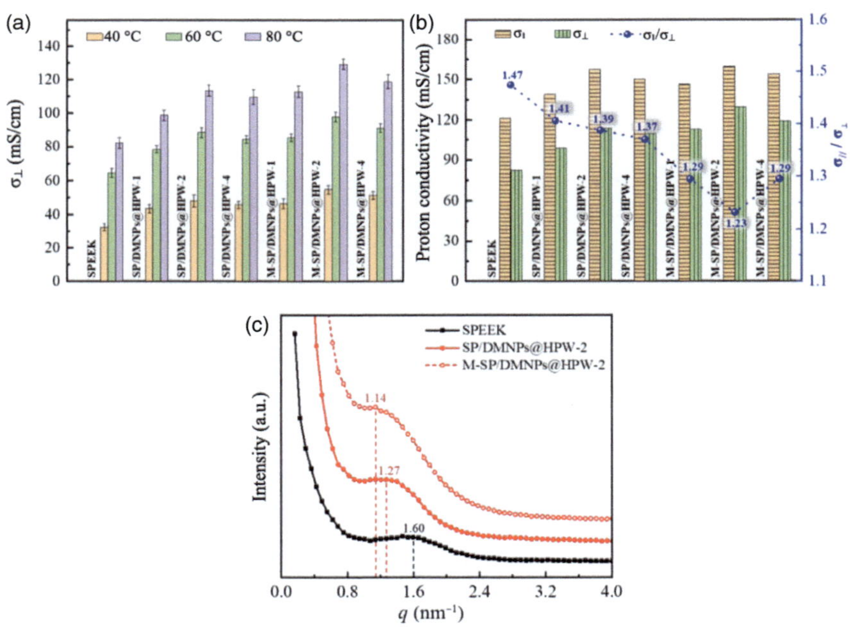

Figure 13.13 (a) Through-plane conductivity and (b) proton conductivity anisotropy coefficient ($\sigma_{//}/\sigma_{\perp}$) at 80 °C of pristine SPEEK and DMNPs@HPW composite membranes. (c) SAXS of pristine SPEEK, SP/DMNPs@HPW-2 and M-SP/DMNPs@HPW-2 composite membranes.

Small-angle X-ray scattering (SAXS) analysis was used to verify the formation of proton transport channels in the prepared membranes. As shown in Figure 13.13c, the pure SPEEK membrane exhibited a weak scattering peak at $q = 1.60\ \text{nm}^{-1}$, corresponding to an ion cluster size of 3.93 nm. This suggests that the ion domains in the pure SPEEK membrane are relatively dispersed and lack well-defined channels. In comparison, the SP/DMNPs@HPW-2 composite membrane demonstrated a stronger scattering peak at $q = 1.27\ \text{nm}^{-1}$, indicating an increase in ion cluster size to 4.95 nm. This shift reflects the impact of DMNPs@HPW, which help interconnect the hydrophilic ion domains and enhance the continuity of the proton-conducting regions.

The M-SP/DMNPs@HPW-2 composite membrane, prepared under a magnetic field, showed an even more pronounced scattering peak at $q = 1.14\ \text{nm}^{-1}$, corresponding to an ion cluster size of 5.51 nm. This further increase in size and scattering intensity confirms the role of the magnetic field in aligning the DMNPs@HPW particles, resulting in well-organized, long-range proton transport channels. These more ordered structures facilitate efficient proton conduction throughout the membrane.

The proton conduction within the membranes after water absorption primarily relies on the Grotthuss and Vehicle mechanisms, as depicted in Figure 13.14. The Arrhenius equation, $\sigma = \sigma_0 \exp(-E_a/\text{RT})$, was applied to determine the activation energy for proton conduction, with values derived from the slope of the Arrhenius plots. The calculated activation energies for the membranes range from

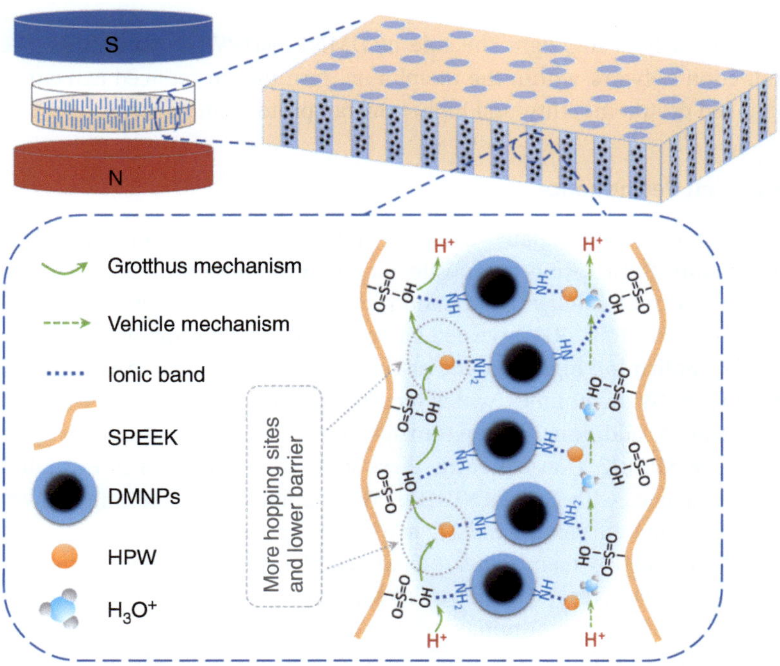

Figure 13.14 Schematic illustration for the proton transportation mechanism of the M-SP/DMNPs@HPW composite membrane.

10.85 to 12.77 kJ/mol, aligning closely with the activation energy associated with the Grotthuss mechanism (~10 kJ/mol). This agreement suggests that proton transport predominantly occurs through hydrogen-bonding networks, where protons "hop" between adjacent water molecules and acidic groups.

DMNP, polydopamine-coated magnetic nanoparticle; HPW, phosphotungstic acid; SPEEK, sulfonated poly(ether ether ketone).

Several factors contribute to the enhanced proton conductivity observed in DMNPs@HPW composite membranes:

1) **Water Adsorption and Proton Sites Provided by HPW:** HPW immobilized on the surface of DMNPs increases water absorption and acts as a superacid, offering additional sites for proton hopping.
2) **Reduced Distances Between Proton Hopping Sites:** The acid–base interactions between amine groups on polydopamine and sulfonic groups in the SPEEK matrix shorten the distance between proton conduction sites, facilitating new proton transfer pathways. The proximity of these sites, approximately 0.376 nm apart, is crucial for efficient proton hopping via the Grotthuss mechanism.
3) **Long-Range Continuous Proton Transport Pathways:** In M-SP/DMNPs@HPW membranes, the alignment of DMNPs@HPW nanoparticles under a magnetic field forms structured, long-range proton transport channels. These channels interconnect previously isolated ionic regions within the SPEEK matrix, significantly improving proton conductivity, particularly in the vertical direction.

Together, these factors highlight the role of DMNPs@HPW in optimizing the proton conduction network within the membrane, resulting in reduced activation energy, enhanced proton mobility, and higher overall proton conductivity.

13.3.8 Fuel Cell Performance

Fuel cell performance was evaluated using membrane electrode assembly (MEA) prepared from pure SPEEK and composite membranes containing 2 and 4 wt% DMNPs@HPW. Current-voltage polarization tests as shown in Figure 13.15, conducted at 80 °C and 100% RH, revealed an open-circuit voltage above 0.98 V for all membranes, confirming high catalytic activity and low hydrogen permeability. The SP/DMNPs@HPW membranes, which incorporate well-dispersed HPW-loaded nanoparticles, showed additional proton transfer sites and reduced ohmic resistance compared to pure SPEEK. Consequently, these composite membranes exhibited improved performance, with SP/DMNPs@HPW-2 achieving a maximum power density of 360 mW/cm^2, surpassing the 277 mW/cm^2 of pure SPEEK.

The M-SP/DMNPs@HPW membranes, prepared under a magnetic field, benefited from the alignment of DMNPs@HPW particles within the SPEEK matrix. This alignment resulted in ordered, continuous proton transport channels, particularly enhancing vertical proton conductivity. The M-SP/DMNPs@HPW-2 membrane reached a peak power density of 433 mW/cm^2, further demonstrating the performance advantages conferred by the magnetic field–induced structure.

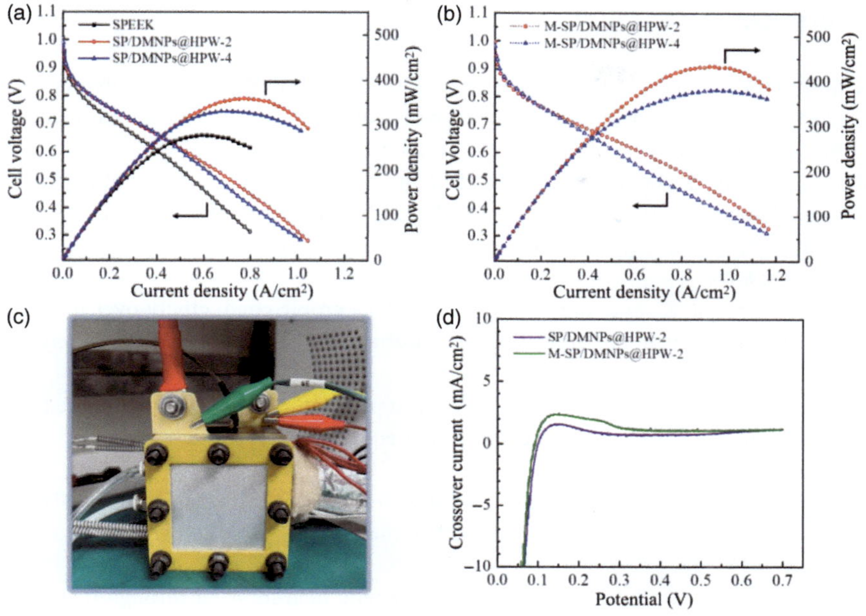

Figure 13.15 (a, b) H_2/O_2 PEMFC single-cell performance of MEAs employing pristine SPEEK, SP/DMNPs@HPW and M-SP/DMNPs@HPW composite membranes at 80 °C, 100% RH. (c) Optical photograph of single-cell equipment for fuel cell test. (d) Linear sweep voltammetry for hydrogen crossover on the MEAs of SP/DMNPs@HPW-2 and M-SP/DMNPs@HPW-2 membranes. Source: Han et al. [14]/with permission of Elsevier.

Additionally, hydrogen permeability was measured under operating conditions at 80 °C and 100% RH. The SP/DMNPs@HPW-2 composite membrane recorded a hydrogen permeation current of 0.73 mA/cm^2 at 0.4 V, while the M-SP/DMNPs@HPW-2 membrane, benefiting from vertical proton pathways, displayed a slightly higher value of 1.07 mA/cm^2. These results confirm the effectiveness of the magnetic field–aligned channels in optimizing fuel cell performance while maintaining acceptable hydrogen permeability.

13.4 Conclusion

In this chapter, we introduced DMNPs@HPW nanoparticles into a SPEEK matrix, with some membranes prepared under a magnetic field to achieve ordered nanoparticle alignment. These composite membranes exhibited improved mechanical strength, dimensional stability, water uptake, and proton conductivity compared to pure SPEEK. The ordered arrangement of nanoparticles in M-SP/DMNPs@HPW membranes created long-range proton transport channels, further enhancing performance. Among the tested membranes, M-SP/DMNPs@HPW-2 demonstrated the best results, achieving 159.3 mS/cm proton conductivity and a maximum power density of 433 mW/cm^2. These findings highlight the potential of magnetic field–induced nanoparticle alignment to advance PEM performance.

References

1 Shin, D.W., Guiver, M.D., and Lee, Y.M. (2017). Hydrocarbon-based polymer electrolyte membranes: Importance of morphology on ion transport and membrane stability. *Chemical Reviews* 117 (6): 4759–4805.

2 Liu, L., Xing, Y., Li, Y. et al. (2022). Enhanced mechanical durability of perfluorosulfonic acid proton-exchange membrane based on a double-layer ePTFE reinforcement strategy. *International Journal of Hydrogen Energy* 47 (67): 29014–29026.

3 Ma, S., Siroma, Z., and Tanaka, H. (2006). Anisotropic conductivity over in-plane and thickness directions in Nafion-117. *Journal of The Electrochemical Society* 153 (12): A2274.

4 Park, M.J. and Balsara, N.P. (2010). Anisotropic proton conduction in aligned block copolymer electrolyte membranes at equilibrium with humid air. *Macromolecules* 43 (1): 292–298.

5 Yun, S.H., Shin, S.H., Lee, J.Y. et al. (2012). Effect of pressure on through-plane proton conductivity of polymer electrolyte membranes. *Journal of Membrane Science* 417–418: 210–216.

6 Hyun, J., Doo, G., Yuk, S. et al. (2020). Magnetic field-induced through-plane alignment of the proton highway in a proton exchange membrane. *ACS Applied Energy Materials* 3 (5): 4619–4628.

7 Lee, J.Y., Lee, J.H., Ryu, S. et al. (2015). Electrically aligned ion channels in cation exchange membranes and their polarized conductivity. *Journal of Membrane Science* 478: 19–24.

8 Gahlot, S. and Kulshrestha, V. (2015). Dramatic improvement in water retention and proton conductivity in electrically aligned functionalized CNT/SPEEK nanohybrid Proton exchange membrane (PEM). *ACS Applied Materials & Interfaces* 7 (1): 264–272.

9 Ryu, S., Kim, J.H., Lee, J.Y. et al. (2018). Investigation of the effects of electric fields on the nanostructure of Nafion and its proton conductivity. *Journal of Materials Chemistry A* 6 (42): 20836–20843.

10 Zhang, B., Cao, Y., Li, Z. et al. (2017). Proton exchange nanohybrid membranes with high phosphotungstic acid loading within metal-organic frameworks for PEMFC applications. *Electrochimica Acta* 240: 186–194.

11 Wu, H., Shen, X., Cao, Y. et al. (2014). Composite proton conductive membranes composed of sulfonated poly(ether ether ketone) and phosphotungstic acid-loaded imidazole microcapsules as acid reservoirs. *Journal of Membrane Science* 451: 74–84.

12 Hasani-Sadrabadi, M.M., Dashtimoghadam, E., Majedi, F.S. et al. (2016). Ionic nanopeapods: Next-generation proton conducting membranes based on phosphotungstic acid filled carbon nanotube. *Nano Energy* 23: 114–121.

13 Wang, S., Zhang, Z., Liu, B. et al. (2013). Silica coated magnetic Fe3O4 nanoparticles supported phosphotungstic acid: A novel environmentally friendly catalyst for the synthesis of 5-ethoxymethylfurfural from 5-hydroxymethylfurfural and fructose. *Catalysis Science & Technology* 3 (8): 2104–2112.

14 Han, D., Sun, J., Ge, J. et al. (2024). Magnetically aligned composite membranes based on sulfonated poly (ether ether ketone) and phosphotungstic acid-loaded magnetic nanoparticles for fuel cell. *International Journal of Hydrogen Energy* 57: 301–314. https://doi.org/10.1016/j.ijhydene.2024.01.023.

15 Masteri-Farahani, M., Movassagh, J., Taghavi, F. et al. (2012). Magnetite–polyoxometalate hybrid nanomaterials: Synthesis and characterization. *Chemical Engineering Journal* 184: 342–346.

16 Zhang, Z., Zhang, F., Zhu, Q. et al. (2011). Magnetically separable polyoxometalate catalyst for the oxidation of dibenzothiophene with H_2O_2. *Journal of Colloid and Interface Science* 360 (1): 189–194.

17 Okuhara, T., Mizuno, N., and Misono, M. (1996). Catalytic chemistry of heteropoly compounds. In: *Advances in Catalysis*, vol. 41 (ed. D.D. Eley, W.O. Haag, and B. Gates), 113–252. Academic Press.

18 Xing, P., Robertson, G.P., Guiver, M.D. et al. (2004). Synthesis and characterization of sulfonated poly(ether ether ketone) for proton exchange membranes. *Journal of Membrane Science* 229 (1-2): 95–106.

14

Chemical Covalent Bonding of Silicotungstic Acid Proton Exchange Membrane

14.1 Introduction

Heteropoly acids (HPAs) are renowned for their superstrong acidity and exceptional water retention, making them appealing candidates for enhancing proton exchange membrane (PEMs). However, their high water solubility poses a significant challenge: direct incorporation into PEMs often leads to HPA leaching, undermining both membrane performance and long-term stability. Preventing HPA leakage is essential to ensure reliable fuel cell operation.

One effective strategy to address the solubility and recyclability challenges of HPAs involves their immobilization through nanostructuring or chemical modification. For instance, Lu et al. [1] utilized vacuum-assisted impregnation to incorporate HPAs into mesoporous silica frameworks, enhancing stability and dispersion. Similarly, Mohammad et al. [2] employed carbon nanotube nanosponges embedded in a polymer matrix as carriers for HPA confinement. More recent advances have emphasized covalent anchoring of HPAs onto functionalized support materials, yielding nanohybrids with significantly reduced water solubility. Zhai et al. [3] synthesized water-insoluble HPA hybrids via solvothermal techniques, incorporating nitrogen-doped carbon and covalent organic frameworks. Other approaches use amine-functionalized nanofillers—such as mesoporous silica, halloysite tubes, and nitrogen-doped carbon nanosheets—which interact with HPAs through hydrogen bonding. These functionalized nanostructures help confine HPA, reducing its solubility and enhancing retention. By employing these advanced immobilization strategies, PEMs maintain high proton conductivity and durability, even under prolonged water exposure, paving the way for more robust and reliable fuel cell technologies.

As discussed in Chapter 9, sulfonated poly(ether ether ketone) (SPEEK) sulfonic acid groups can react with phenyl rings, forming covalent sulfonyl bonds through a dimethylsulfoxide (DMSO)-catalyzed thermal cross-linking process. This reaction builds a three-dimensional network structure that enhances mechanical

Electrospinning for Proton Exchange Membrane Fuel Cells, First Edition.
Yong Liu and Mohideen Meerasahib Mohamedazeem.

strength, thermal stability, and chemical durability. Extending this approach, small molecules with phenyl rings can be similarly immobilized onto SPEEK membranes via heat-activated covalent bonding [4–6].

Herein, phenoxyphenyl-modified silica tungstic acid (POP-SiWA) was blended with high-sulfonation SPEEK in a DMSO/cyclobutanesulfonic anhydride (CBSA) solvent mixture. CBSA played a crucial role in preventing POP-SiWA precipitation and optimizing the cross-linking efficiency. After forming films, a thermal grafting reaction at 180 °C bonded the POP-SiWA molecules to SPEEK chains. This process produced SP/PSiW-*x* membranes with covalently attached silicotungstic acid, significantly enhancing proton conductivity due to the strong acidity and excellent water retention of silicotungstic acid. Moreover, the covalent immobilization prevented leaching, maintained conductivity stability, and conferred improved resistance to free radicals. The results highlight the benefits of integrating phenoxyphenyl-modified silicotungstic acid into SPEEK, demonstrating an effective strategy for developing high-performance PEMs.

14.2 Materials Preparation

14.2.1 Synthesis of 4-Phenoxyphenyl Phosphoric Acid

The synthesis of 4-phenoxyphenyl phosphoric acid (POP-PA) follows the procedure outlined in Figure 14.1 [8, 9]. The process begins with 4-bromodiphenyl ether as the precursor, which undergoes a reaction to form 4-phenoxyphenyl diethylphosphoric acid ester in the following steps.

14.2.2 Synthesis of 4-Phenoxyphenyl Diethylphosphoric Acid Ester

In a controlled reaction setup, 10 g (40.14 mmol, 1 equivalent) of 4-bromodiphenyl ether and 0.7 g (3.21 mmol, 0.08 equivalent) of anhydrous nickel bromide (as the catalyst) were added to 15 mL of xylene under a nitrogen atmosphere. The mixture was heated to 160–180 °C while stirring vigorously to ensure proper dispersion of the catalyst particles. A constant-pressure dropping funnel was used to gradually introduce 10 g (60.21 mmol, 1.5 equivalent) of triethyl phosphite into the system. The reaction was maintained under the same temperature and atmosphere for 2 hours, allowing for efficient synthesis. Following completion, the reaction mixture was cooled to room temperature, and the residual catalyst was removed via multiple extractions with a 5% aqueous hydrogen chloride (HCl) solution. The organic phase was then concentrated under reduced pressure using distillation to obtain the crude 4-phenoxyphenyl diethylphosphoric acid ester.

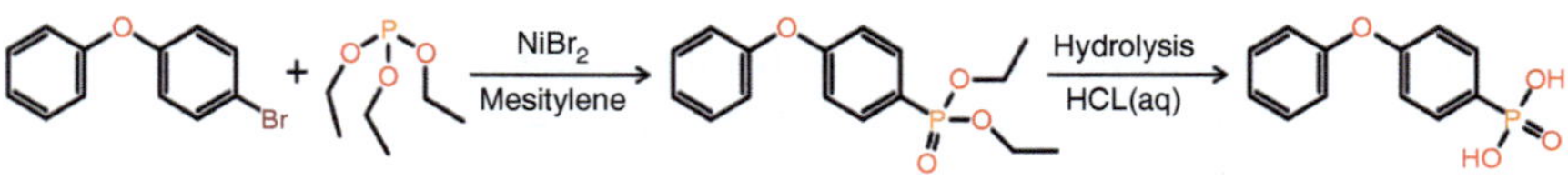

Figure 14.1 Synthetic route of (4-phenoxyphenyl)phosphonic acid.

14.2.3 Hydrolysis to Prepare 4-Phenoxyphenyl Phosphoric Acid

The crude 4-phenoxyphenyl diethylphosphoric acid ester was subjected to direct hydrolysis in 30 mL of concentrated HCl under vigorous stirring and reflux conditions for 24 hours. Upon hydrolysis, a white solid precipitate was observed. The reaction mixture was cooled to room temperature and then filtered to isolate the crude POP-PA product.

Due to its limited solubility in dichloromethane, the crude product was subjected to washing with dichloromethane for purification. The final POP-PA powder was obtained by drying the purified product at 60 °C, ensuring an efficient and reproducible synthesis. This process yields POP-PA, a crucial material for subsequent PEM research and applications.

14.2.4 Preparation of Deficient Silicotungstic Acid Potassium

The synthesis of deficient silicotungstic acid potassium ($K_8SiW_{11}O_{39}$) follows a procedure adapted from literature sources. The preparation involves the following steps:

1) **Dissolution:** Dissolve 33.0 g (100 mmol) of sodium tungstate dihydrate ($Na_2WO_4{\cdot}2H_2O$) in 50 mL of distilled water at room temperature under magnetic stirring.
2) **Silicate Addition:** Introduce 1.11 g (9.09 mmol) of sodium metasilicate (Na_2SiO_3) into the solution and heat until boiling.
3) **Acid Treatment:** While maintaining a boiling state, slowly add 34 mL of 4 M concentrated hydrochloric acid. Continue boiling for 1 hour, then allow the mixture to cool to room temperature.
4) **Precipitation and Isolation:** Once cooled, 12.5 g of solid potassium chloride (KCl) is added, leading to the formation of a white precipitate. The precipitate is filtered, washed twice with 1 M KCl solution, followed by ice water rinsing, and then air-dried to obtain the final white solid $K_8SiW_{11}O_{39}$.

14.2.5 Preparation of POP-SiWA

The process includes:

1) **Reaction Setup:** Dissolve 1.0 g (4 mmol) of POP-PA in 60 mL of *N*,*N*-dimethylacetamide (DMAc) under continuous stirring. Heat the solution to 80 °C to facilitate dissolution.
2) **Reagent Addition and Transformation:** Slowly add 4.0 g (1.34 mmol) of $K_8SiW_{11}O_{39}$, causing the mixture to turn cloudy. Magnetic stirring is employed to prevent precipitation. Introduce 0.8 mL of 12 M concentrated HCl, leading to a transformation from a cloudy mixture to a transparent light yellow solution.
3) **Reaction Completion:** Maintain the reaction at 80 °C for 70 hours. Cool the solution to room temperature and filter to remove KCl crystals.
4) **Purification and Isolation:** The filtrate is concentrated under reduced pressure, yielding a viscous crude product. Since POP-PA is highly soluble in hot water but has limited solubility in cold water, hot water is used for purification.

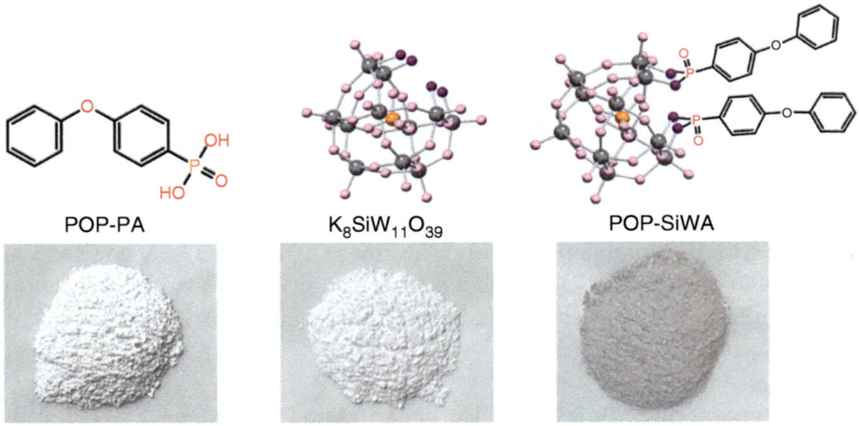

Figure 14.2 Optical photographs of POP-PA, $K_8SiW_{11}O_{39}$ and POP-SiWA powders. Source: Han et al. [7]/with permission of Elsevier.

> The crude product is fully dissolved in hot water and slowly cooled to room temperature. The precipitate (POP-PA) is filtered out, and the remaining filtrate is further concentrated to obtain the white solid POP-SiWA.

The morphological appearances of POP-PA, $K_8SiW_{11}O_{39}$, and POP-SiWA powders are depicted in Figure 14.2.

14.2.6 Sulfonated Poly(Ether Ether Ketone)-Immobilized POP-SiWA

The process of immobilizing POP-SiWA into SPEEK polymer membranes is illustrated in Figure 14.3. The preparation of highly sulfonated SPEEK begins with vacuum-drying poly(ether ether ketone) (PEEK) powder at 80 °C for 24 hours. Following this, 10 g of PEEK is added to 100 mL of concentrated sulfuric acid and stirred at 20 °C for 12 hours to dissolve and react. The temperature is then increased to 50 °C and maintained for an additional 3.5 hours. To halt the reaction, the solution is rapidly cooled using cold water. The reaction mixture is then slowly poured into an excess amount of ice water, resulting in the precipitation of a white fibrous product, SPEEK. This product is washed multiple times with deionized ice water until neutral and then vacuum-dried at 60 °C for 24 hours. The sulfonation degree of SPEEK is determined to be 80% using ^{1}H nuclear magnetic resonance (NMR) analysis.

An improvement to the conventional SPEEK thermal cross-linking method has been implemented. Traditionally, DMSO is used as a solvent to prepare the casting solution, and after film formation, it is subjected to long-term cross-linking at 160–180 °C. In this study, a DMSO/cyclo-butene sulfone–mixed solvent system is used. For instance, 0.3 g of SPEEK polymer is dissolved in a DMSO/cyclo-butene sulfone mixture (3.0 g, mass ratio 6:1) to obtain the casting solution. The solution is filtered, degassed, and then cast onto a clean glass plate before being heated at 60 °C for 12 hours to form a film.

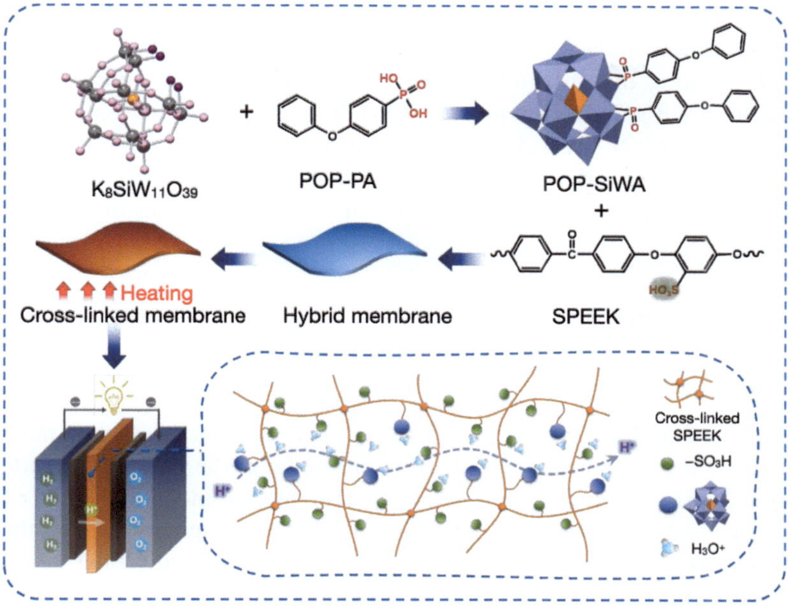

Figure 14.3 The process of covalently immobilized silicotungstic acid in cross-linked SPEEK membrane.

During this process, most of the DMSO solvent evaporates, whereas cyclo-butene sulfone, which has a high boiling point of 285 °C, remains in the film. The film then undergoes a staged thermal treatment, consisting of pretreatment at 120 °C for 2 hours, followed by sequential heating at 140 and 160 °C for 2 hours each, and a final cross-linking step at 180 °C for 6 hours. As the cross-linking reaction progresses, the film changes from a transparent light yellow to a dark brown color.

The thermal cross-linking reaction occurs through an electrophilic substitution mechanism, wherein the sulfonic acid groups of SPEEK react with the phenyl rings of the polymer backbone, forming sulfone bridges that lead to polymer cross-linking. To facilitate this process, small molecules containing phenyl rings are introduced into the casting solution. After film formation, these phenyl rings react with the sulfonic groups of SPEEK, establishing covalent bonds that ensure the immobilization of the small molecules within the membrane structure.

In this study, an enhanced thermal cross-linking method has been employed to graft POP-SiWA onto SPEEK. A 20 wt% POP-SiWA cross-linked grafted film is prepared by dissolving 0.3 g of SPEEK polymer and 0.075 g of POP-SiWA in the DMSO/cyclo-butene sulfone (3.0 g, mass ratio 6:1) mixed solvent to create the casting solution. After filtration and degassing, the solution is cast onto a glass plate to form a film. The temperature treatment process remains identical to that used for the pure SPEEK cross-linked film. The resulting cross-linked grafted membranes are designated as SP/PSiW-x, where $x = 10$, 20, or 30, representing the mass fraction of POP-SiWA in the composite membrane.

Since not all of the incorporated POP-SiWA is successfully immobilized in the film, an accurate determination of the immobilized POP-SiWA content is necessary. To minimize leaching, the cross-linked films undergo a controlled acidification process. The films are first immersed in 2 M HCl at 20 °C for 6 hours. Given the high molecular weight of POP-SiWA, its leaching is slow at low temperatures, and the mass loss during this stage is negligible. After acidification, the films are rinsed with cold water and subsequently dried for further characterization and testing.

14.3 Results and Discussion

14.3.1 Characterization of POP-SiWA

The synthesis of POP-PA begins with 4-bromodiphenyl ether as the starting material, which undergoes a reaction with triethyl phosphite in the presence of $NiBr_2$ as a catalyst. During this process, the bromine atom on 4-bromodiphenyl ether is substituted, leading to the formation of 4-phenoxyphenyl phosphonic acid diethyl ester. This intermediate is then hydrolyzed using a strong acid to yield POP-PA. Based on the method reported by Andrew et al., $K_8SiW_{11}O_{39}$ can react with the phenyl phosphonic acid group, forming a covalent chemical bond that integrates the inorganic heteropoly acid with the organic compound. In this study, POP-PA, which contains the phenyl phosphonic acid group, was reacted with $K_8SiW_{11}O_{39}$ to synthesize POP-SiWA. The reaction products were characterized using ^{1}H NMR and ^{31}P NMR spectroscopy, with the corresponding spectra presented in Figure 14.4a,b.

First, ^{1}H NMR spectrum was analyzed as shown in Figure 14.4a. The hydrogen atoms in the chemical structure of the compounds are labeled alongside their corresponding signal peaks. The protons present on the benzene rings of both POP-PA and POP-SiWA exhibit clear correlations with the signal peaks in the spectrum, with minimal impurities, indicating the high purity of the synthesized products. The chemical shift of the benzene ring protons in POP-PA appears in the range of 7.72 to 7.03 ppm. Since the benzene ring structure in POP-SiWA remains similar to that of POP-PA, the position and shape of the signal peaks are expected to be relatively consistent.

However, the signal peaks of the benzene ring in POP-SiWA exhibit an overall shift, with the chemical shifts of the protons occurring in the range of 7.67 to 6.95 ppm, slightly shifted to the right compared to those in POP-PA. This shift is primarily attributed to the grafting of silicotungstic acid onto POP-SiWA. The difference in electronegativity between silicotungstic acid and the phenyl phosphonic acid group alters the electron cloud distribution within the molecule, causing slight variations in the chemical shifts of the hydrogen atoms on the benzene rings. These shifts arise due to the different inductive effects exerted by the functional groups. Additionally, the differences observed in the ^{1}H NMR spectra of POP-PA and POP-SiWA further confirm the successful synthesis of POP-SiWA.

Figure 14.4b presents the ^{31}P NMR spectra for both POP-PA and POP-SiWA. The chemical shift of the phosphorus atom in POP-PA is observed at 12.81 ppm,

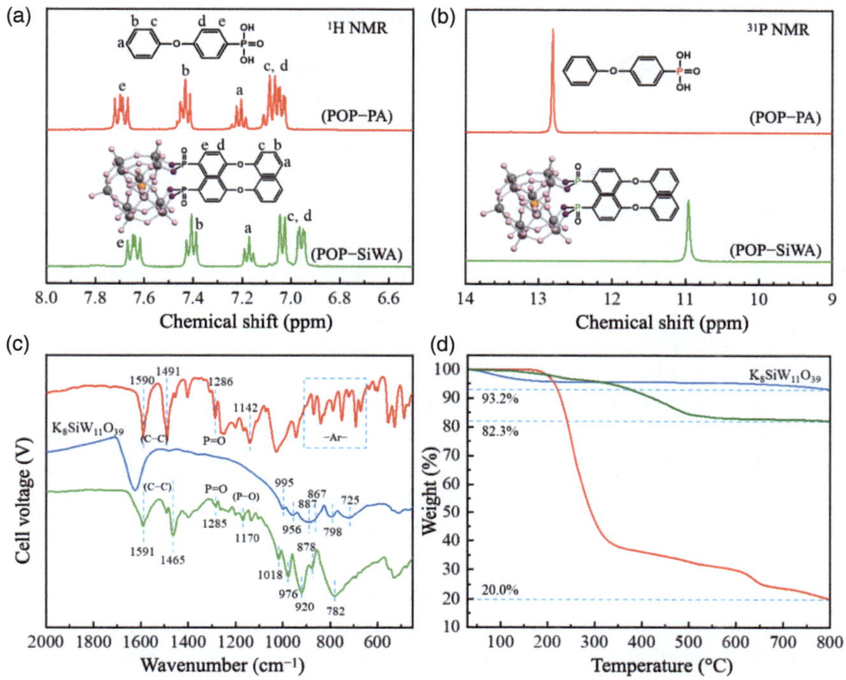

Figure 14.4 (a, b) ^{1}H NMR and ^{31}P NMR spectra of POP-PA and POP-SiWA; (c, d) FT-IR spectra and TGA curves of $K_8SiW_{11}O_{39}$, POP-PA, and POP-SiWA.

while in POP-SiWA, it shifts to 10.95 ppm, indicating a rightward shift. This shift can also be attributed to the difference in electronegativity between silicotungstic acid and phenyl phosphonic acid, further confirming that $K_8SiW_{11}O_{39}$ successfully reacts with POP-PA through the formation of a covalent chemical bond.

Fourier transform infrared (FT-IR) curves of POP-PA, POP-SiWA, and $K_8SiW_{11}O_{39}$ is presented in Figure 14.4c. The infrared absorption spectrum reveals the characteristic peaks corresponding to the phenyl phosphonic acid functional group in POP-PA, which appear at 1286 cm^{-1} for P=O, 1142 cm^{-1} for P–O, and at 1590 and 1491 cm^{-1} for the benzene ring carbon–carbon single bond (C—C) characteristic vibrations. In $K_8SiW_{11}O_{39}$, the four oxygen atoms produce characteristic vibration peaks at 887 cm^{-1} for Si–Oa, 867 cm^{-1} for W–Ob–W, 798 and 725 cm^{-1} for W–Oc–W, and 995 and 956 cm^{-1} for W=Od. Here, Oa represents the oxygen atom in the SiO_4 tetrahedron, Ob represents the shared corner oxygen, Oc represents the shared edge oxygen, and Od corresponds to the terminal oxygen atom.

In POP-SiWA, the characteristic peaks of W=Od appear at 1018, 976, and 920 cm^{-1}, while the Si–Oa peak shifts to 878 cm^{-1} and the W–Oc–W peak appears at 782 cm^{-1}, indicating a noticeable shift in absorption relative to $K_8SiW_{11}O_{39}$. Meanwhile, in the absorption spectrum of POP-SiWA, the characteristic C–C peaks of the benzene ring appear at 1591 and 1465 cm^{-1}, along with the characteristic P=O peak at 1285 cm^{-1} and the P–O vibration peak at 1170 cm^{-1}. These FT-IR results confirm the successful preparation of POP-SiWA.

Thermogravimetric analysis (TGA) was performed on POP-PA, POP-SiWA, and $K_8SiW_{11}O_{39}$ to evaluate their thermal stability. The TGA results are shown in Figure 14.4d. $K_8SiW_{11}O_{39}$ retains 93.2% of its mass in the temperature range of 30 to 800 °C, with the major mass loss occurring around 150 °C, corresponding to the loss of adsorbed and crystalline water, indicating that the inorganic compound $K_8SiW_{11}O_{39}$ exhibits good thermal stability. POP-PA experiences significant mass loss, primarily in the range of 200 to 350 °C, likely due to the sublimation and volatilization of POP-PA during heating. The subsequent mass loss from 350 to 800 °C is attributed to the decomposition of the benzene ring and phosphonic acid groups.

POP-SiWA undergoes major mass loss between 350 and 500 °C, likely due to the rupture of the P—O bond at high temperatures, leading to the formation of small molecular organic compounds that quickly volatilize. The final residual mass of POP-SiWA is 82.3%. The TGA weight loss curve also indicates that POP-SiWA maintains good thermal stability below 250 °C, meeting the requirements for the subsequent SPEEK thermal cross-linking grafting reaction.

The XRD patterns of $K_8SiW_{11}O_{39}$ and POP-SiWA powders are shown in Figure 14.5. The diffraction peak positions of $K_8SiW_{11}O_{39}$ match well with the literature data (JCPDS No. 34-0205), confirming a cubic crystal system. The increased peak width suggests an increase in surface crystallite size. For POP-SiWA, the introduction of the phenoxyphenyl group to $K_8SiW_{11}O_{39}$ via P—O bonds reduces molecular regularity, resulting in poor crystallinity. A sharp diffraction peak appears at 8.3°, corresponding to the characteristic diffraction peak of the Keggin-type structure of silicotungstic acid on POP-SiWA. The presence of broad peaks at other positions indicates that POP-SiWA exhibits an amorphous nature.

14.3.2 Discussion on Thermal Cross-Linking and Grafting

The conventional SPEEK thermal cross-linking method uses DMSO as a solvent for film preparation. After the film is formed, the residual DMSO acts as a catalyst

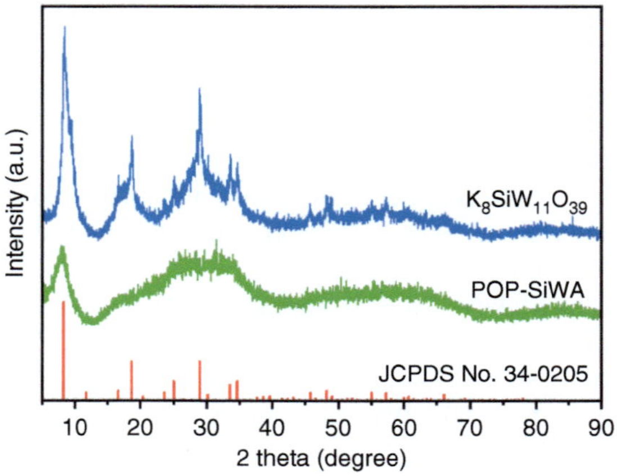

Figure 14.5 XRD patterns of $K_8SiW_{11}O_{39}$ and POP-SiWA.

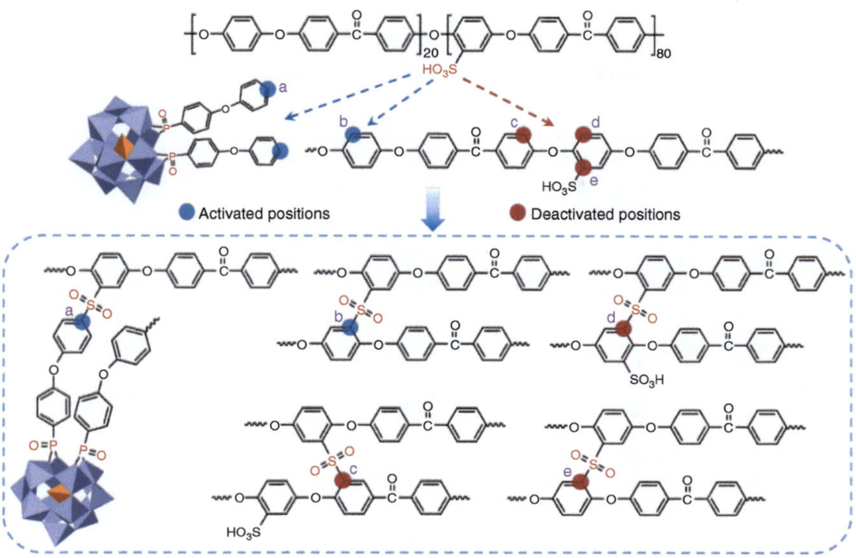

Figure 14.6 Schematic of activation barriers and possible cross-linking reactions.

for thermal cross-linking, promoting the reaction between the sulfonic acid groups on SPEEK and the benzene rings to form sulfone cross-links. Maranesi et al. studied the mechanism of SPEEK thermal cross-linking and pointed out that the polar aprotic solvent DMSO can promote charge separation and the formation of electrophilic agents such as $-SO_2^+$, with the formation of $-SO_2^+$ being the initiation step in the thermal cross-linking reaction.

In Figure 14.6, the active sites on the benzene ring that can react with the sulfonic acid group are indicated in blue and red. The blue sites have a higher density of π-electron cloud on the benzene ring, making them more susceptible to electrophilic substitution reactions with $-SO_2^+$. Therefore, the benzene ring of POP-SiWA reacts more readily with the sulfonic acid groups on SPEEK during the grafting reaction.

In this study, the conventional thermal cross-linking method was first used to prepare films, where DMSO was used to dissolve SPEEK along with a certain amount of POP-SiWA to form the casting solution. The solution was then cast onto a glass plate and heated at 60 °C to allow solvent evaporation and film formation. The film underwent pretreatment at 120 °C for 6 hours, followed by cross-linking at 180 °C to initiate the grafting reaction, leading to the covalent bonding of $-SO_2^-$ to immobilize POP-SiWA.

As shown in Figure 14.7, the photograph of the film prepared using the conventional thermal cross-linking method reveals that after the reaction, the color of the film changes from colorless and transparent to dark yellow. The SP/PSiW-30 film on the right exhibits thicker edges, indicating the presence of more residual DMSO solvent. At high temperatures, the solvent vaporizes, generating bubbles. This suggests that the amount of residual DMSO must be carefully controlled in the conventional thermal cross-linking method to avoid defects.

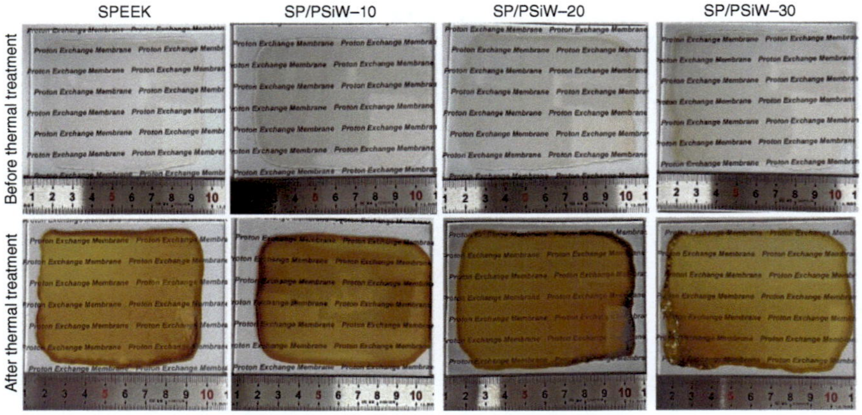

Figure 14.7 Optical images of SPEEK-based membranes before and after thermal treatment. Source: Han et al. [7]/with permission of Elsevier.

For SPEEK with a specific degree of sulfonation, factors influencing its cross-linking degree include cross-linking temperature, cross-linking time, and solvent type. Beyraghi et al. systematically studied the effect of different residual DMSO amounts on the cross-linking degree after film formation [10]. Their findings indicated that a higher residual DMSO content facilitates the cross-linking reaction. Under specific cross-linking temperature and time conditions, a higher initial DMSO residual amount in the SPEEK film results in a greater cross-linking degree.

However, if the residual solvent content is too low, two problems may arise. First, the amount of the electrophilic agent $-SO_2^+$ catalyzed by DMSO decreases, reducing the efficiency of the reaction. Second, a low solvent residue limits the free volume of large molecular chains, restricting their movement and thereby reducing the overall efficiency of the thermal cross-linking process. Conversely, excessive residual solvent also has adverse effects. While higher solvent residues promote the cross-linking reaction, excessive solvent vaporizes at high temperatures, leading to bubble formation and severe film defects. Additionally, an excessive solvent residue increases the free volume of molecular chains, potentially creating a loose, porous structure within the cross-linked and cured film. This porous structure can reduce the film's ability to block gases effectively [10]. To address these issues, the film is typically pretreated at 120 °C to regulate the residual DMSO amount within an optimal range.

The cross-sectional scanning electron microscope (SEM) morphology of the pure SPEEK membrane and SP/PSiW-x crosslinked membranes prepared using the conventional thermal cross-linking method is shown in Figure 14.8. The cross-sectional morphology of the SPEEK membrane and SP/PSiW-10 membrane appears relatively dense and smooth. In contrast, the cross-sections of SP/PSiW-20 and SP/PSiW-30 membranes exhibit a rougher structure, with numerous solid particles embedded within the membrane. This phenomenon is likely due to excess POP-SiWA precipitating inside the membrane as the DMSO solvent content decreases during the thermal cross-linking process.

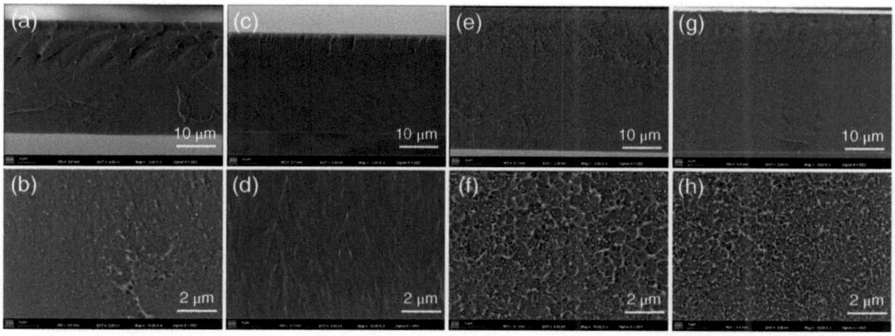

Figure 14.8 SEM images of the cross-section of (a, b) pristine SPEEK, (c, d) SP/PSiWA-10, (e, f) SP/PSiWA-20, and (g, h) SP/PSiWA-30 membranes prepared by the conventional thermal cross-linking method. Source: Han et al. [7]/with permission of Elsevier.

During the initial stage of thermal cross-linking, the membrane contains a high amount of residual DMSO solvent, enabling the cross-linking reaction to proceed rapidly. At this stage, POP-SiWA remains dissolved in DMSO and is uniformly distributed throughout the membrane. It is important to note that the evaporation of DMSO and the progression of the cross-linking reaction occur simultaneously. In the case of the SP/PSiW-10 membrane, where a relatively low amount of POP-SiWA is added, most of the POP-SiWA undergoes grafting during the early phase of thermal cross-linking. Consequently, after the complete evaporation of DMSO, no significant precipitation of solid particles is observed.

However, in SP/PSiW-20 and SP/PSiW-30 membranes, which contain a higher amount of POP-SiWA, the initial thermal cross-linking process is insufficient to complete most of the grafting reactions. As DMSO continues to evaporate, unreacted POP-SiWA precipitates out of the membrane and remains embedded as fine solid particles. This results in two major issues. First, the grafting efficiency of POP-SiWA decreases in SP/PSiW-20 and SP/PSiW-30 membranes. Second, during subsequent washing, the unreacted POP-SiWA readily leaches out, leaving voids where the dissolved particles were removed. These voids negatively impact the proton conductivity and mechanical strength of the membrane and may even lead to increased gas permeability.

To overcome these issues, modifications to the thermal cross-linking grafting method are necessary. One of the primary challenges is the precipitation of POP-SiWA due to low residual DMSO solvent levels. A higher solvent residue can help prevent this precipitation while enhancing the mobility of polymer chains, thereby accelerating the grafting reaction of POP-SiWA. However, as discussed earlier, an excessive residual amount of DMSO leads to bubble formation. Even when reducing the cross-linking temperature to 150 °C, the formation of bubbles remains unavoidable.

To resolve this issue, an alternative approach involves introducing an inert, high-boiling-point solvent that can slow down the evaporation of residual solvent at high temperatures. This solvent should be capable of effectively dissolving both SPEEK and POP-SiWA while ensuring it does not interfere with the cross-linking reaction. After screening various solvents, cyclobutanesulfonic acid was identified as a suitable candidate. As a nonprotonic polar solvent, cyclobutanesulfonic acid exhibits excellent polymer solubility and possesses a high boiling point of 285 °C at atmospheric pressure, ensuring good thermal stability at elevated temperatures. Moreover, experimental results confirm that cyclobutanesulfonic acid does not interfere with the DMSO-catalyzed sulfonic acid cross-linking reaction.

Therefore, in this study, a mixed solvent system of DMSO and cyclobutanesulfonic acid in a 6:1 mass ratio was used to prepare the casting solution. After film formation at 60 °C, the high-boiling-point solvent cyclobutanesulfonic acid remained within the film. During the subsequent cross-linking process at 180 °C, its slower evaporation allowed sufficient time for the completion of the POP-SiWA grafting reaction. This approach effectively prevents POP-SiWA precipitation, enhances grafting efficiency, and maintains the desired film structure.

As shown in Figure 14.9, photographs of the cross-linked membranes prepared using the improved thermal cross-linking method with DMSO/cyclobutanesulfonic acid mixed solvent indicate a significant color change. After the 180 °C thermal cross-linking reaction, the membrane color changed from light yellow and transparent to dark brown. The deeper color corresponds to a higher degree of cross-linking. Notably, under identical cross-linking time and temperature conditions, membranes prepared using the improved method exhibited a significantly higher cross-linking degree than those prepared using the conventional method. For example, the pure SPEEK membrane, after 6 hours of cross-linking at 180 °C using the improved method, exhibited lower water absorption and reduced swelling, which will be discussed in detail later.

In contrast, under the same cross-linking conditions, membranes prepared using the conventional method swelled and deformed when immersed in water at 80 °C, indicating a lower cross-linking degree. The primary reason for this is that the high-boiling-point solvent cyclobutanesulfonic acid remains within the membrane,

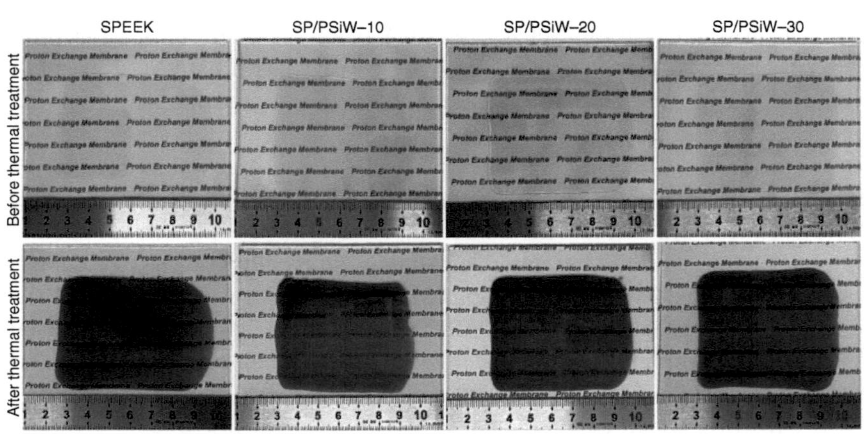

Figure 14.9 Optical images of SPEEK-based membranes before and after thermal treatment using the modified thermal cross-linking method. Source: Han et al. [7]/with permission of Elsevier.

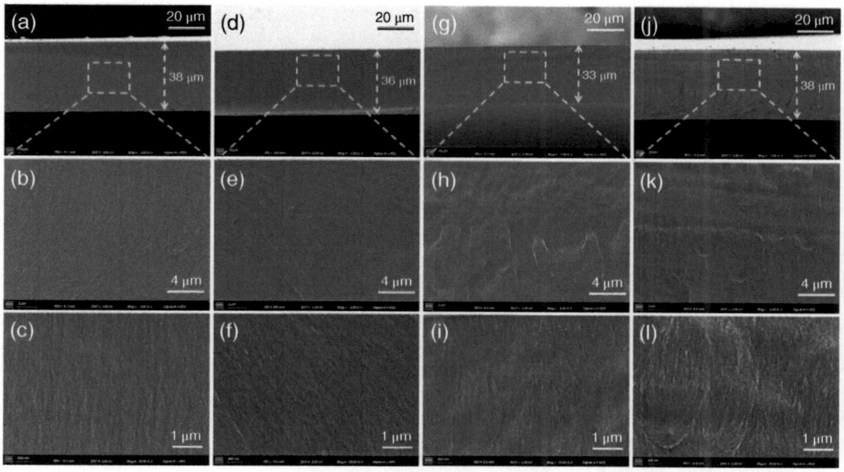

Figure 14.10 SEM images of the cross-section of (a, b, c) pristine SPEEK, (d, e, f) SP/PSiWA-10, (g, h, i) SP/PSiWA-20, and (j, k, l) SP/PSiWA-30 membranes prepared by the modified thermal cross-linking method. Source: Han et al. [7]/with permission of Elsevier.

allowing macromolecular chains to have greater free volume. This enhances the mobility of polymer chain segments, facilitating the cross-linking reaction and achieving a high cross-linking degree in a shorter time. Additionally, the membrane photographs in Figure 14.9 show that all samples exhibit uniform coloration with no visible bubbles, suggesting that the addition of a high-boiling-point solvent enables the preparation of defect-free cross-linked membranes.

As illustrated in Figure 14.10, cross-sectional SEM images of the pure SPEEK membrane and SP/PSiW-*x* cross-linked graft membranes prepared using the improved thermal cross-linking method reveal that all four sample membranes possess a dense and smooth cross-section, with no solid particles visible. This confirms that the use of a high-boiling-point solvent such as cyclobutanesulfonic acid effectively slows down solvent evaporation at elevated temperatures, providing ample time for the complete grafting reaction of POP-SiWA. Following the 180 °C thermal cross-linking process, most of the solvent has evaporated. Even in the SP/PSiW-20 and SP/PSiW-30 membranes, which contain higher POP-SiWA content, no solid particles precipitated out, demonstrating that the improved thermal cross-linking method is effective for grafting small-molecule compounds containing phenyl rings onto sulfonated aromatic polymer membranes.

Figure 14.11 presents the cross-sectional SEM image of the SP/PSiW-20 membrane, along with energy dispersive X-ray spectroscopy (EDS) mapping of element distribution and relative elemental content analysis. The uniform distribution of phosphorus, silicon, tungsten, and other elements indicates that POP-SiWA is well dispersed within the membrane, confirming the effectiveness of the improved grafting method.

As shown in Figure 14.12a, the mass change of the cross-linked membranes after soaking in water at 80 °C for varying durations is presented. The mass of both the

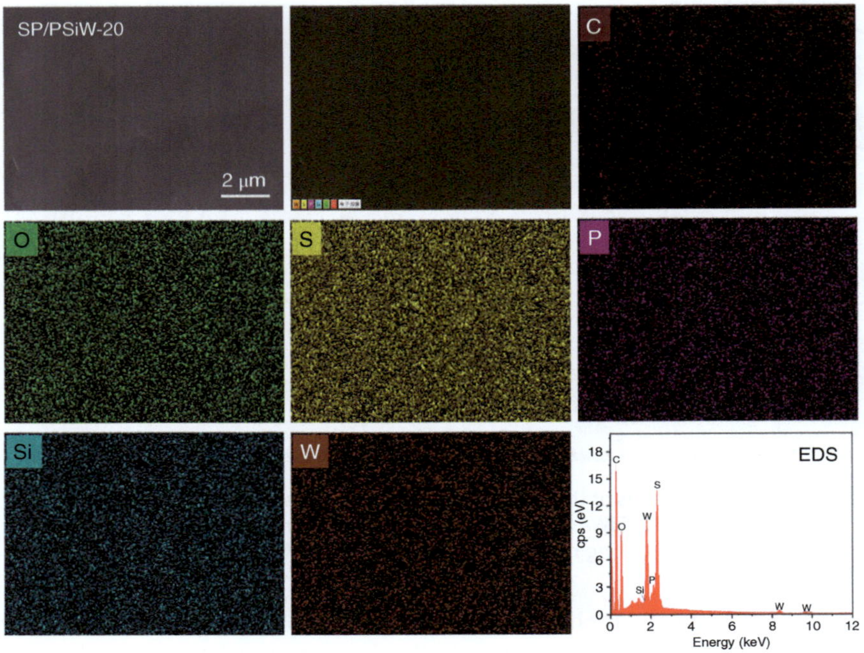

Figure 14.11 Elemental mapping of C, O, S, P, Si, and W in the cross-section of SP/PSiW-20 membrane by SEM-EDS. Source: Han et al. [7]/with permission of Elsevier.

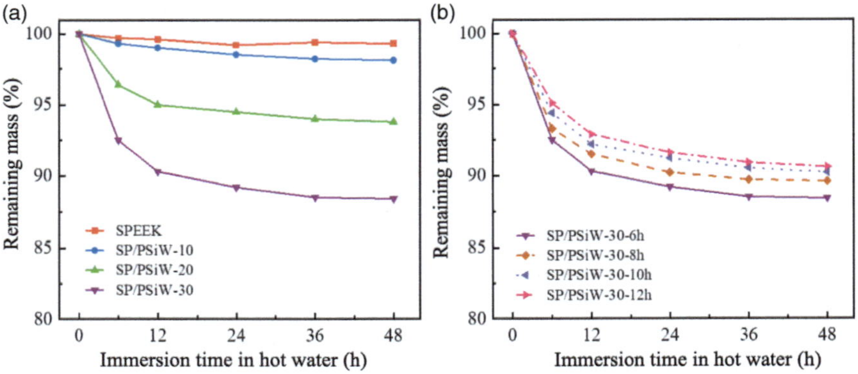

Figure 14.12 Residual mass of membrane after different times of immersion in 80 °C hot water. (a) Remaining mass (%) of SPEEK and SP/PSiW composite membranes with varying POP-SiWA content (10, 20, and 30 wt%) over 48 hours. (b) Remaining mass (%) of SP/PSiW-30 membranes prepared with different POP-SiWA immobilization times (6, 8, 10, and 12 hours) under the same hot water treatment.

Table 14.1 Residual weight of the membranes after 48 hours immersion in water at 80 °C and immobilization ratio of POP-SiWA in the membranes.

Membranes	n(-SO_3H)/n (POP-SiWA)[b)]	Thermal treatment time (h)	Residual weight (%)	POP-SiWA in membrane[c)] (%)	Immobilization ratio (%)
SPEEK	—	6	99.3	—	—
SP/PSiW-10	64.2	6	98.1	9.0	88.0
SP/PSiW-20	28.2	6	93.8	15.4	72.5
SP/PSiW-30	16.4	6	88.4	21.6	63.7
SP/PSiW-30-8h	16.4	8	89.6	22.7	67.6
SP/PSiW-30-10h	16.4	10	90.2	23.2	69.7
SP/PSiW-30-12h	16.4	12	90.6	23.5	71.0
SP50/POP-SiWA-10[a)]	—	—	91.2	2.1	—

POP-SiWA, phenoxyphenyl-modified silica tungstic acid; SPEEK, sulfonated poly(ether ether ketone).

a) Blended membrane containing 10 wt% of POP-SiWA in DS = 50% SPEEK.
b) Molar ratio of sulfonic acid group to POP-SiWA.
c) Calculations include weight lost from SPEEK (residual weight = 99.3%).

pure SPEEK membrane and the SP/PSiW-10 membrane remained relatively stable. After 48 hours of soaking, the mass of the pure SPEEK membrane retained 99.3%, while the SP/PSiW-10 cross-linked membrane exhibited a slight decrease to 98.1%. Table 14.1 presents the POP-SiWA loading and grafting efficiency after 48 hours of soaking. The SP/PSiW-10 membrane showed a POP-SiWA loading of 9.0% with a high grafting efficiency of 88.0%. This high efficiency is attributed to the relatively low amount of added POP-SiWA, with a molar ratio of $n(-SO_3H)/n$(POP-SiWA) of 64.2. Since the number of sulfonic acid groups is significantly greater than that of POP-SiWA, a thorough reaction between POP-SiWA and the sulfonic groups was achieved.

For SP/PSiW-20 and SP/PSiW-30 membranes, mass loss was most significant during the first 6 to 12 hours of soaking. Beyond 12 hours, the rate of mass change gradually decreased and stabilized. After 48 hours, the SP/PSiW-20 and SP/PSiW-30 membranes retained 93.8% and 88.4% of their mass, respectively. Their corresponding POP-SiWA loadings were 15.4% and 21.6%, with grafting efficiencies of 72.5% and 63.7%, respectively. These findings indicate that increasing the POP-SiWA content enhances its overall loading within the membrane, but at the cost of reduced grafting efficiency.

As illustrated in Figure 14.12b and Table 14.1, the effect of cross-linking time on POP-SiWA loading in SP/PSiW-30 membranes was investigated by subjecting the membrane to heat treatment at 180 °C for different durations (6, 8, 10, and 12 hours). As cross-linking time increased, the POP-SiWA loading exhibited a slight increase, from 21.6% at 6 hours to 23.5% at 12 hours. This increase is due to the rapid initial cross-linking and grafting reactions; however, as the solvent evaporates and the cross-linked network forms, molecular chain mobility becomes restricted, leading to a gradual decline in the grafting rate over time. Thus, while extending cross-linking time contributes slightly to increased POP-SiWA loading, its overall effect remains limited.

Additionally, the stability of POP-SiWA in membranes without cross-linking treatment was evaluated. For SPEEK with a sulfonation degree of 80%, the non-cross-linked membrane dissolved in water, demonstrating its lack of stability. To address this issue, a SPEEK membrane with a sulfonation degree of 50% was prepared and tested with 10% POP-SiWA. After 48 hours of soaking in 80 °C water, the membrane retained only 2.1% of the initial POP-SiWA content, as shown in Table 14.1. These findings highlight the importance of cross-linking in preventing the leaching of POP-SiWA and ensuring the long-term stability of the membrane in aqueous environments.

14.3.3 Characterization of Cross-Linked Grafted Membranes

In Figure 14.13a, the FT-IR spectra of the pure SPEEK membrane and SP/PSiW-*x* cross-linked grafted membranes are shown. The characteristic peaks at wavenumbers 1645, 1593, and 1468 cm^{-1} correspond to the C=O stretching vibration of the carbonyl group, C=C stretching of the benzene ring, and C–C stretching vibrations of the SPEEK polymer. The peaks at 1217, 1075, and 1018 cm^{-1} are related to the

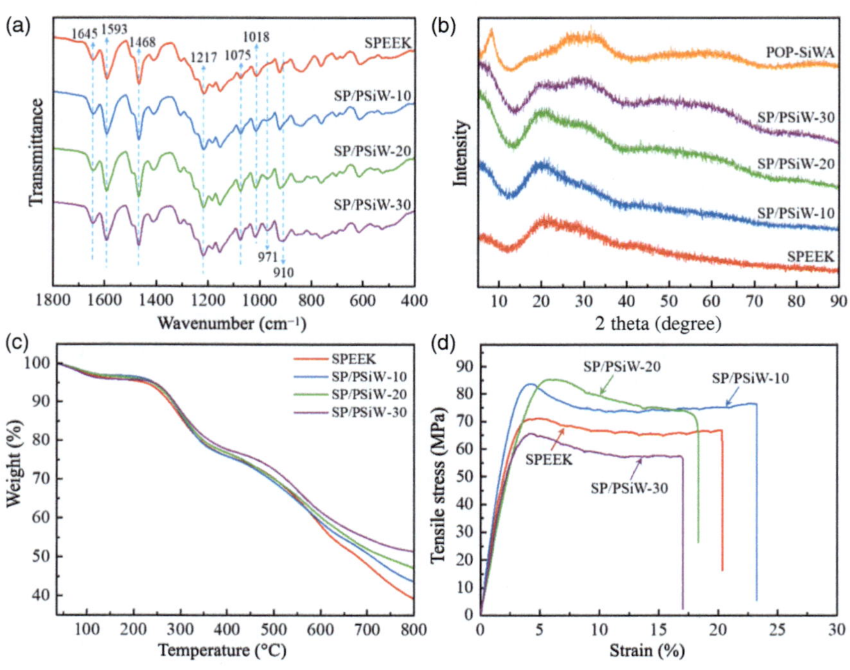

Figure 14.13 (a) FT-IR spectra, (b) XRD patterns, (c) TGA curves, and (d) stress–strain curves of pristine SPEEK and SP/PSiW-*x* membranes.

stretching vibrations of the S=O and S—O bonds in the sulfonic groups of SPEEK. Additionally, the peaks at 971 and 910 cm^{-1} are attributed to the characteristic W=O bond vibrations in the tungsten-silicon acid (POP-SiWA). As the POP-SiWA loading increases, the intensity of these peaks also increases, indicating the successful incorporation of POP-SiWA into the membrane.

Figure 14.13b shows the XRD patterns of pure SPEEK and SP/PSiW-*x* cross-linked grafted membranes. Due to the high sulfonation degree of the SPEEK backbone and the grafting of POP-SiWA, as well as the formation of a three-dimensional network structure during thermal cross-linking, the polymer chains cannot stack orderly. As a result, all the membranes exhibit an amorphous structure with broad diffraction peaks. The pure SPEEK membrane shows a prominent diffraction peak at $2\theta = 21°$, while the POP-SiWA grafted membranes display a broad diffraction peak near $2\theta = 30°$, with the intensity of this peak increasing as the POP-SiWA content rises. This suggests that the increased POP-SiWA content contributes to a more pronounced diffraction pattern, although the overall structure remains amorphous.

The membranes were subjected to TGA to assess their thermal stability and mechanical performance testing to evaluate their strength. The results showed that both the pure SPEEK membrane and the SP/PSiW-*x* grafted membranes exhibited three stages of weight loss, as shown in Figure 14.13c. The first stage of weight loss occurred below 150 °C, primarily due to the evaporation of $-SO_3H$ groups and water absorbed by the tungsten-silica acid. The second stage, between 200 and 400 °C, was mainly attributed to the decomposition of the $-SO_3H$ groups. The third stage, between 450 and 800 °C, involved the degradation of the SPEEK backbone. As previously mentioned, the tungsten-silica acid (POP-SiWA) showed little mass loss upon thermal decomposition, resulting in higher residual mass in the SP/PSiW-*x* grafted membranes compared to the pure SPEEK membrane after heating to 800 °C.

Furthermore, as the POP-SiWA content increased, the residual mass of the grafted membranes also increased. In general, the membranes exhibited good thermal stability up to 200 °C, which is suitable for PEM applications. Figure 14.13d shows the mechanical performance testing results for the pure SPEEK membrane and the SP/PSiW-*x* grafted membranes. The pure SPEEK cross-linked membrane had a tensile strength of 71.1 MPa and a breaking elongation of 20.3%. The SP/PSiW-10 grafted membrane showed an improved tensile strength of 83.7 MPa and a breaking elongation of 23.2%, likely due to the involvement of both phenoxyphenyl groups on the POP-SiWA in the cross-linking reaction, leading to a higher degree of cross-linking.

Similar to SP/PSiW-10, SP/PSiW-20 exhibited a tensile strength of 85.3 MPa, but its breaking elongation decreased to 18.3%. This reduction in elongation was likely due to the increased amount of inorganic tungsten-silica acid, which decreased the flexibility of the membrane. The SP/PSiW-30 membrane showed further decreases in both tensile strength, at 65.7 MPa, and breaking elongation, at 17.0%, which can again be attributed to the higher content of inorganic material, leading to a change in the mechanical properties of the membrane.

14.3.4 Ion Exchange Capacities, Water Uptake, and Swelling Ratio

The ion exchange capacity (IEC) of a PEM reflects the number of functional groups capable of exchanging protons, directly influencing the membrane's water uptake and proton conductivity. Proton conductivity in membranes depends on multiple factors, including water uptake, the number and dissociation ability of acidic functional groups, and the structural morphology of the membrane that forms proton transport channels. Water plays a crucial role in PEMs by dissociating acidic functional groups to release protons and acting as a proton carrier that facilitates proton transport.

Thus, the membrane must absorb a sufficient amount of water to ensure high proton conductivity. However, excessive water uptake can dilute the proton concentration, thereby reducing proton conductivity. Additionally, excessive water absorption can cause significant membrane swelling, which adversely affects mechanical properties. Figures 14.14a and b show the water uptake and swelling ratio of the membranes at different temperatures. After 6 hours of cross-linking at 180 °C, the pure SPEEK membrane exhibited a water uptake of 33.5% and a swelling ratio of 17.6% at 80 °C. In contrast, the 80% sulfonated SPEEK membrane, which was not cross-linked, dissolved completely in hot water. The relatively low proton conductivity of the cross-linked SPEEK membrane is attributed to the partial consumption of sulfonic acid groups during heat cross-linking, reducing its water absorption capacity.

Furthermore, the formation of cross-linked chemical bonds restricts polymer chain movement, reducing water permeability, and consequently limiting both water uptake and swelling. The results indicate that as the temperature increases, both water uptake and swelling ratio increase. This is primarily due to the thermal relaxation of polymer chains, which enhances their mobility, reduces entanglement, and increases free volume, allowing more water penetration.

For the SP/PSiW-x membranes, water uptake was significantly higher than that of the pure SPEEK membrane due to the incorporation of water-absorbing silica tungsten acid (POP-SiWA). For example, at 80 °C, the water uptake of SP/PSiW-10 reached 49.5%, which is 1.48 times higher than that of the pure SPEEK membrane. However, the swelling ratio of the SP/PSiW-10 membrane was only 14.7%, lower

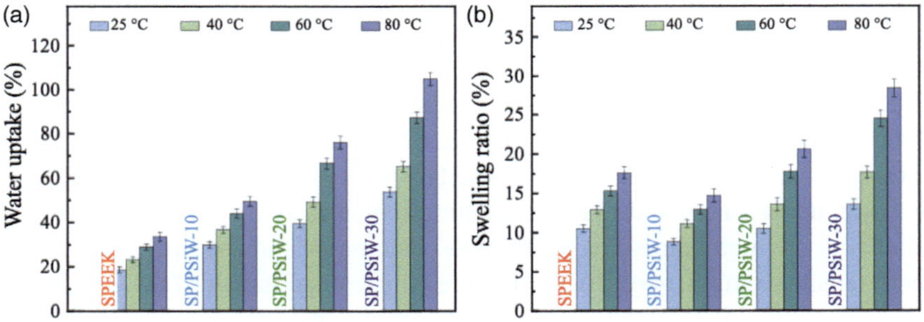

Figure 14.14 (a) Water uptake and (b) swelling ratio of membranes at various temperatures.

than that of the cross-linked pure SPEEK membrane. This could be due to the phenoxyphenyl groups on POP-SiWA participating in the cross-linking reaction, which enhances the degree of cross-linking in SP/PSiW-10 and reduces its swelling ratio.

For the SP/PSiW-20 and SP/PSiW-30 membranes, both water uptake and swelling ratio increased further. This can be attributed to three main factors. First, the higher amount of POP-SiWA grafted into the membranes enables greater absorption of water. Second, more sulfonic acid groups in the SPEEK polymer are involved in POP-SiWA grafting, reducing the extent of cross-linking between SPEEK chains and thereby decreasing the overall cross-linking degree. Third, the increased free volume of the polymer, resulting from the grafted POP-SiWA side chains, further facilitates water absorption. Consequently, at 80 °C, the SP/PSiW-30 membrane exhibited a water uptake of 104.6% and a swelling ratio of 28.3%, which met the requirements for subsequent testing.

The effect of water uptake on proton conductivity can be evaluated using the λ value, which represents the number of water molecules absorbed per sulfonic acid group. It is calculated based on the membrane's water uptake and IEC. Generally, a low λ value indicates insufficient water uptake, making it difficult to form a continuous hydrogen bond network of hydrated ions within the membrane, leading to low proton conductivity. Conversely, a high λ value dilutes the proton concentration inside the membrane, also reducing proton conductivity.

As shown in Figure 14.15, the λ value of the pure SPEEK membrane at 25 °C was only 5.5, and at 80 °C, it reached 10.0, indicating insufficient water uptake and consequently weak proton conductivity. The λ value of the SP/PSiW-10 grafted membrane at 80 °C was 15.1, showing a notable improvement over the pure SPEEK membrane. The λ values for the SP/PSiW-20 and SP/PSiW-30 membranes at 80 °C were 23.5 and 32.8, respectively, suggesting that sufficient water was present near the acidic groups in the membrane. This is beneficial for the formation of hydrogen bond networks and the efficient transport of protons.

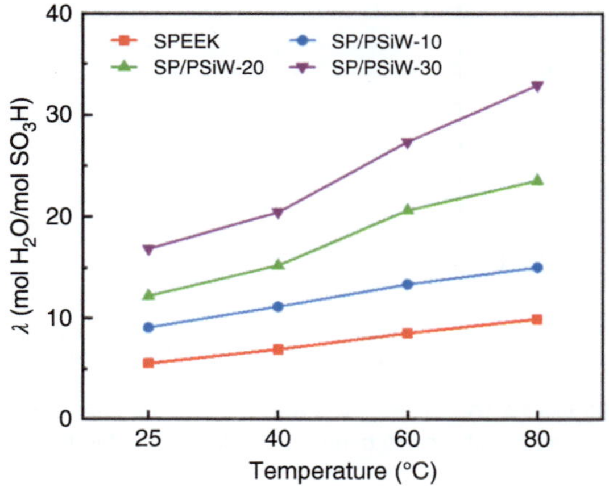

Figure 14.15 The number of water molecules adsorbed per unit of acidic group at different temperatures.

14.3.5 Anti-Free Radical Oxidation Properties

During the operation of proton exchange membrane fuel cells (PEMFCs), peroxide radicals such as HOO• and HO• are generated at the catalyst electrodes. These radicals attack the polymer membrane, leading to membrane degradation and failure, ultimately reducing the lifespan of the fuel cell [11]. To ensure the long-term operation of the fuel cell system, PEMs must exhibit resistance to radical oxidation. Fenton's reagent is commonly used to simulate radical attack on polymer membranes in fuel cells, providing an effective method to evaluate oxidative stability.

As shown in Figure 14.16, the degradation time of the pure SPEEK membrane and SP/PSiW-x cross-linked grafted membranes in Fenton's reagent is presented. The pure SPEEK cross-linked membrane exhibited good oxidative resistance, with a rupture time ($\tau1$) of 165 minutes and complete decomposition time ($\tau2$) of 240 minutes. This enhanced oxidative resistance can be attributed to thermal cross-linking, which improves the membrane's structural stability.

Previous studies have reported that tungstosilicic acid (POP-SiWA) contributes positively to enhancing the radical resistance of membranes [12]. The SP/PSiW-10 and SP/PSiW-20 membranes, which contain grafted POP-SiWA, exhibited slightly improved radical resistance compared to the pure SPEEK membrane. Specifically, the $\tau1$ values for SP/PSiW-10 and SP/PSiW-20 increased to 210 and 190 minutes, respectively, indicating that the introduction of tungstosilicic acid enhances radical resistance.

However, the radical resistance of the SP/PSiW-30 membrane was reduced. This decline is primarily due to its relatively low cross-linking density, as well as its higher water absorption and swelling rate. These factors allow more peroxide radicals to penetrate the membrane along with water, accelerating membrane degradation [13].

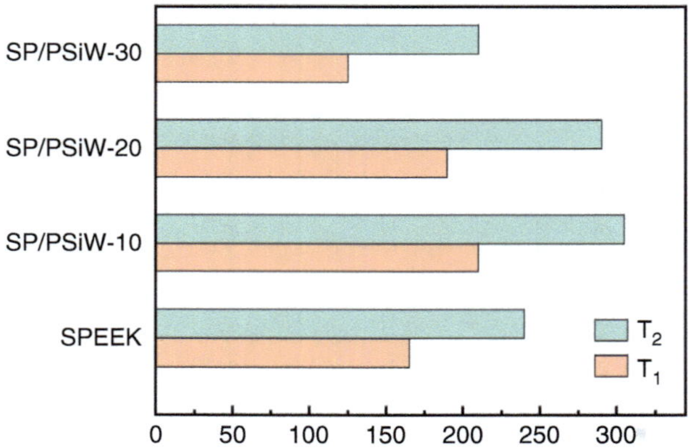

Figure 14.16 Oxidative stability of the prepared cross-linked membranes. τ_1: Elapsed time for the membrane to begin to break down in Fenton's reagent at 80 °C; τ_2: Elapsed time for the membrane to completely dissolve in Fenton's reagent.

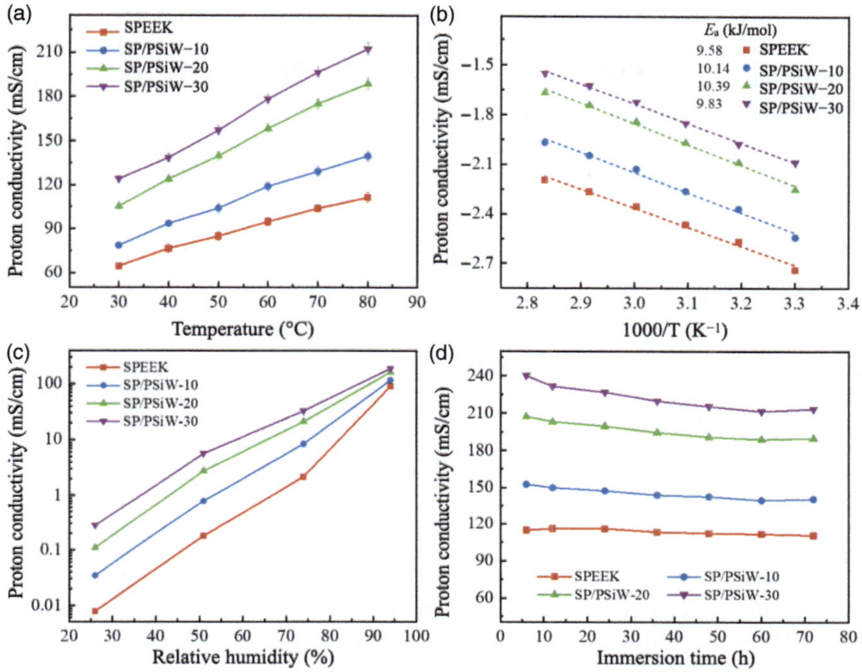

Figure 14.17 Proton conductivity properties of pristine SPEEK and SP/PSiW-*x* membranes: (a) measured at 30–80 °C in water; (b) Arrhenius plots of the membranes; (c) measured at different relative humidities; (d) measured at different stages of immersion time.

14.3.6 Proton Conductivity

Figure 14.17a presents the proton conductivity of the pure SPEEK and SP/PSiW-*x* cross-linked grafted membranes in pure water at different temperatures. As temperature increases, the polymer chains become more flexible, and the mobility of water molecules within the membrane improves, leading to a gradual increase in proton conductivity for all membranes. Figure 14.17b illustrates the relationship between proton conductivity and temperature, which follows the Arrhenius equation. From the slope of the fitted line, the activation energy for proton transport (E_a) was calculated. The results show that the activation energy for proton transport in the prepared membranes ranges from 9.58 to 10.39 kJ/mol. This suggests that proton transport primarily occurs via the Grotthuss mechanism, as E_a values below 10 kJ/mol indicate a dominant hydrogen bond-assisted proton hopping process.

This mechanism relies on the hydrogen bond network formed by sulfonic acid groups, tungstosilicic acid, and water molecules. Protons migrate by "hopping" along proton carriers through repeated formation and dissociation of hydrogen bonds. Since this process requires a relatively low energy barrier, proton transport exhibits minimal temperature dependence. Furthermore the test results indicate that SP/PSiW-*x* cross-linked grafted membranes exhibit higher proton conductivity than do pure SPEEK cross-linked membranes. Although the pure SPEEK membrane has a high IEC (1.86 mmol/g), its proton conductivity in the 30–80 °C range is relatively low, varying between 64.7 and 111.6 mS/cm. This reduced proton conductivity can be attributed to the high degree of cross-linking, which restricts polymer chain movement. Additionally, the sulfonic acid groups are directly attached to the polymer backbone, limiting their mobility. The formation of cross-linked structures further constrains the proton transport capacity of the sulfonic acid groups.

Moreover, proton conductivity in polymer membranes is highly dependent on water content. As discussed earlier, the water uptake of the pure SPEEK membrane at 80 °C is only 33.5%, with a corresponding λ value of 10.0. This suggests that the number of water molecules associated with the $-SO_3H$ groups is relatively low, leading to insufficient water availability for H^+ dissociation and a lack of water molecules as proton carriers, making proton transport inefficient. For SP/PSiW-x membranes, which are grafted with POP-SiWA, proton conductivities at 80 °C were measured as 140, 189, and 212 mS/cm for SP/PSiW-10, SP/PSiW-20, and SP/PSiW-30, respectively, demonstrating a significant improvement in proton conductivity. This enhancement can be attributed to the following factors:

1) Tungstosilicic acid (SiWA) is a superstrong proton acid, capable of rapid and complete dissociation within the membrane. Each SiWA molecule can provide four H^+ ions, and compared to sulfonic acid groups, SiWA enables a more concentrated H^+ distribution.
2) SiWA has a higher water absorption capacity than do sulfonic acid groups, ensuring that SP/PSiW-x membranes exhibit greater water uptake, thereby facilitating the formation of continuous water-hydrated proton transport pathways.
3) SiWA is linked to the polymer backbone via short phenoxyphenyl side chains, offering increased mobility compared to sulfonic acid groups that are directly bound to the polymer backbone. In the Grotthuss mechanism, the mobility of SiWA along these side chains allows H^+ ions to "hop" more efficiently between SiWA sites, accelerating proton transport.

Figure 14.17c presents the proton conductivity of pure SPEEK and SP/PSiW-x cross-linked grafted membranes at 80 °C under varying relative humidity (RH) conditions. At low RH (26, 51, and 74% RH), all membranes exhibit reduced proton conductivity. However, as the humidity increases to 94% RH, a significant increase in proton conductivity is observed, indicating the strong dependence of proton conductivity on water content. Among the tested membranes, SP/PSiW-30 exhibited relatively higher proton conductivity even at lower humidity levels. This can be attributed to the high concentration of SiWA in the membrane, which functions as an excellent water-absorbing material. This property enables SP/PSiW-30 to retain moisture even at low humidity levels, maintaining the hydration required for effective proton transport.

Figure 14.17d shows the proton conductivity of the membranes at different stages of soaking and washing. The results indicate that the proton conductivity of both pure SPEEK and SP/PSiW-10 membranes remained stable after 72 hours of soaking and washing, suggesting that the grafting of POP-SiWA in SP/PSiW-10 is relatively

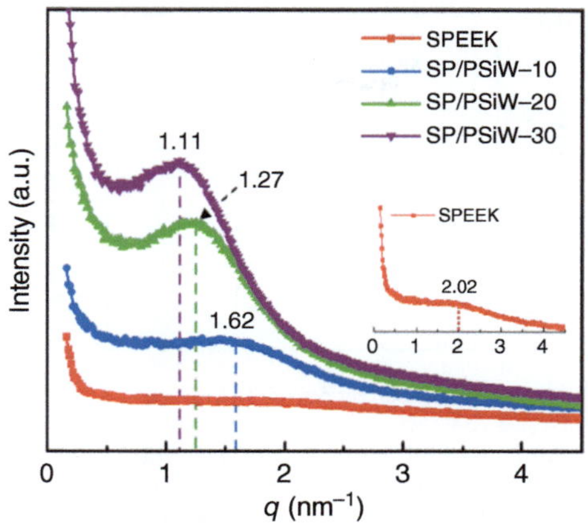

Figure 14.18 SAXS of pristine SPEEK and SP/PSiW membranes.

stable, with minimal loss of unreacted POP-SiWA. However, for SP/PSiW-20 and SP/PSiW-30 membranes, a significant amount of unreacted POP-SiWA was initially present in a free state within the membrane. This unreacted material contributed to higher initial proton conductivity. Over time, as the unreacted POP-SiWA leached out during soaking and washing, proton conductivity gradually decreased until stabilizing after approximately 48 hours.

To further investigate the formation of proton transport channels, small-angle X-ray scattering (SAXS) analysis was conducted on the membranes. As shown in Figure 14.18, the pure SPEEK cross-linked membrane exhibited a weak scattering peak at $2.02\,nm^{-1}$, corresponding to an average ion cluster size of 3.11 nm. This suggests that the ion domains within the pure SPEEK membrane are relatively dispersed, leading to lower proton conductivity. In contrast, SP/PSiW-10 showed a stronger scattering peak at $1.62\,nm^{-1}$, corresponding to an ion cluster size of 3.88 nm, indicating the formation of continuous proton transport channels that slightly improve proton conductivity.

For SP/PSiW-20 and SP/PSiW-30 membranes, which contain higher SiWA content, strong scattering peaks appeared at q values of 1.27 and $1.11\,nm^{-1}$, corresponding to ion cluster sizes of 4.95 and 5.66 nm, respectively. This indicates the presence of larger, well-defined proton transport channels, which significantly enhance proton conductivity.

14.3.7 Fuel Cell Performance

The pure SPEEK membrane and SP/PSiW-x membranes were pressed into membrane electrodes, and the single-cell performance of the membrane electrode assembly was tested at 80 °C and 100% RH under H_2/O_2 conditions. Figure 14.19a shows the current–voltage polarization curves and power density curves of the membrane electrodes based on the SPEEK membrane and SP/PSiW-x membranes. It can be observed that the open circuit voltage (OCV) of the prepared membrane

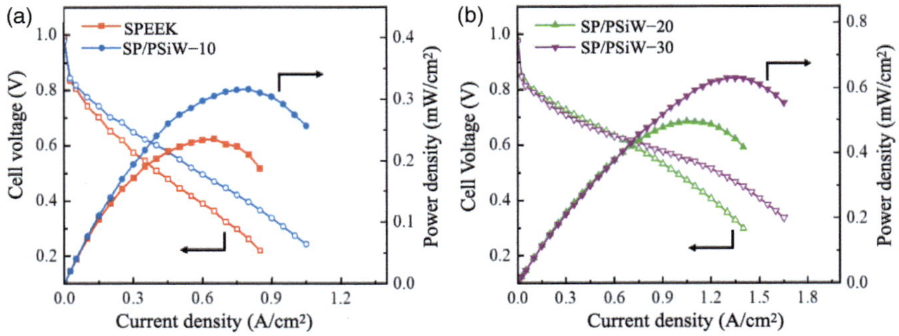

Figure 14.19 H_2/O_2 PEMFC single cell performance of MEAs employing pristine SPEEK and SP/PSiW hybrid membranes at 80 °C, 100% RH. (a) Polarization and power density curves of the pristine SPEEK membrane and the SP/PSiW-10 composite membrane. (b) Comparison of SP/PSiW-20 and SP/PSiW-30 composite membranes.

electrodes exceeds 0.97 V, indicating high catalytic activity and low H_2 permeability of the membrane electrodes. In the electrode overpotential-dominant region, where the current density is below 100 mA/cm^2, there is no significant difference in the battery performance. However, in the ohmic resistance-dominant region, the performance differences become more apparent, as the ohmic resistance is primarily influenced by the proton conductivity of the membrane.

Figure 14.19b shows that the operating voltage of the membrane electrode using SP/PSiW-10 is noticeably higher than that of the membrane electrode using SPEEK, mainly because the membrane resistance of SPEEK is much higher than that of SP/PSiW-10. The maximum output power for the two membranes is 236 mW/cm^2 for the SPEEK membrane and 317 mW/cm^2 for the SP/PSiW-10 membrane. Figure 14.19b shows that SP/PSiW-20 and SP/PSiW-30 membrane electrodes exhibit relatively better battery performance. The polarization curve indicates that as the current density increases, the voltage of SP/PSiW-20 and SP/PSiW-30 membranes decreases more gradually compared to the SPEEK and SP/PSiW-10 membranes. These membranes can output higher power densities of 497 and 629 mW/cm^2, respectively.

At current densities below 600 mA/cm^2, the voltage drop trends of SP/PSiW-20 and SP/PSiW-30 membranes are similar. However, at high current densities greater than 600 mA/cm^2, the voltage of the SP/PSiW-20 membrane electrode decreases more rapidly. This is likely because, at high current densities, protons drag some of the water from the anode side to the cathode side through "electro-osmosis." The SP/PSiW-20 membrane lacks sufficient water back-diffusion ability from the cathode to the anode, leading to dehydration on the anode side and an increase in membrane resistance. In contrast, for the SP/PSiW-30 membrane, water is better transported within the membrane, facilitating water back-diffusion at high current densities and reducing membrane resistance. Therefore, under operating conditions, the membrane electrode using SP/PSiW-30 exhibits better battery performance.

14.4 Conclusion

In this study, we developed POP-SiWA-grafted SPEEK membranes using an improved thermal cross-linking method with a DMSO/cyclohexane sulfone (6:1 by mass) solvent system. The optimized approach enhanced grafting efficiency (88.0–63.7%), prevented POP-SiWA precipitation, and produced uniform, dense membranes with controlled water absorption and swelling. The SP/PSiW-30 membrane demonstrated high proton conductivity (212 mS/cm at 80 °C) and superior fuel cell performance, achieving a peak power density of 629 mW/cm^2 at 80 °C and 100% RH, significantly outperforming pure SPEEK (236 mW/cm^2). These findings establish a promising strategy for high-performance PEMs with enhanced stability, water retention, and conductivity.

References

1 Lu, S., Wang, D., Jiang, S.P. et al. (2010). HPW/MCM-41 phosphotungstic acid/mesoporous silica composites as novel proton-exchange membranes for elevated-temperature fuel cells. *Advanced Materials* 22 (9): 971–976.

2 Hasani-Sadrabadi, M.M., Dashtimoghadam, E., Majedi, F.S. et al. (2016). Ionic nanopeapods: Next-generation proton conducting membranes based on phosphotungstic acid filled carbon nanotube. *Nano Energy* 23: 114–121.

3 Zhai, S., Lu, Z., Ai, Y. et al. (2023). High performance nanocomposite proton exchange membranes based on the nanohybrids formed by chemically bonding phosphotungstic acid with covalent organic frameworks. *Journal of Power Sources* 554: 232332.

4 Di Vona, M.L., Sgreccia, E., Licoccia, S. et al. (2009). Analysis of temperature-promoted and solvent-assisted cross-linking in sulfonated poly(ether ether ketone) (SPEEK) proton-conducting membranes. *The Journal of Physical Chemistry B* 113 (21): 7505–7512.

5 Di Vona, M.L., Alberti, G., Sgreccia, E. et al. (2012). High performance sulfonated aromatic ionomers by solvothermal macromolecular synthesis. *International Journal of Hydrogen Energy* 37 (10): 8672–8680.

6 Knauth, P., Pasquini, L., Maranesi, B. et al. (2013). Proton mobility in sulfonated polyetheretherketone (SPEEK): Influence of thermal crosslinking and annealing. *Fuel Cells* 13 (1): 79–85.

7 Han, D., Sun, J., Ge, J. et al. (2024). Cross-linked proton exchange membrane covalently bonded with silicotungstic acid for enhanced proton conductivity. *International Journal of Hydrogen Energy* https://doi.org/10.1016/j.ijhydene.2024.10.17.

8 Lejeune, N., Dez, I., Jaffrès, P. et al. (2008). Synthesis, crystal structure and thermal properties of phosphorylated cyclotriphosphazenes. *European Journal of Inorganic Chemistry* 2008 (1): 138–143.

9 Grabiak, R.C., Miles, J.A., and Schwenzer, G.M. (1980). Synthesis of phosphonic dichlorides and correlation of their P-31 chemical shifts. *Phosphorus Sulfur and Related Elements* 9 (2): 197–202.

10 Beyraghi, F., Mirfarsi, S.H., Rowshanzamir, S. et al. (2020). Optimal thermal treatment conditions for durability improvement of highly sulfonated poly(ether ether ketone) membrane for polymer electrolyte fuel cell applications. *International Journal of Hydrogen Energy* 45 (24): 13441–13458.

11 Rui, Z. and Liu, J. (2020). Understanding of free radical scavengers used in highly durable proton exchange membranes. *Progress in Natural Science: Materials International* 30 (6): 732–742.

12 Motz, A.R., Kuo, M.C., Horan, J.L. et al. (2018). Heteropoly acid functionalized fluoroelastomer with outstanding chemical durability and performance for vehicular fuel cells. *Energy & Environmental Science* 11 (6): 1499–1509.

13 Borup, R., Meyers, J., Pivovar, B. et al. (2007). Scientific aspects of polymer electrolyte fuel cell durability and degradation. *Chemical Reviews* 107 (10): 3904–3951.

Index

Electrospinning for Proton Exchange Membrane Fuel Cells, First Edition.
Yong Liu and Mohideen Meerasahib Mohamedazeem.

p

r

s

t